ATLANTIS STUDIES IN MATHEMATICS FOR ENGINEERING AND SCIENCE
VOLUME 1

SERIES EDITOR: **C.K. CHUI**

Atlantis Studies in Mathematics for Engineering and Science

Series Editor:

C.K. Chui, Stanford University, USA

(ISSN: 1875-7642)

Aims and scope of the series

The series 'Atlantis Studies in Mathematics for Engineering and Science'(AMES) publishes high quality monographs in applied mathematics, computational mathematics, and statistics that have the potential to make a significant impact on the advancement of engineering and science on the one hand, and economics and commerce on the other. We welcome submission of book proposals and manuscripts from mathematical scientists worldwide who share our vision of mathematics as the engine of progress in the disciplines mentioned above.

All books in this series are co-published with World Scientific.

For more information on this series and our other book series, please visit our website at:

www.atlantis-press.com/publications/books

ATLANTIS PRESS
AMSTERDAM – PARIS

World Scientific

Continued Fractions

Second edition

Volume 1: Convergence Theory

Lisa Lorentzen
Haakon Waadeland

Department of Mathematics
Norwegian University of Science and Technology
Trondheim
Norway

ATLANTIS PRESS
AMSTERDAM – PARIS

World Scientific

Atlantis Press
29 avenue Laumière
75019 Paris, France

For information on all Atlantis Press publications, visit our website at: www.atlantis-press.com.

ISBN: 978-90-78677-07-9
ISSN: 1875-7642

Preface to the second edition

15 years have passed since the first edition of this book was written. A lot has happened since then – also in continued fraction theory. New ideas have emerged and some old results have gotten new proofs. It was therefore time to revise our book "Continued Fractions with Applications" which appeared in 1992 on Elsevier.

The interest in using continued fractions to approximate special functions has also grown since then. Such fractions are easy to program, they have impressive convergence properties, and their convergence is often easy to accelerate. They even have good and reliable truncation error bounds which makes it possible to control the accuracy of the approximation. The bounds are both of the *a posteriori type* which tells the accuracy of a done calculation, and of the *a priori type* which can be used to determine the number of terms needed for a wanted accuracy. This important aspect is treated in this first volume of the second edition, along with the basic theory.

In the second volume we focus more on continued fraction expansions of analytic functions. There are several beautiful connections between analytic function theory and continued fraction expansions. We can for instance mention orthogonal polynomials, moment theory and Padé approximation.

We have tried to give credit to people who have contributed to the continued fraction theory up through the ages. But some of the material we believe to be new, at least we have found no counterpart in the literature. In particular, we believe that tail sequences play a more fundamental role in this book than what is usual. This way of looking at continued fractions is very fruitful.

Each chapter is still followed by a number of problems. This time we have marked the more theoretic ones by ♠. We have also kept the appendix from the first edition. This list of continued fraction expansions of special functions was so well received that we wanted it to stay as part of the book. Finally, we have kept the informality in the sense that the first chapter consists almost entirely of examples which show what continued fractions are good for. The more serious theory starts in Chapter 2.

Lisa Lorentzen carries the main responsibility for the revisions in this second edition. Through the first year of its making, Haakon Waadeland was busy writing a handbook on continued fractions, together with an international group of people. This left Lisa Lorentzen with quite free hands to choose the contents and the way of presentation. Still, he has played an important part in the later phases of the work on volume 1. For volume 2 Lisa Lorentzen bears the blame alone.

Trondheim, 14 February 2008

Lisa Lorentzen Haakon Waadeland

Preface

The name

Shortly before this book was finished, we sent out a number of copies of Chapter 1, under the name "A Taste of Continued Fractions". Now, in the process of working our way through the chapters on a last minute search for errors, unintended omissions and overlaps, or other unfortunate occurrences, we feel that this title might have been the right one even for the whole book. In most of the chapters, in particular in the applications, a lot of work has been put into the process of cutting, canceling and "non-writing". In many cases we are just left with a "taste", or rather a glimpse of the role of the continued fractions within the topic of the chapter. We hope that we thereby can open some doors, but in most cases we are definitely not touring the rooms.

The chapters

Each chapter starts with some introductory information, "About this chapter". The purpose is not to tell about the contents in detail. That has been done elsewhere. What we want is to tell about the intention of the chapter, and thereby also to adjust the expectations to the right (moderate) level. Each chapter ends with a reference list, reflecting essentially literature used in preparing that particular chapter. As a result, books and papers will in many cases be referred to more than once in the book. On the other hand, those who look for a complete, updated bibliography on the field will look in vain. To present such a bibliography has not been one of the purposes of the book.

The authors

The two authors are different in style and approach. We have not made an effort to hide this, but to a certain extent the creative process of tearing up each other's drafts and telling him/her to glue it together in a better way (with additions and omissions) may have had a certain disguising effect on the differences. This struggling type of cooperation leaves us with a joint responsibility for the whole book. The way we then distribute blame and credit between us is an internal matter.

The treasure chest

Anybody who has lived with and loved continued fractions for a long time will also have lived with and loved the monographs by Perron, Wall and Jones/Thron. Actually the love for continued fractions most likely has been initiated by one or more of these books. This is at least the case for the authors of the present book, and more so: these three books have played an essential role in our lives. The present book is in no way an attempt to replace or compete with these books. To the contrary, we hope to urge the reader to go on to these sources for further information.

For whom?

We are aiming at two kinds of readers: On the one hand people *in or near* mathematics, who are curious about continued fractions; on the other hand senior-graduate level students who would like an introduction (and a little more) to the analytic theory of continued fractions. Some basic knowledge about functions of a complex variable, a little linear algebra, elementary differential equations and occasionally a little dash of measure theory is what is needed of mathematical background. Hopefully the students will appreciate the problems included and the examples. They may even appreciate that some examples precede a properly established theory. (Others may dislike it.)

Words of gratitude

We both owe a lot to Wolf Thron, for what we have learned from him, for inspiration and help, and for personal friendship. He has read most of this book, and his remarks, perhaps most of all his objections, have been of great help for us. Our gratitude also extends to Bill Jones, his closest coworker, to Arne Magnus, whose recent death struck us with sadness, and to all other members of the Colorado continued fraction community. Here in Trondheim Olav Njåstad has been a key person in the field, and we have on several occasions had a rewarding cooperation with him.

Many people, who had received our Chapter I, responded by sending friendly and encouraging letters, often with valuable suggestions. We thank them all for their interest and kind help.

The main person in the process of changing the hand-written drafts to a camera-ready copy was Leiv Arild Andenes Jacobsen. His able mastering of LaTeX, in combination with hard work, often at times when most people were in bed, has left us with a great debt of gratitude. We also want to thank Arild Skjølsvold and Irene Jacobsen for their part of the typing job. We finally thank Ruth Waadeland, who made all the drawings, except the LaTeX-made ones in Chapter XI.

The Department of Mathematics and Statistics, AVH, The University of Trondheim generously covered most of the typing expenses. The rest was covered by Elsevier Science Publishers. We are most grateful to Claude Brezinski and Luc Wuytack for urging us to write this book, and to Elsevier Science Publishers for publishing it.

Trondheim, December 1991,

Lisa Lorentzen Haakon Waadeland

Contents

Chapter 1

Introductory examples

We have often been asked questions, by students as well as by established mathematicians, about continued fractions: what they are and what they can be used for. Sometimes the questions have been raised under circumstances where a quick answer is the only alternative to no answer: in the discussion after a talk or lecture, by a cup of coffee in a short break, in an airplane cabin or on a mountain hike. In responding to these questions we have often been pleased by the sparks of interest we have seen, indicating that we had managed to transmit a glimpse of new and apparently appealing knowledge. In quite a few cases this led to further contact and "follow-up activities".

This introductory chapter is to a large extent inspired by the questions we have received and governed by the answers we have given. There is of course a great danger: A quick answer is often a wrong answer. It may tell the truth and nothing but the truth, but it definitely does not tell the whole truth. This may lead to false guesses. This danger is in particular great in cases where observations and experiments are used to create and support guesses. But we still wanted to keep this often non-accepted, but highly useful aspect of mathematics as part of the introductory chapter. We have tried to reduce the danger, partly by the way things are phrased, partly by indicating briefly how wrong such guesses can be, and finally by referring to a more careful treatment later in the book. Still, we have chosen to introduce some basic notation and definitions in the first section of the first chapter.

In this new edition it has been important not only to maintain the intention of the introductory chapter, but also to increase the number of examples. We have therefore moved the convergence theorems to Chapter 3 to make some more space. In doing this we are violating rules and traditions for presentation of mathematics, namely to present the basic theory first, and then illustrate it by examples. This is done on purpose, in the belief that what is lost in mathematical style and structure is gained in glimpses of what it is all about, and in wetting the appetite and curiosity.

1.1 Basic concepts

1.1.1 Prelude to a definition

Let $\{a_n\}$ be a sequence of complex numbers. When we talk about the series

$$\sum_{n=1}^{\infty} a_n = a_1 + a_2 + \cdots + a_n + \cdots , \tag{1.1.1}$$

we have in mind the sequence $\{\sigma_n\}$ of *partial sums*

$$\sigma_n := \sum_{k=1}^{n} a_k .$$

(An empty sum $\sum_{k=m}^{n} a_k := 0$ for $n < m$.) *Convergence* of the series (1.1.1) means convergence of $\{\sigma_n\}$ to a complex number σ, in which case we write

$$\sum_{n=1}^{\infty} a_n = \sigma . \tag{1.1.2}$$

Similarly we are familiar with infinite *products*

$$\prod_{n=1}^{\infty} a_n = a_1 \cdot a_2 \cdots a_n \cdots . \tag{1.1.3}$$

$\{p_n\}$ is the sequence of *partial products*

$$p_n := \prod_{k=1}^{n} a_k .$$

(An empty product $\prod_{k=m}^{n} a_k := 1$ for $n < m$.) *Convergence* of the infinite product (1.1.3) means convergence of the sequence $\{p_n\}$ to a complex number $p \neq 0$, in which case we write

$$\prod_{n=1}^{\infty} a_n = p . \tag{1.1.4}$$

Let $\{f_n\}$ be the sequence

$$f_1 := a_1 , \qquad f_2 := \frac{a_1}{1 + a_2} , \qquad f_3 := \cfrac{a_1}{1 + \cfrac{a_2}{1 + a_3}} ,$$

and generally

$$f_n := \cfrac{a_1}{1 + \cfrac{a_2}{1 + \cfrac{a_3}{1 + \ddots + a_n}}} .$$

For simplicity we assume that all $a_n \neq 0$, and we allow $f_n = \infty$. Then $\{f_n\}$ is well defined in $\widehat{\mathbb{C}} := \mathbb{C} \cup \{\infty\}$ where $\mathbb{C}$ is the set of complex numbers. Similarly to what we have for sums and products, this also leads to a concept, having to do with the nonterminating continuation of the process, in this case the concept of a *continued fraction*

$$\mathop{\mathrm{K}}_{n=1}^{\infty} \frac{a_n}{1} := \cfrac{a_1}{1 + \cfrac{a_2}{1 + \cfrac{a_3}{1 + \ddots}}} \,. \tag{1.1.5}$$

(The letter K is chosen since *Kettenbruch* is the German name for a continued fraction.) *Convergence* of (1.1.5) means convergence of the sequence $\{f_n\}$ of *approximants*. We also accept convergence to ∞, as suggested by Pringsheim ([Prin99a]). The limit $f := \lim f_n$ is the *value* of the *convergent* continued fraction when it exists, and then we adopt the tradition from series and infinite products and write

$$f = \mathop{\mathrm{K}}_{n=1}^{\infty} \frac{a_n}{1} = \mathbf{K}_{n=1}^{\infty}(a_n/1) = \mathbf{K}(a_n/1) \,. \tag{1.1.6}$$

(An empty continued fraction $\mathbf{K}_{k=m}^{n}(a_k/1) := 0$ for $m > n$.) For simplicity we shall write the continued fraction (1.1.5) as

$$\mathop{\mathrm{K}}_{n=1}^{\infty} \frac{a_n}{1} =: \frac{a_1}{1} \, + \, \frac{a_2}{1} \, + \, \frac{a_3}{1} \, + \cdots \,.$$

Note where we place the plus signs to indicate the fraction structure. This distinguishes the continued fraction (1.1.5) from the series (1.1.1).

Example 1. For the continued fraction

$$\mathop{\mathrm{K}}_{n=1}^{\infty} \frac{6}{1} = \cfrac{6}{1 + \cfrac{6}{1 + \cfrac{6}{1 + \cfrac{6}{1 + \ddots}}}} = \frac{6}{1} \, + \, \frac{6}{1} \, + \, \frac{6}{1} \, + \cdots$$

we find

$$f_1 = 6 \,, \qquad f_2 = \frac{6}{7} \,, \qquad f_3 = \frac{42}{13} \,, \cdots \,.$$

By induction (see Problem 13 on page 49 with $x = 2$, $y = 3$) it follows that generally

$$f_n = 6 \frac{3^n - (-2)^n}{3^{n+1} - (-2)^{n+1}} \,.$$

Therefore the continued fraction converges to 2. ◇

Quite similarly we can construct, from any sequence $\{b_n\}$ of complex numbers, a continued fraction

$$\overset{\infty}{\underset{n=1}{\mathrm{K}}} \frac{1}{b_n} = \cfrac{1}{b_1 + \cfrac{1}{b_2 + \cfrac{1}{b_3 + \ddots}}} = \frac{1}{b_1} \, + \, \frac{1}{b_2} \, + \, \frac{1}{b_3} \, + \cdots , \tag{1.1.7}$$

or from *two* sequences, $\{a_n\}$ and $\{b_n\}$ of complex numbers, where all $a_n \neq 0$, a continued fraction

$$\overset{\infty}{\underset{n=1}{\mathrm{K}}} \frac{a_n}{b_n} = \cfrac{a_1}{b_1 + \cfrac{a_2}{b_2 + \cfrac{a_3}{b_3 + \ddots}}} = \frac{a_1}{b_1} \, + \, \frac{a_2}{b_2} \, + \, \frac{a_3}{b_3} \, + \cdots . \tag{1.1.8}$$

We write $\mathbf{K}(1/b_n)$ and $\mathbf{K}(a_n/b_n)$ for these structures.

In all the three cases the nth approximant f_n is what we get by truncating the continued fraction after n *fraction terms* a_k/b_k, and convergence means convergence of $\{f_n\}$. (1.1.5) and (1.1.7) are obviously special cases of (1.1.8). In the particular case when in (1.1.7) all b_n are natural numbers, we get the *regular continued fraction*, well known in number theory.

Let us take a look at the common pattern in the three cases: series, products and continued fractions (and other constructions for that matter). In all three cases the construction can be described in the following way: We have a sequence $\{\phi_k\}$ of mappings from $\widehat{\mathbb{C}}$ to $\widehat{\mathbb{C}}$. By composition we construct a new sequence $\{\Phi_n\}$ of mappings

$$\Phi_1 := \phi_1 , \qquad \Phi_n := \Phi_{n-1} \circ \phi_n = \phi_1 \circ \phi_2 \circ \cdots \circ \phi_n . \tag{1.1.9}$$

For *series* we have

$$\phi_k(w) := w + a_k ,$$

and

$$\Phi_n(w) = \phi_1 \circ \phi_2 \circ \cdots \circ \phi_n(w) = a_1 + a_2 + \cdots + a_n + w .$$

For *products* we have

$$\phi_k(w) := w \cdot a_k ,$$

and

$$\Phi_n(w) = \phi_1 \circ \phi_2 \circ \cdots \circ \phi_n(w) = a_1 \cdot a_2 \cdots a_n \cdot w .$$

For *continued fractions* (1.1.8) we have

$$\phi_k(w) := \frac{a_k}{b_k + w} ,$$

and

$$\Phi_n(w) = \phi_1 \circ \phi_2 \circ \cdots \circ \phi_n(w) = \frac{a_1}{b_1} \! + \! \frac{a_2}{b_2} \! + \! \cdots \! + \! \frac{a_n}{b_n + w} . \tag{1.1.10}$$

These infinite structures can be regarded formally, but in applications the question of convergence often comes up. Convergence of a series is defined as convergence of $\{\Phi_n(0)\}$, convergence of an infinite product is defined as convergence of $\{\Phi_n(1)\}$, and convergence of a continued fraction is defined as convergence of $\{\Phi_n(0)\}$.

1.1.2 Definitions

A linear fractional transformation (or Möbius transformation) τ is a mapping of form

$$\tau(w) = \frac{aw+b}{cw+d}; \quad a,b,c,d \in \mathbb{C} \quad \text{with } ad-bc \neq 0.$$

This representation of τ is not unique since we can multiply the coefficients a, b, c and d with any fixed non-zero constant without changing the mapping. However, we identify all these representations as one single transformation (mapping) τ.

We shall denote this family of mappings by $\mathcal{M}$. Sometimes one emphasizes that $ad-bc \neq 0$ by saying that τ is *non-singular*, as opposed to the *singular* transformations with $ad-bc=0$ which do not belong to $\mathcal{M}$. A singular transformation is constant wherever it is meaningful, whereas $\tau \in \mathcal{M}$ maps $\widehat{\mathbb{C}}$ univalently onto $\widehat{\mathbb{C}}$.

We define a continued fraction in terms of linear fractional transformations:

Definition 1.1. *A continued fraction $b_0 + \mathbf{K}(a_n/b_n)$ is an ordered pair $((\{a_n\}, \{b_n\}), \{S_n\})$ where $\{a_n\}$ and $\{b_n\}$ are sequences of complex numbers with all $a_n \neq 0$ and $\{S_n\}$ is the sequence from $\mathcal{M}$ given by*

$$S_n(w) := b_0 + \frac{a_1}{b_1} \genfrac{}{}{0pt}{}{}{+} \frac{a_2}{b_2} \genfrac{}{}{0pt}{}{}{+\cdots+} \frac{a_n}{b_n+w}. \tag{1.2.1}$$

Remark: That $\{S_n\}$ is a sequence from $\mathcal{M}$ is easily seen by the fact that

$$\begin{aligned} S_n &= s_0 \circ s_1 \circ \cdots \circ s_n \quad \text{where} \\ s_0(w) &:= b_0 + w\,, \quad s_n(w) := \frac{a_n}{b_n+w} \quad \text{for } n \in \mathbb{N}. \end{aligned} \tag{1.2.2}$$

Obviously $s_n \in \mathcal{M}$ when $a_n \neq 0$, and thus $S_n \in \mathcal{M}$ since compositions of linear fractional transformations are again linear fractional transformations (which is easily verified). We owe it to Weyl ([Weyl10]) for this very useful connection between continued fractions and the class $\mathcal{M}$. Normally we set $b_0 := 0$, in which case $\phi_k = s_k$ and $\Phi_k = S_k$ in (1.1.10). Still, it is useful to have the possibility to set $b_0 \neq 0$.

The numbers a_n and b_n are called the *elements* of $b_0 + \mathbf{K}(a_n/b_n)$, and $\frac{a_n}{b_n}$ is called a *fraction term* of $b_0 + \mathbf{K}(a_n/b_n)$. Evaluations $S_n(w)$ of S_n are called nth *approximants*. The name "convergents" has also been used in the literature. Historically, the word "approximant" always meant

$$f_n := S_n(0) \tag{1.2.3}$$

whereas $S_n(w_n)$ was referred to as a "modified approximant" where w_n was the "modifying factor". *Classical approximants* $f_n = S_n(0)$ play a special role in the theory. In particular, the concept of convergence is based on $\{f_n\}$:

Definition 1.2. *A continued fraction converges (in the classical sense) to a value* $f \in \widehat{\mathbb{C}}$ *if* $\lim f_n = f$.

If $\mathbf{K}(a_n/b_n)$ fails to converge, we say that it *diverges.*

A classical approximant f_n is obtained by truncating the continued fraction after n fraction terms. The part we cut away,

$$f^{(n)} := \frac{a_{n+1}}{b_{n+1}} + \frac{a_{n+2}}{b_{n+2}} + \frac{a_{n+3}}{b_{n+3}} + \cdots \tag{1.2.4}$$

is called the nth *tail* of $b_0 + \mathbf{K}(a_n/b_n)$. This is also a continued fraction, and it converges if and only if $b_0 + \mathbf{K}(a_n/b_n)$ converges. Indeed, (1.2.4) converges to $f^{(n)}$ if and only if $b_0 + \mathbf{K}(a_n/b_n)$ converges to

$$f = S_n(f^{(n)}). \tag{1.2.5}$$

The sequence $\{f^{(n)}\}$ is then called *the sequence of tail values* for $b_0 + \mathbf{K}(a_n/b_n)$. This sequence will be important in our investigations.

Lemma 1.1. *Let* S_n *be given by (1.2.1). Then*

$$S_n(w) = \frac{A_{n-1}w + A_n}{B_{n-1}w + B_n} \quad \text{for } n = 1, 2, 3, \ldots \tag{1.2.6}$$

where

$$A_n = b_n A_{n-1} + a_n A_{n-2}, \qquad B_n = b_n B_{n-1} + a_n B_{n-2} \tag{1.2.7}$$

with initial values $A_{-1} = 1$, $A_0 = b_0$, $B_{-1} = 0$ *and* $B_0 = 1$.

Proof : It is clear that

$$S_0(w) = b_0 + w = \frac{b_0 + w}{1 + 0w}, \qquad S_1(w) = b_0 + \frac{a_1}{b_1 + w} = \frac{b_0 b_1 + a_1 + b_0 w}{b_1 + w},$$

so (1.2.7) holds for $n = 1$. To see that it holds for general $n \in \mathbb{N}$, we observe that

$$S_n(w) = S_{n-1}(s_n(w)) = \frac{A_{n-1} + A_{n-2}\dfrac{a_n}{b_n + w}}{B_{n-1} + B_{n-2}\dfrac{a_n}{b_n + w}}\,.$$

□

A_n and B_n are called the nth *canonical numerator* and *denominator* of $b_0 + \mathbf{K}(a_n/b_n)$, or just its nth *numerator* and *denominator* for short. These names are quite natural since

$$f_n = S_n(0) = A_n/B_n\,. \tag{1.2.8}$$

The useful *determinant formula*

$$\Delta_n := A_{n-1}B_n - A_nB_{n-1} = \prod_{k=1}^{n}(-a_k) \tag{1.2.9}$$

is a consequence of (1.2.7) (it follows by induction). Therefore also

$$A_{n-1}B_{n+1} - A_{n+1}B_{n-1} = b_{n+1}\Delta_n. \tag{1.2.10}$$

Another simple, but important observation follows from the fact that

$$S_n(w_n) = S_0(w_0) + \sum_{k=1}^{n}(S_k(w_k) - S_{k-1}(w_{k-1})). \tag{1.2.11}$$

By the recurrence relations (1.2.7) and the determinant formula (1.2.9)

$$S_k(w_k) - S_{k-1}(w_{k-1}) = \frac{\lambda_k \prod_{m=1}^{k-1}(-a_m)}{(B_{k-1}w_k + B_k)(B_{k-2}w_{k-1} + B_{k-1})} \quad \text{where } \lambda_k := a_k - w_{k-1}(b_k + w_k), \tag{1.2.12}$$

and thus

$$S_n(w_n) = b_0 + w_0 + \sum_{k=1}^{n} \frac{\lambda_k \prod_{m=1}^{k-1}(-a_m)}{(B_{k-1}w_k + B_k)(B_{k-2}w_{k-1} + B_{k-1})}\,. \tag{1.2.13}$$

For the case $w_k := 0$ for all k, we get the well known *Euler-Minding formula*

$$f_n = \frac{A_n}{B_n} = b_0 - \sum_{k=1}^{n} \frac{(-1)^k a_1 a_2 \cdots a_k}{B_k B_{k-1}} = b_0 - \sum_{k=1}^{n} \frac{\Delta_k}{B_k B_{k-1}}\,, \tag{1.2.14}$$

where Δ_k is given by (1.2.9). This formula was used by Euler ([Euler48]) and rediscovered by Minding ([Mind69]). Of course, Euler considered classical approximants A_n/B_n, and the recurrence relations (1.2.7) are normally attributed to him ([Euler48], Chapter 18).

Remark: For a given continued fraction $b_0 + \mathbf{K}(a_n/b_n)$, we shall use the notation s_n, S_n, A_n, B_n, f_n, $f^{(n)}$ and Δ_n throughout the book.

In Lemma 1.1 we saw that the canonical numerators $\{A_n\}$ and denominators $\{B_n\}$ for a given continued fraction $b_0 + \mathbf{K}(a_n/b_n)$ are uniquely determined. In fact, Daniel Bernoulli ([Berno75]) proved that also the converse holds true if all $\Delta_n \neq 0$:

Theorem 1.2. *The given sequences $\{A_n\}_{n=0}^{\infty}$ and $\{B_n\}_{n=0}^{\infty}$ of complex numbers are the canonical numerators and denominators of some continued fraction $b_0 + \mathbf{K}(a_n/b_n)$ if and only if $\Delta_n \neq 0$ for $n \geq 1$ and $B_0 = 1$. Then $b_0 + \mathbf{K}(a_n/b_n)$ is uniquely determined by*

$$\begin{aligned} &b_0 := A_0\,, \qquad b_1 := B_1\,, \qquad a_1 := A_1 - A_0 B_1\,, \\ &a_n := -\frac{\Delta_n}{\Delta_{n-1}}\,, \quad b_n := \frac{A_{n-2}B_n - B_{n-2}A_n}{\Delta_{n-1}} \quad \text{for } n \geq 2\,. \end{aligned} \tag{1.2.15}$$

Proof : Let $\{A_n\}$ and $\{B_n\}$ be given with all $\Delta_n \neq 0$ and $B_0 = 1$. Then the elements a_n and b_n must be solutions of the system

$$\begin{aligned} b_n A_{n-1} + a_n A_{n-2} &= A_n\,, \\ b_n B_{n-1} + a_n B_{n-2} &= B_n \end{aligned} \tag{1.2.16}$$

of linear equations. The determinant of this system is $-\Delta_{n-1} \neq 0$. Hence the solution is given by (1.2.15), and it is unique. □

This theorem allows us to construct as many continued fraction identities

$$f = \mathbf{K}(a_n/b_n)$$

as we may possibly want.

Example 2. We shall find the continued fraction $b_0 + \mathbf{K}(a_n/b_n)$ for which $A_n := n^2$ and $B_n := n^2 + 1$ for $n = 0, 1, 2, \ldots$. In this case

$$\Delta_n = A_{n-1}B_n - B_{n-1}A_n = 1 - 2n$$

and

$$A_{n-2}B_n - B_{n-2}A_n = 4 - 4n\,,$$

so by Theorem 1.2

$$\begin{aligned} &b_0 = 0\,, \quad b_1 = 2\,, \quad a_1 = 1\,, \\ &a_n = -\frac{2n-1}{2n-3}\,, \quad b_n = \frac{4n-4}{2n-3} \quad \text{for } n = 2, 3, 4, \ldots. \end{aligned}$$

Hence, the continued fraction

$$\frac{1}{2+}\,\frac{-3/1}{4/1}\,+\,\frac{-5/3}{8/3}\,+\,\frac{-7/5}{12/5}+\frac{-9/7}{16/7}+\,\frac{-11/9}{20/9}\,+\cdots,$$

which also can be written

$$\frac{1}{2}\,-\,\frac{3/1}{4/1}\,-\,\frac{5/3}{8/3}\,-\,\frac{7/5}{12/5}\,-\,\frac{9/7}{16/7}\,-\,\frac{11/9}{20/9}\,-\cdots,$$

is the one with $A_n = n^2$ and $B_n = n^2 + 1$. Since $\lim A_n/B_n = 1$, this continued fraction converges to 1, and we have proved the continued fraction identity

$$\frac{1}{2}\,-\,\frac{3/1}{4/1}\,-\,\frac{5/3}{8/3}\,-\,\frac{7/5}{12/5}\,-\,\frac{9/7}{16/7}\,-\,\frac{11/9}{20/9}\,-\cdots = 1.$$

◇

Since $f_n = A_n/B_n$, we also have, with $\mathbb{N} := \{\text{natural numbers}\}$:

Theorem 1.3.
A. The sequence $\{f_n\}_{n=0}^{\infty}$ from $\widehat{\mathbb{C}}$ is a sequence of classical approximants for some continued fraction $\mathbf{K}(a_n/b_n)$ if and only if $f_0 = 0$, $f_1 \neq 0, \infty$ and $f_n \neq f_{n-1}$ for all $n \in \mathbb{N}$.
B. If $\{f_n\}$ is a sequence of classical approximants for $\mathbf{K}(a_n/b_n)$, then $f_n \neq f_{n-2}$ if and only if $b_n \neq 0$.

Proof : A. Let first $f_0 = 0$, $f_1 \neq 0, \infty$ and $f_n \neq f_{n-1}$ for all n. Let $\{A_n\}_{n=0}^{\infty}$ and $\{B_n\}_{n=0}^{\infty}$ be given by

$$A_n := \begin{cases} f_n & \text{if } f_n \neq \infty, \\ 1 & \text{if } f_n = \infty, \end{cases} \qquad B_n := \begin{cases} 1 & \text{if } f_n \neq \infty, \\ 0 & \text{if } f_n = \infty. \end{cases}$$

Then $f_n = A_n/B_n$, $\Delta_n \neq 0$ for $n \geq 1$ and $B_0 = 1$, and thus the existence of $\mathbf{K}(a_n/b_n)$ follows from Theorem 1.2. Conversely, let $\{f_n\}$ be the classical approximants for $\mathbf{K}(a_n/b_n)$. Then $f_0 = 0$, $B_0 = 1$ and $f_n = S_n(0) \neq S_n(\infty) = f_{n-1}$ since S_n is a univalent mapping of $\widehat{\mathbb{C}}$ to $\widehat{\mathbb{C}}$.

B. This follows since $f_n = S_n(0)$ and $f_{n-2} = S_n(-b_n)$. □

As already mentioned, the nth tail $\mathbf{K}_{m=n+1}^{\infty}(a_m/b_m)$ of $\mathbf{K}(a_n/b_n)$ is also a continued fraction. We shall let $A_k^{(n)}$ and $B_k^{(n)}$ denote its kth canonical numerator and denominator. Then the following follows by manipulating the recurrence relations:

Lemma 1.4.
$$A_k^{(n)} = a_{n+1} B_{k-1}^{(n+1)}. \tag{1.2.17}$$
$$B_k^{(n)} = b_{n+1} B_{k-1}^{(n+1)} + a_{n+2} B_{k-2}^{(n+2)}. \tag{1.2.18}$$

Proof : Let $n \in \mathbb{N} \cup \{0\}$ be arbitrarily chosen. The first identity holds trivially for $k = 0$ and $k = 1$. Hence it holds for all k since $\{A_k^{(n)}\}$ and $\{B_{k-1}^{(n+1)}\}$ are solutions of the same recurrence relation

$$X_k = b_{n+k} X_{k-1} + a_{n+k} X_{k-2}. \tag{1.2.19}$$

The second identity holds trivially for all n for $k = 1$ and $k = 2$ since $B_{-1}^{(n)} = 0$, $B_0^{(n)} = 1$, $B_1^{(n)} = b_{n+1}$ and $B_2^{(n)} = b_{n+1} b_{n+2} + a_{n+2}$. Assume it holds for $2 \leq k \leq \nu - 1$. Then it also holds for $k = \nu$ since

$$\begin{aligned} B_k^{(n)} &= b_{n+k} B_{k-1}^{(n)} + a_{n+k} B_{k-2}^{(n)} \\ &= b_{n+k} \left(b_{n+1} B_{k-2}^{(n+1)} + a_{n+2} B_{k-3}^{(n+2)} \right) + a_{n+k} \left(b_{n+1} B_{k-3}^{(n+1)} + a_{n+2} B_{k-4}^{(n+2)} \right) \\ &= b_{n+1} \left(b_{n+k} B_{k-2}^{(n+1)} + a_{n+k} B_{k-3}^{(n+1)} \right) + a_{n+2} \left(b_{n+k} B_{k-3}^{(n+2)} + a_{n+k} B_{k-4}^{(n+2)} \right) \end{aligned}$$

which is equal to the right hand side of (1.2.18). □

Our definition of a continued fraction leads to an infinite structure. In some applications one only needs a continued fraction-like structure which terminates after n terms,

$$b_0 + \frac{a_1}{b_1} + \frac{a_2}{b_2} + \cdots + \frac{a_n}{b_n}.$$

This will be called a *terminating continued fraction* even though it is not a continued fraction by our definition. A related situation occurs if $a_n = 0$ for some or all $n \in \mathbb{N}$. Let n_0 be the first index for which $a_n = 0$. Then S_n is singular for $n \geq n_0$, and the effect is essentially the same as if $b_0 + \mathbf{K}(a_n/b_n)$ was truncated after $(n_0 - 1)$ fraction terms. (The only thing that can complicate the situation is that the n_0th tail converges to $-b_{n_0}$. In such cases it is a matter of definition to determine the action.) This is in particular relevant if the elements a_n and b_n are functions of a complex variable z.

1.1.3 Computation of approximants

There are several algorithms to compute approximants

$$S_n(w_n) = b_0 + \frac{a_1}{b_1} + \frac{a_2}{b_2} + \cdots + \frac{a_n}{b_n + w_n} = \frac{A_{n-1} w_n + A_n}{B_{n-1} w_n + B_n}.$$

We shall only mention the most obvious ones:

1. The *forward recurrence algorithm* consists of computing A_n and B_n by the recurrence relation (1.2.7).

2. The *backward recurrence algorithm* starts by setting $q_n := w_n$ and then work backwards by setting
$$q_{k-1} := \frac{a_k}{b_k + q_k}$$
for $k = n, n-1, \ldots, 1$. Then $S_n(w_n) = b_0 + q_0$. (Dirichlet [Diri63], p 46.)

3. *Euler-Minding summation* consists of computing B_n by the recurrence relation (1.2.7) and then finding f_n or $S_n(w_n)$ by means of the Euler-Minding formula (1.2.14) or by (1.2.13).

The *arithmetic complexity* $\omega(f_n)$ or $\omega(f_1, f_2, \ldots, f_n)$ of an algorithm is the minimal number of operations ($+$, $\times$, $:$) needed to compute f_n or $(f_1, f_2, \ldots, f_n)$. Straightforward counting shows:

1. The arithmetic complexity of the forward algorithm is
$$\omega(f_1) = 3, \quad \omega(f_n) = 6n - 5, \quad \omega(f_1, \ldots, f_n) = 7n - 5 \text{ for } n \geq 2.$$

2. The arithmetic complexity of the backward algorithm is
$$\omega(f_1) = 2, \quad \omega(f_n) = 2n, \quad \omega(f_1, \ldots, f_n) = n^2 + n.$$

3. The arithmetic complexity of the Euler-Minding summation is $\omega(f_1, \ldots, f_n) = 6n - 3$.

Note that the work involved is essentially independent of what we choose for w_n. Method 1 and 3 has the advantage that if you have found $S_n(w_n)$, you can use this to find $S_{n+1}(w_{n+1})$, whereas you must start again from scratch in the second method. On the other hand, the backward recurrence algorithm is in general more stable. We return to this claim in Section 5.3.1 on page 260. The computations in this book are done by means of the backward recurrence algorithm. Approximants are computed with high precision, and the value of a convergent continued fraction is estimated by high order approximants with reliable truncation error bounds.

1.1.4 Approximating the value of $\mathbf{K}(a_n/b_n)$

Example 3. The continued fraction
$$\mathop{\mathrm{K}}_{n=1}^{\infty} \frac{30 + 0.9^n}{1} \tag{1.4.1}$$
is known to converge. We want to find its value f. If we cannot find the exact value, we can use an approximant $f_n = S_n(0)$ as an approximation, since $f_n \to f$. However, the nth tail that we then cut away, looks more and more like $\mathbf{K}(30/1)$ as n increases. Since $\mathbf{K}(30/1)$ converges to 5 (Problem 13 on page 49 with $x = -5$, $y = 6$), it is reasonable to believe that $S_n(5)$ is a better approximation than $S_n(0)$ to the value of (1.4.1). The table below indicates strongly that this is true.

In fact, we can do even better! The tail values

$$f^{(n)} := \frac{30+0.9^{n+1}}{1} \;+\; \frac{30+0.9^{n+2}}{1} \;+\; \frac{30+0.9^{n+3}}{1} \;+\cdots$$

satisfy the recurrence

$$f^{(n)} = \frac{30+0.9^{n+1}}{1+f^{(n+1)}}, \quad \text{i.e.,} \quad f^{(n)}(1+f^{(n+1)}) = 30+0.9^{n+1}.$$

Assume that $f^{(n)} = 5 + cr^{n+1} + o(r^{n+1})$ for some $c \in \mathbb{R}$ and $|r| < 1$, where $o(r^{n+1})$ is the usual "little o-notation" defined by

$$u_n = o(r^{n+1}) \quad \Longleftrightarrow \quad \lim_{n\to\infty} \frac{u_n}{r^{n+1}} = 0.$$

That is, we assume that c is chosen such that

$$\lim_{n\to\infty} \frac{f^{(n)} - (5 + cr^{n+1})}{r^{n+1}} = 0.$$

Since then

$$\begin{aligned} f^{(n)}(1+f^{(n+1)}) &= (5+cr^{n+1})(6+cr^{n+2}) + o(r^{n+1}) \\ &= 30 + c(6+5r)r^{n+1} + o(r^{n+1}), \end{aligned}$$

we must have $r = 0.9$ and $c(6+5r) = 1$; i.e., $c = 2/21$. Therefore the approximants $S_n(w_n)$ with $w_n = 5 + \frac{2}{21} \cdot 0.9^{n+1}$ should be an even better choice than $S_n(5)$, in particular for large n. This is in fact true. We shall return to this idea in Section 5.1.5 on page 232. The table below gives a very clear indication that this is so. The value f is in this case $f = 5.085066164924$ correctly rounded to 12 decimal places.

n	$S_n(0)$	$f - S_n(0)$	$f - S_n(5)$	$f - S_n(w_n)$
1	30.9000	-25.8	$-6.5 \cdot 10^{-2}$	$4.4 \cdot 10^{-4}$
2	0.97139	4.1	$4.8 \cdot 10^{-2}$	$-3.0 \cdot 10^{-4}$
3	15.6770	-10.6	$-3.7 \cdot 10^{-2}$	$2.0 \cdot 10^{-4}$
4	1.85765	3.2	$2.7 \cdot 10^{-2}$	$-1.4 \cdot 10^{-4}$
35	5.10127	$-1.6 \cdot 10^{-2}$	$-3.8 \cdot 10^{-6}$	$7.3 \cdot 10^{-10}$
36	5.07160	$1.3 \cdot 10^{-2}$	$2.8 \cdot 10^{-6}$	$-4.9 \cdot 10^{-10}$
37	5.09631	$-1.1 \cdot 10^{-2}$	$-2.1 \cdot 10^{-6}$	$3.3 \cdot 10^{-10}$
38	5.07571	$9.3 \cdot 10^{-3}$	$1.6 \cdot 10^{-6}$	$-2.2 \cdot 10^{-10}$
85	5.08507	$-1.8 \cdot 10^{-6}$	$-2.1 \cdot 10^{-12}$	$2.1 \cdot 10^{-18}$
86	5.08506	$1.5 \cdot 10^{-6}$	$1.6 \cdot 10^{-12}$	$-1.4 \cdot 10^{-18}$
87	5.08507	$-1.2 \cdot 10^{-6}$	$-1.2 \cdot 10^{-12}$	$9.7 \cdot 10^{-19}$
88	5.08507	$1.0 \cdot 10^{-6}$	$9.0 \cdot 10^{-13}$	$-6.5 \cdot 10^{-19}$

◇

Example 4. For the continued fraction

$$\frac{3+1/1^2}{1} + \frac{4+3/2^2}{1} + \frac{3+1/3^2}{1} + \frac{4+3/4^2}{1} + \frac{3+1/5^2}{1} + \cdots \tag{1.4.2}$$

the tails "look more and more like"

$$\frac{3}{1} + \frac{4}{1} + \frac{3}{1} + \frac{4}{1} + \cdots \tag{1.4.3}$$

and

$$\frac{4}{1} + \frac{3}{1} + \frac{4}{1} + \frac{3}{1} + \cdots . \tag{1.4.4}$$

We take convergence for granted, since it will be obvious later. The value of the continued fraction (1.4.3) is then clear: it is the positive root of the quadratic equation

$$u = \cfrac{3}{1 + \cfrac{4}{1+u}},$$

which is 1. The value of the continued fraction (1.4.4) is therefore $v = 4/(1+1) = 2$, and it seems reasonable to choose approximants $S_n(w_n)$ for (1.4.2) with $w_{2n} := 1$ and $w_{2n+1} := 2$. It will be proved later that $\{S_n(w_n)\}$ converges faster than $\{S_n(0)\}$ to the value of (1.4.2), something that is clearly indicated by the table below. In this example the value of the continued fraction is $f = 1.1873788917921$, correctly rounded.

n	$S_n(0)$	$f - S_n(0)$	$S_n(w_n)$	$f - S_n(w_n)$
1	4.00000000	$-2.8 \cdot 10^{0}$	1.33333333	$-1.5 \cdot 10^{-1}$
2	0.69565217	$4.9 \cdot 10^{-1}$	1.18518519	$2.2 \cdot 10^{-3}$
3	1.85579937	$-6.7 \cdot 10^{-1}$	1.20054570	$-1.3 \cdot 10^{-2}$
4	1.00774785	$1.8 \cdot 10^{-1}$	1.18752336	$-1.4 \cdot 10^{-4}$
10	1.18032950	$7.0 \cdot 10^{-3}$	1.18738410	$-5.2 \cdot 10^{-6}$
11	1.19450818	$-7.1 \cdot 10^{-3}$	1.18739871	$-2.0 \cdot 10^{-5}$
12	1.18502153	$2.4 \cdot 10^{-3}$	1.18738030	$-1.4 \cdot 10^{-6}$
13	1.18975231	$-2.4 \cdot 10^{-3}$	1.18738379	$-4.9 \cdot 10^{-6}$
23	1.18738866	$-9.8 \cdot 10^{-6}$	1.18737890	$-7.1 \cdot 10^{-9}$
24	1.18737564	$3.3 \cdot 10^{-6}$	1.18737889	$-7.1 \cdot 10^{-10}$
25	1.18738215	$-3.3 \cdot 10^{-6}$	1.18737889	$-2.0 \cdot 10^{-9}$

◇

This idea of carefully choosing the approximants goes way back, at least to Sylvester ([Sylv69]) who in 1769 claimed that this possibility was one of the main advantages of continued fractions. But it requires that $\{w_n\}$ is properly picked. We can easily make "improper choices": Take *any* sequence $\{\beta_n\}$ from $\widehat{\mathbb{C}}$ and choose

$$w_n := S_n^{-1}(\beta_n).$$

Then $\beta_n = S_n(w_n)$. This shows that we can make $\{S_n(w_n)\}$ converge to anything we want, or diverge, regardless of the convergence behavior of the continued fraction itself. There is no reason for panic, though. This warning looks more serious than it really is. It is one of the wonders of continued fractions that if $S_n(0) \to f$, then $S_n(w_n) \to f$ for almost every sequence $\{w_n\}$. (Section 2.1.2.)

Example 3, and more so Example 4, may perhaps look artificial. But there is a huge family of functions, including some hypergeometric functions and ratios of hypergeometric functions, for which this method of convergence acceleration can be used. (Section 1.3.2 on page 27.) Another matter is that such acceleration is not much use for practical purposes unless one has fast converging, easy to use truncation error bounds. This will be a recurrent theme in this book.

1.2 Regular continued fractions

1.2.1 Introduction

Regular continued fractions are continued fractions of the form $b_0 + \mathbf{K}(1/b_n)$ with $b_n \in \mathbb{N}$, although we also allow $b_0 = 0$. They play an important role in number theory. In this section we shall give a few samples of their power. We first note that A_n and B_n are positive integers and relatively prime since by the determinant formula (1.2.9)

$$\Delta_n := A_{n-1}B_n - A_nB_{n-1} = (-1)^n. \tag{2.1.1}$$

A regular continued fraction always converges, and we have some very useful truncation error bounds:

Theorem 1.5. *A regular continued fraction* $\mathbf{K}(1/b_n)$ *(with* $b_n \in \mathbb{N}$ *for* $n \in \mathbb{N}$*) converges to a value* $0 < f < 1$*. Moreover,*

$$f_2 < f_4 < f_6 < \cdots f_5 < f_3 < f_1\,, \tag{2.1.2}$$

and

$$|f - f_n| < |f_{n+1} - f_n| = \frac{1}{B_nB_{n+1}} \tag{2.1.3}$$

where the canonical denominators $\{B_n\}_{n=1}^{\infty}$ *are strictly increasing natural numbers.*

Proof : The monotonicity of $\{B_k\}$ follows from the recurrence formula $B_n = b_nB_{n-1} + B_{n-2}$. The Euler-Minding formula (1.2.14) on page 7 therefore shows that $\{f_n\}$ are the partial sums for an alternating series with monotonely decreasing terms $1/B_kB_{k-1} \to 0$. The result follows therefore from standard properties of alternating series. □

The alternating character of $(f_{n+1} - f_n)$ also suggests the rational approximation

$$f \approx f_n^* := \tfrac{1}{2}(f_n + f_{n+1}) \tag{2.1.4}$$

which guarantees that

$$|f - f_n^*| < \tfrac{1}{2}|f_n - f_{n+1}|\,. \tag{2.1.5}$$

This is also a good approximation, with a better truncation error bound.

We now turn to the question of how to find the regular continued fraction $b_0 + \mathbf{K}(1/b_n)$ which converges to a given positive number x; i.e., the *regular continued fraction expansion* of x. The most natural method is probably the one we use for $x := \pi$ in the following example.

Example 5. We shall expand $\pi = 3.1415926535897932385\ldots$ in a regular continued fraction:

$$\pi = 3.14159265358979323\ldots = 3 + \cfrac{1}{\cfrac{1}{0.14159265358979323\ldots}}$$

$$= 3 + \frac{1}{7.0625133305931045\ldots} = 3 + \cfrac{1}{7 + \cfrac{1}{\cfrac{1}{0.0625133305931045\ldots}}}$$

$$= 3 + \cfrac{1}{7 + \cfrac{1}{15.996594406685\ldots}} = 3 + \frac{1}{7} {}_{+} \cfrac{1}{15 + \cfrac{1}{\cfrac{1}{0.996594406685\ldots}}}$$

$$= \cdots = 3 + \frac{1}{7} {}_{+} \frac{1}{15} {}_{+} \frac{1}{1} {}_{+} \frac{1}{292} {}_{+} \frac{1}{1} {}_{+} \cdots\,.$$

Of course, we only get the start of the continued fraction in this way since we in essence started with an approximation to π. $\diamond$

The method applied in this example can be described as follows: for a positive number u let $\lfloor u \rfloor$ denote the largest integer $\leq u$. Starting with a positive number u_0 we write

$$u_0 = \lfloor u_0 \rfloor + (u_0 - \lfloor u_0 \rfloor) = \lfloor u_0 \rfloor + \cfrac{1}{\cfrac{1}{u_0 - \lfloor u_0 \rfloor}}.$$

We proceed with

$$u_1 := \frac{1}{u_0 - \lfloor u_0 \rfloor} = \lfloor u_1 \rfloor + (u_1 - \lfloor u_1 \rfloor) = \lfloor u_1 \rfloor + \cfrac{1}{\cfrac{1}{u_1 - \lfloor u_1 \rfloor}},$$

and repeat this process until it stops (if $u_n \in \mathbb{N}$) or to get an infinite regular continued fraction.

Theorem 1.6. *To a given $x > 0$ there exists an essentially unique regular continued fraction $b_0 + \mathbf{K}(1/b_n)$ which converges to x. This continued fraction is terminating if and only if $x > 0$ is a rational number.*

The convergence to x in the non-terminating case follows since

$$x = \lfloor u_0 \rfloor + \frac{1}{\lfloor u_1 \rfloor} + \cdots + \frac{1}{\lfloor u_n \rfloor + (u_n - \lfloor u_n \rfloor)} = S_n(u_n - \lfloor u_n \rfloor),$$

so $x - f_{n-1} = S_n(u_n - \lfloor u_n \rfloor) - S_{n-1}(0) \to 0$ by (1.2.12). The only problem with the uniqueness occurs in the terminating case: the two regular continued fractions

$$b_0 + \frac{1}{b_1} + \frac{1}{b_2} + \cdots + \frac{1}{b_n} + \frac{1}{1} \quad \text{and} \quad b_0 + \frac{1}{b_1} + \frac{1}{b_2} + \cdots + \frac{1}{b_n + 1}$$

are expansions of the same rational number. The next example demonstrates why expanding a rational number in a regular continued fraction by this method terminates, and why the process is equivalent to the euclidean algorithm.

Example 6. The euclidean algorithm ([Eucl56], book 7) attributed to the Greek mathematician Euclid (325 - 265 BC), is a tool to find the greatest common divisor of two integers. Applied to the pair $(71, 47)$ it works as follows:

$$\begin{aligned} 71 &= 1 \cdot 47 + 24, \\ 47 &= 1 \cdot 24 + 23, \\ 24 &= 1 \cdot 23 + 1, \\ 23 &= 23 \cdot 1. \end{aligned}$$

The greatest common divisor is therefore 1. That is, (71, 47) are relatively prime. A rewriting of these equations gives

$$\begin{aligned} \frac{71}{47} &= 1 + \frac{1}{47/24}, \\ \frac{47}{24} &= 1 + \frac{1}{24/23}, \\ \frac{24}{23} &= 1 + \frac{1}{23/1}. \end{aligned}$$

Linking these together we get the (terminating) regular continued fraction expansion of $71/47$:

$$\frac{71}{47} = 1 + \frac{1}{1} + \frac{1}{1} + \frac{1}{23}.$$

Its classical approximants are $f_1 = \frac{2}{1}$, $f_2 = \frac{3}{2}$ and $f_3 = \frac{71}{47}$. That the process has to terminate follows since the left hand side in the euclidean algorithm consists of decreasing positive integers. ◇

1.2.2 Best rational approximation

Example 7. In Example 5 we found the first terms of the regular continued fraction expansion of π. This expansion can be used to produce rational approximations to π. We compute the first canonical numerators and denominators:

n	-1	0	1	2	3	4	5	$\cdots$
b_n	—	3	7	15	1	292	1	$\cdots$
A_n	1	3	22	333	355	103993	104348	$\cdots$
B_n	0	1	7	106	113	33102	33215	$\cdots$

This gives the rational approximations

$$f_0 = 3, \quad f_1 = \frac{22}{7} = 3.142857\ldots, \quad f_2 = \frac{333}{106} = 3.14150943\ldots,$$
$$f_3 = \frac{355}{113} = 3.14159292\ldots, \quad f_4 = \frac{103993}{33102} = 3.1415926530119\ldots$$

to π. They are actually quite good. Indeed, even for such low order approximants we have

$$|\pi - f_3| < 3 \cdot 10^{-7} \quad \text{and} \quad |\pi - f_4| < 6 \cdot 10^{-10}.$$

$\diamond$

It is no coincidence that the approximants are good. Indeed, as the heading of this section indicates, the rational approximants we obtain from regular continued fractions are best in a certain sense:

Theorem 1.7. *Let $\mathbf{K}(1/b_n)$ be a regular continued fraction with value f, and let p and q be two natural numbers such that*

$$\left| f - \frac{p}{q} \right| \le \left| f - \frac{A_n}{B_n} \right|. \tag{2.2.1}$$

Then $q \ge B_n$. If $q = B_n$, then $p = A_n$.

Proof : Assume that $q < B_n$. We shall first prove that then

$$|qf - p| \ge |B_{n-1}f - A_{n-1}| > |B_n f - A_n|. \tag{2.2.2}$$

Let M and N be such that

$$\begin{aligned} A_n M + A_{n-1} N &= p, \\ B_n M + B_{n-1} N &= q. \end{aligned} \tag{2.2.3}$$

Since the determinant of this 2×2-system of linear equations with unknowns M, N is

$$A_n B_{n-1} - B_n A_{n-1} = (-1)^{n-1},$$

such numbers M, N exist uniquely, and they are integers. If $N = 0$, then $A_n/B_n = p/q$, which means that $A_n = p$ and $B_n = q$ since A_n and B_n are relatively prime. Let $N \neq 0$. Then M is either $= 0$ or has opposite sign of N, since otherwise $q > B_n$, contradicting our assumption.

With M and N as given above we get the identity

$$qf - p = M(B_n f - A_n) + N(B_{n-1} f - A_{n-1}) . \tag{2.2.4}$$

Here the two expressions in parentheses have opposite signs by Theorem 1.5, and M, N also have opposite signs (unless $M = 0$). Hence

$$|qf - p| = |M(B_n f - A_n)| + |N(B_{n-1} f - A_{n-1})| ,$$

and, since N is an integer $\neq 0$, we have

$$|qf - p| \geq |B_{n-1} f - A_{n-1}| . \tag{2.2.5}$$

Since always $|B_{n-1} f - A_{n-1}| > |B_n f - A_n|$ (see Problem 18 on page 50), (2.2.2) follows. Simultaneous division, left by q and right by B_n, gives

$$\left| f - \frac{p}{q} \right| > \left| f - \frac{A_n}{B_n} \right| . \tag{2.2.6}$$

This contradicts (2.2.1), and thus $q \geq B_n$.

Finally, assume that $q = B_n$. If $N \neq 0$, then (2.2.4) still holds, and thus, by the same argument as above (2.2.5) holds, which leads to the contradiction (2.2.6). Therefore $N = 0$, and thus $A_n = p$ and $B_n = q$. □

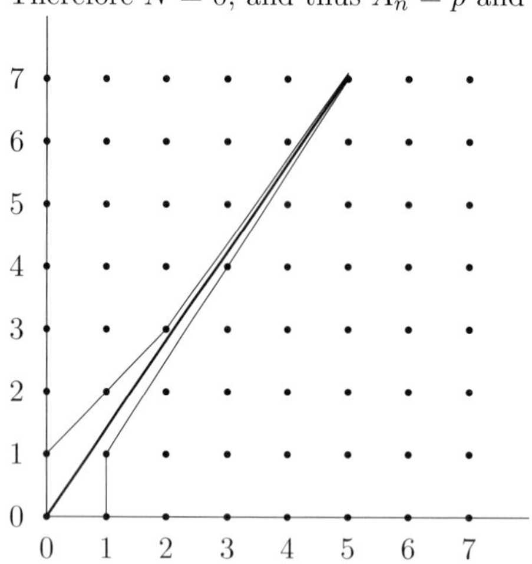

This was first proved by H.J.S. Smith ([Smith60]), but the present proof is due to Lagrange ([Lagr98]). The theorem shows that in order to find a better rational approximation to f than A_n/B_n, we have to increase the denominator. Smith also gave a very nice geometric interpretation of this fact. A description of this interpretation can also be found in a paper by Felix Klein ([Klein95]): Let a be a positive irrational number. We shall let this number a be represented by the ray $y = ax$ from the origin into the first quadrant of a cartesian coordinate system. The lattice point (n, m) with integer n and m represents the fraction m/n. Assume there is a nail in every lattice point, and a rubber band fastened in the points $(1, 0)$ and $(0, 1)$ on the figure. Stretch the rubber band with a pencil following the ray $y = ax$.

The ray $y = ax$ does not pass through any lattice points, but the rubber band will create a polygon with corners at certain nails, say the nails (n_2, m_2), $(n_4, m_4), \ldots$ above the ray and the nails (n_1, m_1), $(n_3, m_3), \ldots$ below the ray, numbered with increasing distance from the origin. Then

$$\frac{m_{2k-2}}{n_{2k-2}} < \frac{m_{2k}}{n_{2k}} < a < \frac{m_{2k+1}}{n_{2k+1}} < \frac{m_{2k-1}}{n_{2k-1}} \quad \text{for all } k.$$

Since there are no nails inside the polygon, these fractions are the best rational approximations to a in our meaning of "best", and therefore they are the approximants of the regular continued fraction expansion of a. That is,

$$m_k/n_k = A_k/B_k.$$

In our figure we have used $a = \sqrt{2}$. The points $(1,1)$, $(2,3)$, $(5,7)$ on the figure corresponds exactly to the first approximants $1/1$, $3/2$, $7/5$ of the regular continued fraction expansion $1 + \mathbf{K}(2/1)$ of $\sqrt{2}$ found in our next example.

Example 8. We shall find rational numbers which approximate the irrational number $\sqrt{2}$. From the equality

$$\sqrt{2} - 1 = \frac{1}{2 + (\sqrt{2} - 1)} \tag{2.2.7}$$

follow the equalities

$$\begin{aligned} \sqrt{2} - 1 &= \frac{1}{2} \, \frac{1}{+\, 2 + (\sqrt{2} - 1)} = \frac{1}{2} \, \frac{1}{+\, 2} \, \frac{1}{+\, 2 + (\sqrt{2} - 1)} \\ &= \frac{1}{2} \, \frac{1}{+\, 2} \, \frac{1}{+\, 2} \, \frac{1}{+\, 2 + (\sqrt{2} - 1)}, \end{aligned}$$

and so on, as far out as we want. Since obviously

$$\sqrt{2} = 1 + \frac{1}{2} \, \frac{1}{+ 2} \, {+ \cdots +} \, \frac{1}{2} \, \frac{1}{+ 2 + (\sqrt{2} - 1)}$$

for any length of the row of dots, it seems to be a good idea to take a look at the approximants of the regular continued fraction

$$1 + \frac{1}{2} \, \frac{1}{+ 2} \, \frac{1}{+ 2} \, {+ \cdots +} \, \frac{1}{2} \, {+ \cdots}.$$

Indeed, by Problem 13 with $x = 1 - \sqrt{2}$ and $y = 1 + \sqrt{2}$, this continued fraction converges to $\sqrt{2}$. To determine its first approximants we can use the recurrence relations $A_n = b_n A_{n-1} + A_{n-2}$, $B_n = b_n B_{n-1} + B_{n-2}$ to obtain

n	-1	0	1	2	3	4	5	$\cdots$
b_n	—	1	2	2	2	2	2	$\cdots$
A_n	1	1	3	7	17	41	99	$\cdots$
B_n	0	1	2	5	12	29	70	$\cdots$

This gives

$$
\begin{aligned}
1+\frac{1}{2} &= 1.5,\\
1+\frac{1}{2}\,\frac{1}{+2} &= \frac{7}{5}=1.4,\\
1+\frac{1}{2}\,\frac{1}{+2}\,\frac{1}{+2} &= \frac{17}{12}=1.4166\ldots,\\
1+\frac{1}{2}\,\frac{1}{+2}\,\frac{1}{+2}\,\frac{1}{+2} &= \frac{41}{29}=1.41379\ldots,\\
1+\frac{1}{2}\,\frac{1}{+2}\,\frac{1}{+2}\,\frac{1}{+2}\,\frac{1}{+2} &= \frac{99}{70}=1.4142857\ldots,
\end{aligned}
$$

which seems to approach $\sqrt{2}$ pretty quickly. Already the fifth one, the last one listed, has an error less than .00008. Indeed, by Theorem 1.5

$$f_4 = 41/29 < \sqrt{2} < f_5 = 99/70,$$

or alternatively

$$
\begin{aligned}
&\sqrt{2} \approx \frac{1}{2}(f_4+f_5) = \frac{5741}{4060} \approx 1.41404,\\
&\left|\sqrt{2}-\frac{5741}{4060}\right| < \frac{1}{2}(f_5-f_4) = \frac{1}{4060} \approx 2.5\cdot 10^{-4}
\end{aligned}
$$

which agrees with the value $\sqrt{2} = 1.41421356\ldots$. ◇

This is a good place for a warning: using identities like (2.2.7) to derive a continued fraction as in this example, must *always* be followed by a check: does the continued fraction really converge to the expected value? The equality

$$-\sqrt{2}-1 = \frac{1}{2+(-\sqrt{2}-1)}$$

will for instance lead to

$$-\sqrt{2} = 1+\frac{1}{2}\,\frac{1}{+2}+\cdots\frac{1}{+2+(-\sqrt{2}-1)},$$

but the continued fraction (still) converges to $\sqrt{2}$. More generally, if we start with any number p, the recurrence

$$p_0 := p, \quad p_{n-1} = a_n/(b_n+p_n) \quad \text{for } n = 1,2,3,\ldots \tag{2.2.8}$$

defines a sequence $\{p_n\}$. So how do we know which p is the value of $\mathbf{K}(a_n/b_n)$, if this continued fraction converges at all? This question will be resolved in the next chapter. (Definition 2.9 on page 63, combined with Corollary 2.7 on page 68, gives one possible answer.)

1.2.3 Solving linear diophantine equations

For regular continued fractions $b_0 + \mathbf{K}(1/b_n)$ the determinant in the determinant formula (1.2.9) on page 7 is

$$A_{n-1}B_n - A_nB_{n-1} = (-1)^n. \tag{2.3.1}$$

This property is essential when terminating regular continued fractions are used to solve linear diophantine equations of the form

$$Mx + Ny = P, \tag{2.3.2}$$

where M, N, P are integers, and we seek integer solutions x, y. The Greek mathematician Diophantos (ca 250 AD) studied such equations, but solving them by means of terminating regular continued fractions, as we shall do in our next example, seems to have been introduced by the Indian mathematician Aryabhata the elder (ca 475 - 550 AD).

Example 9. A shipowner wants to update and upgrade her fleet of ships. After having received bids from different shipyards, she made up her mind and placed a contract for her ships at Beakersheep Shipyard where they offered two types of ships, X-type for $71 \cdot 10^8$ (7.1 billion) US-dollars a piece and Y-type for $47 \cdot 10^8$ (4.7 billion) US-dollars a piece. The number and types of ships were meant to be confidential until further notice. However, an industrial spy finds information on the total price paid, $449 \cdot 10^8$ US-dollars. He also has a price list from the ship yard. How can he determine how many ships of the two types were contracted?

Let x and y be the number of X-ships and Y-ships, respectively. The problem is then to solve the diophantine equation

$$71x + 47y = 449. \tag{2.3.3}$$

We expand the rational number $71/47$ in a terminating regular continued fraction, as done in Example 6, and find that the expansion is

$$\frac{71}{47} = 1 + \frac{1}{1} + \frac{1}{1} + \frac{1}{23}, \quad \begin{array}{ll} A_3 = 71 & A_2 = 3 \\ B_3 = 47 & B_2 = 2 \end{array}$$

and thus, by (2.3.1) with $n := 3$

$$A_3B_2 - A_2B_3 = 71 \cdot 2 + 47 \cdot (-3) = 1.$$

To get 449 on the right hand side of this identity, we multiply by 449

$$71 \cdot 898 + 47 \cdot (-1347) = 449.$$

From this follows that for all integer values t, the pair

$$x = 898 + 47t, \quad y = -1347 - 71t$$

is a solution of (2.3.3). Actually, these pairs are the only solutions. Since we want non-negative solutions we must have

$$-19.10\cdots = -\frac{898}{47} \leq t \leq -\frac{1347}{71} = -18.97\ldots .$$

The only t-value for which this holds is $t = -19$, leading to the solution

$$x = 5, \qquad y = 2.$$

That is, the shipowner had placed a contract for 5 ships of X-type and 2 ships of Y-type. ◇

Of course, to find the integer solution $x = 5$, $y = 2$ in the example above would only take a few seconds of trial and error. But that is only because we can expect the solution to be relatively small. For large solutions the method illustrated by Example 9 is a clear winner.

1.2.4 Grandfather clocks

If cogwheel M with m teeth interlace with cogwheel N with n teeth, then N has rotated m/n times around when M has rotated once. This is the idea of grandfather clocks where cogwheels transmit the movement from the shaft of the second hand to the shaft of the minute hand, and ditto from the minute hand to the hour hand.

But some of these old treasures also show the faces of the moon. The moon takes about 29.53059 nights and days to go from full moon to full moon. The cogwheel for the moon's movement should therefore have $2 \cdot 29.53059$ times as many teeth as the wheel for the hour hand. Therefore $m = 5906118$ teeth and $n = 100000$ teeth could be a solution if it was at all possible and practical to make durable cogwheels with so many teeth. A slightly better solution would be to use an extra cogwheel in between, such that

$$\frac{5906118}{100000} = \frac{6}{10} \cdot \frac{984353}{10000}$$

describes ratios in the two transmissions. But the number of teeth is still unrealistically high.

A better solution still is to find a good rational approximation to the rational number 59.06118 (which is also a rational approximation to the true number). And here the regular continued fractions come into play:

$$\frac{5906118}{100000} = 59 + \cfrac{1}{\cfrac{100000}{6118}} = 59 + \cfrac{1}{16 + \cfrac{2112}{6118}}$$

$$= 59 + \cfrac{1}{16 + \cfrac{1}{\cfrac{6188}{2112}}} = 59 + \cfrac{1}{16 + \cfrac{1}{2 + \cfrac{1894}{2112}}}$$

$$= \cdots = 59 + \frac{1}{16+}\,\frac{1}{2+}\,\frac{1}{1+}\,\frac{1}{8+}\,\frac{1}{1+}\,\frac{1}{2+}\,\frac{1}{4+}\,\frac{1}{1+}\,\frac{1}{6}\,.$$

The choice

$$\frac{m}{n} = 59 + \frac{1}{16} = \frac{945}{16} = 59.0625$$

gives a more practical solution to the problem. Huygens ([Huyg95]) used this method for designing cogwheels for his planetarium.

1.2.5 Musical scales

In a piano the fixed tones with their frequencies f_n are essentially chosen such that

(i) the ratio f_{n+1}/f_n between two neighboring tones is fixed, no matter where they are chosen on the piano. This makes a melody independent of where it was started on the piano.

(ii) the consonant ratios

2 : 1	octave
3 : 2	fifth
4 : 3	fourth
5 : 4	major third
6 : 5	minor third

are present; i.e. there are frequencies such that their ratios give these fractions. This is done because such tones sound beautiful together.

The problem is that it is impossible to achieve both (i) and (ii) exactly, simultaneously. In a piano the octave is divided into 12 intervals, so we get the twelve tones

C, #C, D, #D, E, F, #F, G, #G, A, #A, H

with corresponding frequencies $f_1,\ f_2, \ldots, f_{12}$, and $f_{13} := 2f_1$ is the next C. If $q := f_{n+1}/f_n$ is constant, this means that

$$\frac{f_{13}}{f_1} = 2 = \frac{f_{13}}{f_{12}} \cdot \frac{f_{12}}{f_{11}} \cdots \frac{f_2}{f_1} = q^{12}$$

and thus $q = 2^{1/12}$ which is an *irrational* number! We can therefore not get the consonant ratios we asked for exactly, except for the octave. We only get approximations to the consonant ratios. For instance

$$\begin{aligned}
\frac{f_8}{f_1} &= q^7 = 2^{7/12} \approx 1.498\cdots \approx \frac{3}{2} = 1.5 \\
\frac{f_6}{f_1} &= q^5 = 2^{5/12} \approx 1.335\cdots \approx \frac{4}{3} = 1.333\ldots \\
\frac{f_5}{f_1} &= q^4 = 2^{4/12} \approx 1.260\cdots \approx \frac{5}{4} = 1.250 \\
\frac{f_4}{f_1} &= q^3 = 2^{3/12} \approx 1.189\cdots \approx \frac{6}{5} = 1.200
\end{aligned}$$

For a mathematician, a natural question is then: given our two concerns (i) and (ii), what is an optimal number of tones in a octave? That is, for which $N \in \mathbb{N}$ will there exist integers k such that $(2^{1/N})^k$ are good approximations to the consonant ratios? This is a question of rational approximation: we want to choose N such that $2^{1/N}$ is "almost rational". In particular we want, say,

$$(2^{1/N})^k = 2^{k/N} \approx \frac{3}{2}; \quad \text{i.e.,} \quad \frac{k}{N} \approx \log_2 \frac{3}{2} = \frac{\operatorname{Ln} 3/2}{\operatorname{Ln} 2}$$

for some $k, N \in \mathbb{N}$ where $\operatorname{Ln} x = \log_e x$ is the natural logarithm. In other words, k/N should be a good rational approximation to the irrational number $\log_2 \frac{3}{2}$. Therefore we expand $\log_2 \frac{3}{2} \approx 0.5849625007$ in a regular continued fraction ([Lore06]). Of course, the regular continued fraction expansion of $\log_2 \frac{3}{2}$ never stops, whereas the regular continued fraction expansion of the rational number 0.5849625007 terminates. However, we only want to use the first few terms of the continued fraction anyway, since we do not want to have thousands of tones in one octave. We get

$$\log_2 \frac{3}{2} \approx 0.5846925007 = \frac{1}{1}\frac{}{+}\frac{1}{1}\frac{}{+}\frac{1}{2}\frac{}{+}\frac{1}{2}\frac{}{+}\frac{1}{3}\frac{}{+}\frac{1}{1}\frac{}{+}\cdots .$$

The approximation

$$\log_2 \frac{3}{2} \approx \frac{1}{1}\frac{}{+}\frac{1}{1}\frac{}{+}\frac{1}{2} = \frac{3}{5} = 0.60000$$

says: 5 equally spaced tones in an octave with (f_1, f_4) as the fifth is reasonably good. Taking one more term of the continued fraction gives

$$\log_2 \frac{3}{2} \approx \frac{1}{1}\frac{}{+}\frac{1}{1}\frac{}{+}\frac{1}{2}\frac{}{+}\frac{1}{2} = \frac{7}{12} \approx 0.58333 .$$

Hah!! That is exactly what we have on a piano. 12 tones in an octave, and for instance C–G makes up a fifth. What is the next good choice? We let the continued fraction come up with an answer:

$$\log_2 \frac{3}{2} \approx \frac{1}{1}\frac{}{+}\frac{1}{1}\frac{}{+}\frac{1}{2}\frac{}{+}\frac{1}{2}\frac{}{+}\frac{1}{3} = \frac{24}{41} \approx 0.58537 .$$

Interesting! A piano with 41 equally spaced tones in an octave will give very good approximations to the consonant ratio 3:2, with (f_1, f_{25}) as the fifth.

1.3 Rational approximation to functions

1.3.1 Expansions of functions

Example 10. In this book $\sqrt{\ldots}$ shall always denote the principal branch of the square root; that is, $-\frac{\pi}{2} < \arg\sqrt{\ldots} \le \frac{\pi}{2}$. For $z \in D := \{z \in \mathbb{C};\, |\arg(1+z)| < \pi\}$, the identity

$$\sqrt{1+z}-1 = \frac{z}{2+(\sqrt{1+z}-1)}$$

leads to the identities

$$\sqrt{1+z}-1 = \frac{z}{2} \frac{z}{+2} + \cdots + \frac{z}{2} + \frac{z}{2+(\sqrt{1+z}-1)},$$

just as in Example 8. This suggests the continued fraction

$$\frac{z}{2} \frac{z}{+2} + \cdots + \frac{z}{2} + \cdots. \tag{3.1.1}$$

But the same warning still applies: we do not know off-hand that this continued fraction converges to $(\sqrt{1+z}-1)$. However, by Problem 13 on page 49 with $x = 1+\sqrt{1+z}$ and $y = 1-\sqrt{1+z}$ we get its classical approximants $f_n(z)$ on closed form. They show that (3.1.1) converges to $\sqrt{1+z}-1$ for all $z \in D$. The approximants of (3.1.1) are rational functions,

$$f_1(z) = \frac{z}{2}, \quad f_2(z) = \frac{z}{2} \frac{z}{+2} = \frac{2z}{z+4}, \quad f_3(z) = \frac{z}{2} \frac{z}{+2} \frac{z}{+2} = \frac{z^2+4z}{4z+8},$$

and so on. Hence $f_n(z)$ is a rational approximation to the function $\sqrt{1+z}-1$ in D. Strictly speaking, the continued fraction (3.1.1) is not defined for $z=0$, but we can make the additional definition that it has the value 0 for $z=0$. ◇

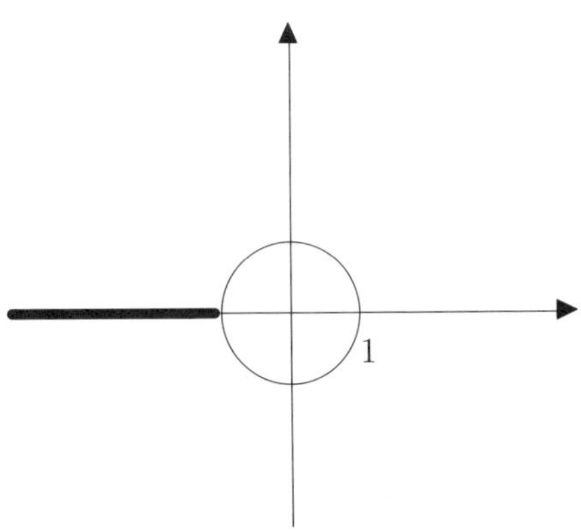

Continued fraction expansions are less known than power series expansions. The Taylor series expansion of $\sqrt{1+z}-1$ at 0 is

$$\sqrt{1+z}-1 = \sum_{k=1}^{\infty} \binom{\frac{1}{2}}{k} z^k.$$

It converges for $|z| < 1$ and diverges for $|z| > 1$. Its partial sums are

$$\sigma_1(z) = \frac{1}{2}z, \; \sigma_2(z) = \frac{1}{2}z - \frac{1}{8}z^2,$$
$$\sigma_3(z) = \frac{1}{2}z - \frac{1}{8}z^2 + \frac{1}{16}z^3, \ldots,$$

and they are polynomial approximations to $\sqrt{1+z}\,-1$ for z in the unit disk $\mathbb{D} := \{z \in \mathbb{C};\ |z| < 1\}$. The approximants $f_n(z)$ in Example 10 approximate $\sqrt{1+z}-1$ in a much larger region, $D := \{z \in \mathbb{C};\ |\arg(1+z)| < \pi\}$. In our next example we compare their degree of approximation.

Example 11. We compute the two types of approximants $\sigma_n(z)$ and $f_n(z)$ for $\sqrt{1+z}-1$ for a given z–value to see what happens. We choose $z := 0.96$, in which case the value of the function is exactly 0.4. The table below shows $\sigma_n(0.96)$ and $f_n(0.96)$, all correctly rounded in the 4th decimal place:

n	1	2	3	4	5	6	7
σ_n	.4800	.3648	.4201	.3869	.4092	.3932	.4053
f_n	.4800	.3871	.4022	.3996	.4001	.4000	.4000

Or, if we rather list the index $n(k)$ which is the smallest index for which the approximation is correct, rounded to k decimals, for all $n \geq n(k)$:

	$n(3)$	$n(6)$	$n(12)$	$n(20)$
σ_n	23	129	424	849
f_n	4	8	16	26

This of course does not prove anything, but it certainly suggests that the continued fraction is better (converges faster) than the power series expansion. (It is, however, only fair to say, that such a comparison, based merely upon the order n of the approximant, does not always give the correct picture. Essential in the comparison is the resources needed for the computation.)

Even more flattering for the continued fraction expansion is the choice $z = 3$ for which $\sqrt{1+z}-1 = 1$. In this case it does not make sense to compute power series approximants, since we know that the power series diverges. In the next table the first seven continued fraction approximants are listed, correctly rounded in the 4th decimal place.

n	1	2	3	4	5	6	7
f_n	1.5000	.8571	1.0500	.9836	1.0055	.9982	1.0006

◇

Example 12. We use the notation $\operatorname{Ln} z$ or $\operatorname{Ln}(z)$ to denote the principal branch of the natural logarithm; that is, $-\pi < \operatorname{Im} \operatorname{Ln} z \leq \pi$. The continued fraction

$$\frac{z}{1+}\,\frac{z/2}{1}\,+\,\frac{z/6}{1}\,+\,\frac{2z/6}{1}\,+\,\frac{2z/10}{1}\,+\,\frac{3z/10}{1}\,+\cdots \tag{3.1.2}$$

converges to $\operatorname{Ln}(1+z)$ for all z in the cut plane $D := \{z \in \mathbb{C};\ |\arg(1+z)| < \pi\}$ when we set its value to 0 for $z = 0$. More precisely,

$$\operatorname{Ln}(1+z) = \frac{z}{1+}\,\frac{a_2 z}{1}\,+\,\frac{a_3 z}{1}\,+\cdots \qquad \text{for } z \in D,$$

where for all $k \geq 1$

$$a_{2k} := \frac{k}{2(2k-1)}\,, \qquad a_{2k+1} := \frac{k}{2(2k+1)}\,.$$

Therefore its classical approximants $f_n(z)$ (which are rational functions) approximate the value of $\mathrm{Ln}(1+z)$ for $z \in D$. Right now we shall not worry about *how* one gets this continued fraction expansion or prove its convergence.

We shall compare this rational approximation to the polynomial approximation obtained from the power series expansion

$$\mathrm{Ln}(1+z) = z - \frac{z^2}{2} + \frac{z^3}{3} - \frac{z^4}{4} + \cdots$$

which converges in the unit disk and diverges for $|z| > 1$. Let us take $z = 1$. The series then converges to Ln(2), but very slowly. The first seven continued fraction approximants are listed in the table below, correctly rounded. The value is $\mathrm{Ln}(2) = .69314718$, correctly rounded.

n	1	2	3	4	5	6	7
f_n	1.000000	.666667	.700000	.692308	.693333	.693121	.693152

In order to get the polynomial approximation with the same accuracy we need $n > 10^5$ terms of the power series.

Let us also try a z-value where the series does not converge, for instance $z = 3$. In the next table the first 7 classical approximants f_n are listed, correctly rounded. The value is now $\mathrm{Ln}(4) = 2\mathrm{Ln}(2)$, so a better idea is to approximate Ln(2), and then multiply by 2. But as an experiment we try $\mathrm{Ln}(1+3) \approx 1.38629436$.

Since $a_n \to \frac{1}{4}$, the approximants $S_n(w(z))$ (probably) works better if $w(z)$ is the value of the periodic continued fraction $\mathbf{K}(\frac{z}{4}/1)$; i.e., $w(z) = \frac{1}{2}(\sqrt{1+z} - 1)$. The approximants are no longer rational, but they give good approximations. For $z := 3$, $w(3) = \frac{1}{2}$, and $f_n^* := S_n(\frac{1}{2})$, correctly rounded, is shown in the next line of the table.

n	1	2	3	4	5	6	7
f_n	3.00000	1.20000	1.50000	1.36363	1.39726	1.38367	1.38744
f_n^*	2.00000	1.50000	1.41176	1.39286	1.38806	1.38679	1.38644

Even better approximants $S_n(w_n(z))$ will be suggested in Chapter 5. ◇

1.3.2 Hypergeometric functions

For $n \in \mathbb{N}$ and $a \in \mathbb{C}$, the *Pochhammer symbol* $(a)_n$ is defined to be

$$(a)_0 := 1, \quad (a)_n := (a)_{n-1} \cdot (a+n-1) = \prod_{k=0}^{n-1} (a+k)\,. \tag{3.2.1}$$

For given a, b, $c \in \mathbb{C}$ with $c \neq 0, -1, -2, \ldots$, the *hypergeometric function* ${}_2F_1(a,b;c;z)$ is the analytic function with power series expansion

$$\sum_{n=0}^{\infty} \frac{(a)_n (b)_n}{(c)_n} \frac{z^n}{n!} = 1 + \frac{ab}{c}\frac{z}{1!} + \frac{a(a+1)b(b+1)}{c(c+1)}\frac{z^2}{2!} + \cdots \tag{3.2.2}$$

at 0. Many useful special functions are special cases of ${}_2F_1(a,b;c;z)$. If we assume that also a and b are different from 0 and the negative integers, then (3.2.2) is an infinite power series whose radius of convergence is 1. The following formal identities can be established from (3.2.2) by comparing the power series on both sides term by term:

$$\begin{aligned}
{}_2F_1(a,b;c;z) &= {}_2F_1(a,b+1;c+1;z) \\
&\qquad - \frac{a(c-b)}{c(c+1)} z\, {}_2F_1(a+1,b+1;c+2;z) \\
{}_2F_1(a,b+1;c+1;z) &= {}_2F_1(a+1,b+1;c+2;z) \\
&\qquad - \frac{(b+1)(c-a+1)}{(c+1)(c+2)} z\, {}_2F_1(a+1,b+2;c+3;z).
\end{aligned}$$

Assuming that we avoid zeros in the denominators, this can be written

$$\begin{aligned}
\frac{{}_2F_1(a,b;c;z)}{{}_2F_1(a,b+1;c+1;z)} &= 1 + \cfrac{\cfrac{-a(c-b)z}{c(c+1)}}{\cfrac{{}_2F_1(a,b+1;c+1;z)}{{}_2F_1(a+1,b+1;c+2;z)}} \\
\frac{{}_2F_1(a,b+1;c+1;z)}{{}_2F_1(a+1,b+1;c+2;z)} &= 1 + \cfrac{\cfrac{-(b+1)(c-a+1)z}{(c+1)(c+2)}}{\cfrac{{}_2F_1(a+1,b+1;c+2;z)}{{}_2F_1(a+1,b+2;c+3;z)}}.
\end{aligned}$$

Observe that the denominator on the right hand side of the first equality is equal to the left hand side of the second equality. Furthermore, the denominator on the right hand side of the second equality coincides with the left hand side of the first one, if in the former a is replaced by $a+1$, b is replaced by $b+1$ and c by $c+2$ in all places. Hence, we have

$$\begin{aligned}
\frac{{}_2F_1(a,b;c;z)}{{}_2F_1(a,b+1;c+1;z)} &= 1 + \cfrac{a_1 z}{\cfrac{{}_2F_1(a,b+1;c+1;z)}{{}_2F_1(a+1,b+1;c+2;z)}} \\
&= 1 + \cfrac{a_1 z}{1 + \cfrac{a_2 z}{\cfrac{{}_2F_1(a+1,b+1;c+2;z)}{{}_2F_1(a+1,b+2;c+3;z)}}} \\
&= 1 + \frac{a_1 z}{1} \,{+}\, \frac{a_2 z}{1} \,{+}\, \cfrac{a_3 z}{\cfrac{{}_2F_1(a+1,b+2;c+3;z)}{{}_2F_1(a+2,b+2;c+4;z)}}
\end{aligned}$$

and so on, where

$$a_{2n+1} = \frac{-(a+n)(c-b+n)}{(c+2n)(c+2n+1)} \quad \text{for } n \geq 0, \qquad a_{2n} = \frac{-(b+n)(c-a+n)}{(c+2n-1)(c+2n)} \quad \text{for } n \geq 1, \tag{3.2.3}$$

and we get the continued fraction

$$1 + \frac{a_1 z}{1} + \frac{a_2 z}{1} + \frac{a_3 z}{1} + \cdots. \tag{3.2.4}$$

Again we do not have any guarantee off-hand that this continued fraction converges, let alone that it converges to this ratio of hypergeometric functions. However, since $a_n \to -\frac{1}{4}$, it is a consequence of Theorem 4.13 on page 188 that (3.2.4) converges in the cut plane $D := \{z \in \mathbb{C};\, 0 < \arg(z-1) < 2\pi\}$. That its value indeed is

$$\frac{{}_2F_1(a,b;c;z)}{{}_2F_1(a,b+1;c+1;z)} \tag{3.2.5}$$

will be proved in volume 2. This means that its classical approximants provide rational approximation to this ratio. This continued fraction expansion was developed by Gauss ([Gauss13]).

A particularly interesting example occurs for the choice $b = 0$. Since $F(a,0;c;z) \equiv 1$, we get

$${}_2F_1(a,1;c+1;z) = \frac{1}{1} + \frac{a_1 z}{1} + \frac{a_2 z}{1} + \frac{a_3 z}{1} + \cdots \tag{3.2.6}$$

where

$$a_{2n+1} = \frac{-(a+n)(c+n)}{(c+2n)(c+2n+1)}, \quad a_{2n} = \frac{-n(c-a+n)}{(c+2n-1)(c+2n)} \tag{3.2.7}$$

for $n = 0, 1, 2, \ldots$. Examples of such functions are for instance

$$\begin{aligned} {}_2F_1(1,1;2;z) &= -z^{-1}\mathrm{Ln}(1-z) \\ {}_2F_1(\tfrac{1}{2},1;\tfrac{3}{2};-z^2) &= z^{-1}\mathrm{Arctan}\, z \quad \text{(principal part)} \\ z \cdot {}_2F_1(\tfrac{1}{n},1;1+\tfrac{1}{n};-z^n) &= \int_0^z \frac{dt}{1+t^n}. \end{aligned} \tag{3.2.8}$$

Since $a_n \to -\frac{1}{4}$, the nth tail of (3.2.4) looks more and more like the periodic continued fraction

$$\frac{-z/4}{1} + \frac{-z/4}{1} + \frac{-z/4}{1} + \cdots$$

which converges to $w(z) := (\sqrt{1-z} - 1)/2$ for $z \in D$. Therefore, we are not surprised to find than $S_n(w(z))$ converges considerably faster to the value of (3.2.4) that $f_n = S_n(0)$. In Chapter 5 we also suggest faster converging approximants $S_n(w_n)$.

1.4 Correspondence between power series and continued fractions

1.4.1 From power series to continued fractions

A continued fraction of the form

$$b_0 + \frac{a_1 z}{1} \underset{+}{} \frac{a_2 z}{1} \underset{+\cdots+}{} \frac{a_n z}{1} \underset{+\cdots}{} ; \qquad 0 \neq a_n \in \mathbb{C} \tag{4.1.1}$$

is called a *regular C-fraction.* (This concept has no particular connection to regular continued fractions.) Regular C-fractions often work better than power series as far as *speed of convergence* and *domain of convergence* are concerned. Hence it is of interest to go from a power series to a continued fraction of this particular form. One way of doing this was demonstrated in the previous section. A more primitive way is the method of *successive substitutions* ([Lamb61]) due to Lambert (1728 - 1777), a colleague of Euler and Lagrange in Berlin. We illustrate this method by an example:

Example 13. We shall compute the circumference L of the ellipse

$$\frac{x^2}{a^2} + \frac{y^2}{b^2} = 1\,, \qquad a \geq b \geq 0\,, \quad a > 0\,. \tag{4.1.2}$$

The well known arc length formula leads to the elliptic integral

$$L = 4a \int_0^{\pi/2} \sqrt{1 - \varepsilon^2 \sin^2 \theta}\, d\theta$$

where $\varepsilon := \sqrt{a^2 - b^2}/a$ is the eccentricity of the ellipse, and thus $b^2 = a^2(1-\varepsilon^2)$. By setting

$$t := \left(\frac{a-b}{a+b}\right)^2, \tag{4.1.3}$$

and expanding the integrand in a series and integrate, we get

$$L = \pi(a+b) \sum_{n=0}^{\infty} \binom{1/2}{n}^2 t^n = \pi(a+b)\left(1 + \frac{t}{2^2} + \frac{t^2}{2^6} + \frac{t^3}{2^8} + \frac{25t^4}{2^{14}} + \cdots\right), \tag{4.1.4}$$

([Hütte55], [LoWa85]). One way of finding approximate values for L is to truncate the series. But we can also transform the series into a continued fraction:

$$1 + \frac{t}{2^2} + \frac{t^2}{2^6} + \frac{t^3}{2^8} + \frac{25\,t^4}{2^{14}} + \cdots = 1 + \frac{t/2^2}{\left(1 + \frac{t}{2^4} + \frac{t^2}{2^6} + \frac{25\,t^3}{2^{12}} + \cdots\right)^{-1}}$$

$$= 1 + \frac{t/4}{1 - \frac{t}{16} - \frac{3\,t^2}{256} - \frac{9\,t^3}{2048} + \cdots} = 1 + \frac{t/4}{1} + \frac{-t/16}{\left(1 + \frac{3\,t}{16} + \frac{9\,t^2}{128} + \cdots\right)^{-1}}$$

$$= 1 + \frac{t/4}{1} + \frac{-t/16}{1 - \frac{3t}{16} - \frac{9t^2}{256} + \cdots} = 1 + \frac{t/4}{1} + \frac{-t/16}{1} + \frac{-3t/16}{\left(1 + \frac{3t}{16} + \cdots\right)^{-1}}$$

$$= 1 + \frac{t/4}{1} + \frac{-t/16}{1} + \frac{-3t/16}{1 - 3t/16 + \cdots}.$$

That is, we have obtained the first fraction terms of a continued fraction expansion

$$1 + \frac{t/4}{1} + \frac{-t/16}{1} + \frac{-3t/16}{1} + \frac{-3t/16}{1} + \cdots \tag{4.1.5}$$

of $L/(\pi(a+b))$. We shall prove in volume 2 that this continued fraction actually converges to $L/(\pi(a+b))$. Therefore its classical approximants provide rational approximations to L:

$$L \approx \pi(a+b) f_n(t)\,.$$

This approximation is surprisingly accurate, even for moderate n. For $n = 2$, simple as it is, it has an error less than 3 mm for an ellipse with size and eccentricity as the orbit of the planet Mercury. For $n = 3$ it has for the same ellipse an error roughly $= 1/10$ of the wave length of blue light. In the "flat" case $b = 0$, which is likely to be the "worst" case ($t = 1$), the exact value of L is $4a$. The approximate formulas give

$$L \approx \begin{cases} \pi a\, f_0(1) & = 3.1416\, a \\ \pi a\, f_1(1) & = 3.9270\, a \\ \pi a\, f_2(1) & = 3.9794\, a \\ \pi a\, f_3(1) & = 3.9924\, a \\ \pi a\, f_4(1) & = 3.9964\, a\,, \end{cases}$$

correctly rounded in the 4th decimal place.

However, we can do better. In the continued fraction (4.1.5) we pretend that *all* fraction terms from the second one are equal to

$$\frac{-t/16}{1},$$

i.e. we replace the continued fraction (4.1.5) by

$$1 + \frac{t/4}{1} + \frac{-t/16}{1} + \frac{-t/16}{1} + \frac{-t/16}{1} + \cdots.$$

(This is of course only an experiment, but one can prove that it will lead to a good approximation.) This periodic continued fraction has the value

$$S_1(w) = 1 + \frac{t/4}{1 + w}, \tag{4.1.6}$$

where w is the value of its tail

$$w = \frac{-t/16}{1} \underset{+}{} \frac{-t/16}{1} \underset{+}{} \frac{-t/16}{1} \underset{+}{} \cdots = \frac{1}{2}\left(\sqrt{1-t/4}-1\right). \qquad (4.1.7)$$

Here we have used that the continued fraction $\mathbf{K}(\frac{z}{4}/1)$ has the value $(\sqrt{1+z}-1)/2$, and replaced z by $z := -t/4$ with $0 < t \leq 1$. Since therefore

$$S_1(w) = 1 + \frac{t/4}{1+(\sqrt{1-t/4}-1)/2} = 3 - \sqrt{4-t}\,,$$

we get the formula

$$L \approx \pi(a+b)(3-\sqrt{4-t}). \qquad (4.1.8)$$

This formula was first found by Ramanujan ([Bern78]). For an ellipse of size and eccentricity as the orbit of Mercury this formula has an error less than 2 mm. For the degenerate case ("flat" case) it gives the value $3.9834a$.

The next formula uses the observation that the continued fraction (4.1.5) has two equal fraction terms

$$\frac{-3t/16}{1}.$$

If we pretend that all subsequent ones also are $(-3t/16)/1$ (again merely an experiment), the first tail of (4.1.5) would be

$$w = \frac{1}{3}\left(\frac{-3t/16}{1} \underset{+}{} \frac{-3t/16}{1} \underset{+}{} \frac{-3t/16}{1} \underset{+}{} \frac{-3t/16}{1} \underset{+}{} \cdots\right),$$

which by (4.1.7) has the value

$$w = \frac{1}{6}(\sqrt{1-3t/4}-1).$$

This suggests the approximate formula

$$L \approx \pi(a+b)\left(1 + \frac{t/4}{1+(\sqrt{1-3t/4}-1)/6}\right),$$

or, rewritten in a nicer form

$$L \approx \pi(a+b)\left(1 + \frac{3t}{10+\sqrt{4-3t}}\right). \qquad (4.1.9)$$

For the "flat" case it gives the value $3.9984a$. This was published in [JaWa85]. To our delight it was pointed out by Almquist and Berndt ([AlBe88]) that also this was one of Ramanujan's formulas, although it was not known *how* he established it. It may therefore well be assumed that the method shown in the present section was the one he used. ◇

1.4.2 From continued fractions to power series

Example 14. Sometimes we want to go in the opposite direction, i.e. from continued fraction to power series. We shall use our continued fraction expansion for $\mathrm{Ln}(1+z)$ as an example; i.e.,

$$\mathrm{Ln}(1+z) = \frac{z}{1} \underset{+}{} \frac{z/2}{1} \underset{+}{} \frac{z/6}{1} \underset{+}{} \frac{2z/6}{1} \underset{+}{} \frac{2z/10}{1} \underset{+}{} \frac{3z/10}{1} \underset{+}{} \cdots .$$

We look at its first classical approximants and their power series expansions at 0:

$$f_1(z) = \frac{z}{1} = \underline{z} + 0z^2 + 0z^3 + 0z^4 + \cdots$$

$$f_2(z) = \frac{z}{1+z/2} = \underline{z - \frac{z^2}{2}} + \frac{z^3}{4} - \frac{z^4}{8} + \cdots$$

$$f_3(z) = \frac{z}{1 + \cfrac{z/2}{1+z/6}} = \underline{z - \frac{z^2}{2} + \frac{z^3}{3}} - \frac{2}{9}z^4 + \cdots$$

$$f_4(z) = \frac{z}{1 + \cfrac{z/2}{1 + \cfrac{z/6}{1+z/3}}} = \underline{z - \frac{z^2}{2} + \frac{z^3}{3} - \frac{z^4}{4}} + \cdots$$

The underlined terms coincide with terms in the power series for $\mathrm{Ln}(1+z)$. Observe that the agreement increases with the order of the approximants. It can be proved (and it will be proved in volume 2) that this continues. It is called *correspondence* between the power series and the continued fraction expansion at 0. ◇

1.4.3 One fraction, two series; analytic continuation

Example 15. The identity

$$z = \frac{z}{1-z+z}$$

used repeatedly leads to the identities

$$z = \frac{z}{1-z} \underset{+}{} \frac{z}{1-z+z}, \qquad z = \frac{z}{1-z} \underset{+}{} \frac{z}{1-z} \underset{+}{} \frac{z}{1-z+z},$$

and generally

$$z = \frac{z}{1-z} \underset{+}{} \frac{z}{1-z} \underset{+}{} \cdots \underset{+}{} \frac{z}{1-z+z}.$$

Similarly, the identity

$$-1 = \frac{z}{1-z-1}, \qquad z \neq 0,$$

leads to

$$-1 = \frac{z}{1-z} + \frac{z}{1-z} + \cdots + \frac{z}{1-z-1} .$$

Inspired by these two identities we look at the continued fraction

$$\frac{z}{1-z} + \frac{z}{1-z} + \cdots + \frac{z}{1-z} + \cdots . \tag{4.3.1}$$

For simplicity we assume that $z \neq -1$. Then its classical approximants $f_n(z)$ can be written

$$\begin{aligned} f_1(z) &= \frac{z}{1-z} = \frac{z(1+z)}{1-z^2} , \\ f_2(z) &= \frac{z}{1-z} + \frac{z}{1-z} = \frac{z(1-z)}{1-z+z^2} = \frac{z(1-z^2)}{1+z^3} , \\ f_3(z) &= \frac{z}{1-z} + \frac{z}{1-z} + \frac{z}{1-z} = \frac{z(1+z^3)}{1-z^4} , \end{aligned}$$

and by Problem 13 on page 49 with $x = -z$ and $y = 1$,

$$f_n(z) = \frac{z(1-(-z)^n)}{1-(-z)^{n+1}} = \frac{z+(-z)^{n+1}}{1-(-z)^{n+1}} .$$

We therefore distinguish between two cases:

$\mathbf{0 < |z| < 1}$: The continued fraction converges to z. Since

$$f_n(z) = z + (-z)^{n+1} + \text{higher powers of } z$$

it corresponds at 0 to the series

$$z + 0z^2 + 0z^3 + \cdots$$

$\mathbf{|z| > 1}$: The continued fraction converges to -1. Since

$$f_n(z) = -1 + (-z)^{-n} + \text{higher powers of } z^{-1}$$

it corresponds at ∞ to the series

$$-1 + 0z^{-1} + 0z^{-2} + 0z^{-3} + \cdots$$

◇

This example shows that one and the same continued fraction expansion may converge to two different functions in two different regions and correspond to two different series at two different points (here 0 and ∞). There also exist non-trivial analogues, where one continued fraction simultaneously represents two different analytic functions by convergence and correspondence. But we can say more:

1. The approximants $S_n(z)$ in Example 15 are all equal to z. This implies two things: The convergence to z in $|z| < 1$ is *accelerated* (bull's eye, the value is hit right away), and the convergence to z is extended also to the region $|z| \geq 1$, i.e. we have an *analytic continuation* of the limit function in $|z| < 1$ to the whole plane.

2. The approximants $S_n(-1)$ are all equal to -1. This implies two things: Acceleration of convergence to -1 in $|z| > 1$ (again bull's eye), and the convergence to -1 is extended to $0 < |z| \leq 1$, i.e. we have an analytic continuation of the limit function in $|z| > 1$ to the whole plane, minus the origin.

Also these properties have their non-trivial analogues. For instance, if

$$a_n(z) \to z \quad \text{and} \quad b_n(z) =\!\!\to 1 - z \tag{4.3.2}$$

then the continued fraction $\mathbf{K}(a_n(z)/b_n(z))$ converges to one function for $|z| < 1$ and to another function for $|z| > 1$. We have already seen that $S_n(z)$ (probably) converges faster to its value than its classical approximants $S_n(0)$. If the convergence in (4.3.2) is fast enough, then $S_n(z)$ actually converges in a larger domain, and thus provides analytic continuation under proper conditions. So also for $S_n(-1)$. This idea was presented and developed further in [Waad66], [Waad67], [ThWa80b], [Jaco88], [Lore93]. We return to this idea in volume 2.

1.4.4 Padé approximation

Problem: For a given (formal) power series

$$L(z) := \sum_{k=0}^{\infty} c_k z^k$$

and given non-negative integers m and n, find the rational function

$$R_{m/n}(z) := \frac{P_{m/n}(z)}{Q_{m/n}(z)}$$

whose Taylor series at $z = 0$ agrees with $L(z)$ as far out as possible, when $P_{m/n}$ and $Q_{m/n}$ are polynomials of degree $\leq m$ and $\leq n$ respectively.

Solutions to this problem are essentially what is called *Padé approximants* (in one of the two possible definitions). The idea is due to Stirling ([Stir30]) who probably based his work on a paper by Bernoulli ([Berno78]). For finer distinctions we refer to the extensive literature on Padé approximation, such as for instance [BaGM96]. Padé approximants for a given series $L(z)$ are usually presented in an array called a *Padé table*:

$m \backslash n$	0	1	2	3	4	$\cdots$
0	$R_{0/0}$	$R_{0/1}$	$R_{0/2}$	$R_{0/3}$	$R_{0/4}$	$\cdots$
1	$R_{1/0}$	$R_{1/1}$	$R_{1/2}$	$R_{1/3}$	$R_{1/4}$	$\cdots$
2	$R_{2/0}$	$R_{2/1}$	$R_{2/2}$	$R_{2/3}$	$R_{2/4}$	$\cdots$
3	$R_{3/0}$	$R_{3/1}$	$R_{3/2}$	$R_{3/3}$	$R_{3/4}$	$\cdots$
4	$R_{4/0}$	$R_{4/1}$	$R_{4/2}$	$R_{4/3}$	$R_{4/4}$	$\cdots$
$\vdots$	$\vdots$	$\vdots$	$\vdots$	$\vdots$	$\vdots$	$\ddots$

The 0-column $\{R_{n/0}\}$ is just the partial sums of $L(z)$. (Some authors have chosen to use the transposed table where the numerator-degree increases in each row and the denominator-degree increases column-wise.) For practical computation of Padé approximants different algorithms are available. We refer to ([CuWu87], sect 2.3) and references therein. Some key words deserve to be mentioned: the *qd*-algorithm, the Viskovatov algorithm, Gragg's algorithm and the ε-algorithm.

Example 16. Let

$$L(z) := 1 + \sum_{n=1}^{\infty}(-1)^{n+1}\frac{z^n}{2n-1} = 1 + z - \frac{z^2}{3} + \frac{z^3}{5} - \frac{z^4}{7} + \frac{z^5}{9} - \frac{z^6}{11} + \cdots .$$

The upper left corner of the Padé table is then

$m \backslash n$	0	1	2	$\cdots$
0	1	$\dfrac{1}{1-z}$	$\dfrac{3}{3-3z+4z^2}$	$\cdots$
1	$1+z$	$\dfrac{3+4z}{3+z}$	$\dfrac{15+21z}{15+6z-z^2}$	$\cdots$
2	$\dfrac{3+3z-z^2}{3}$	$\dfrac{15+24z+4z^2}{15+9z}$	$\dfrac{105+195z+64z^2}{105+90z+9z^2}$	$\cdots$
$\vdots$	$\vdots$	$\vdots$	$\vdots$	$\ddots$

For instance

$$R_{2/1}(z) = \frac{15+24z+4z^2}{15+9z} \sim 1 + z - \frac{z^2}{3} + \frac{z^3}{5} - \frac{3z^4}{25} + \frac{9z^5}{125} + \cdots$$

which agrees with $L(z)$ up to and including the term of order 3. It is quite easy to see that $R_{m/n}(z)$ always agrees with $L(z)$ up to and including at least the term of order $m+n$.

Now, in our particular case $L(z)$ can be written

$$L(z) = 1 + z\,{}_2F_1(\tfrac{1}{2}, 1; \tfrac{3}{2}; -z)$$

where the hypergeometric series ${}_2F_1(\frac{1}{2}, 1; \frac{3}{2}; \; z)$ has a continued fraction expansion given by (3.2.6) - (3.2.7) on page 29; that is,

$$ {}_2F_1(\tfrac{1}{2}, 1; \tfrac{3}{2}; -z) = \frac{1}{1+} \frac{\frac{1^2 z}{1\cdot 3}}{1} + \frac{\frac{2^2 z}{3\cdot 5}}{1} + \frac{\frac{3^2 z}{5\cdot 7}}{1} + \frac{\frac{4^2 z}{7\cdot 9}}{1} + \cdots . $$

Therefore L(z) has the continued fraction expansion

$$ 1 + \operatorname*{K}_{n=1}^{\infty} \frac{a_n z}{1} \qquad \text{where} \quad a_1 = 1 \text{ and } a_{n+1} = \frac{n^2}{4n^2 - 1} \qquad \text{for } n \geq 1. $$

As in the previous example, the correspondence shows up when we compare $L(z)$ to the Taylor expansions of the classical approximants

$$ f_1(z) = 1 + z $$

$$ f_2(z) = 1 + \frac{z}{1+} \frac{\frac{1}{3}z}{1} = \frac{3 + 4z}{3 + z} $$

$$ f_3(z) = 1 + \frac{z}{1+} \frac{z/3}{1} + \frac{4z/15}{1} = \frac{15 + 24z + 4z^2}{15 + 9z} $$

$$ f_4(z) = 1 + \frac{z}{1+} \frac{z/3}{1} + \frac{4z/15}{1} + \frac{9z/35}{1} = \frac{105 + 195z + 64z^2}{105 + 90z + 9z^2} $$

and so on. We recognize these approximants from the Padé table. The approximants of the regular C-fraction are indeed the staircase diagonal Padé approximants

$$ R_{0/0}, \; R_{1/0}, \; R_{1/1}, \; R_{2/1}, \; R_{2/2}, \; R_{3/2}, \; \ldots $$

as illustrated below.

$m \backslash n$	0	1	2	3	4	5	6	$\cdots$
0	$R_{0/0}$							$\cdots$
1	$R_{1/0}$	$R_{1/1}$						$\cdots$
2		$R_{2/1}$	$R_{2/2}$					$\cdots$
3			$R_{3/2}$	$R_{3/3}$				$\cdots$
4				$R_{4/3}$	$R_{4/4}$			$\cdots$
5					$R_{5/4}$	$R_{5/5}$		$\cdots$
6						$R_{6/5}$	$R_{6/6}$	$\cdots$
$\vdots$	$\vdots$	$\vdots$	$\vdots$	$\vdots$	$\vdots$	$\vdots$	$\vdots$	$\ddots$

This means that we can

- use regular C-fraction expansions to compute Padé approximants.
- use convergence theory for continued fractions to prove convergence of Padé approximants.

◇

1.5 More examples of applications

1.5.1 A differential equation

One can solve certain differential equations by means of continued fractions. We include a very simple example here. (More substantial examples can for instance be found in [Khov63], [Steen73], [Ince26], [Waad83].)

Example 17. To solve the differential equation

$$y = 2y' + y''$$

we first differentiate the equation repeatedly to get

$$\begin{aligned} y' &= 2y'' + y''', \\ &\vdots \\ y^{(n)} &= 2y^{(n+1)} + y^{(n+2)}. \end{aligned}$$

Assuming that we do not divide by 0, this gives

$$\begin{aligned} \frac{y}{y'} &= 2 + \frac{1}{y'/y''}, \\ \frac{y'}{y''} &= 2 + \frac{1}{y''/y'''}, \\ &\vdots \\ \frac{y^{(n)}}{y^{(n+1)}} &= 2 + \frac{1}{y^{(n+1)}/y^{(n+2)}}. \end{aligned}$$

From this it follows that

$$\frac{y}{y'} = \underbrace{2 + \frac{1}{2} + \frac{1}{2} + \cdots + \frac{1}{2}}_{n+1} + \frac{1}{y^{(n+1)}/y^{(n+2)}}.$$

This suggests the continued fraction

$$2 + \frac{1}{2} + \frac{1}{2} + \cdots + \frac{1}{2} + \cdots.$$

We know from Example 8 on page 19 that it converges to $\sqrt{2} + 1$, which suggests that the differential equation has a solution y satisfying

$$\frac{y}{y'} = \sqrt{2} + 1,$$

that is,

$$\frac{y'}{y} = \sqrt{2} - 1\,,$$

from which it would follow that

$$y = C \exp\left[(\sqrt{2} - 1)x\right].$$

This is actually a solution of the given differential equation.

By the warning on page 20, the value $1 - \sqrt{2}$ is also associated with the continued fraction above. Since the recursion more than the asymptotics is the important part of the investigation so far, we try

$$\frac{y}{y'} = 1 - \sqrt{2}\,.$$

This leads to

$$\frac{y'}{y} = -(\sqrt{2} + 1),$$

and thus $y = C_1 \exp(-(\sqrt{2} + 1)x)$. A quick check shows that this also is a solution of the differential equation. The method therefore gives the general solution

$$y = C_1 \exp[(\sqrt{2} - 1)x] + C_2 \exp[(-\sqrt{2} - 1)x]\,.$$

There is of course no good reason to use this "method" for the present differential equation, since there exists a perfectly good, simple method taught in elementary calculus. But there are cases, where this idea leads to non-trivial results. ◇

1.5.2 Moment problems and divergent series

In Example 10 on page 25 and Example 12 on page 26 a series with convergence radius 1 corresponded to a continued fraction which converged to the right value in a much larger region D containing the convergence disk of the series. Therefore, transforming the series to the continued fraction in question worked as a method to sum the divergent series. The effect is even more spectacular in the next example where the power series diverges for all $z \in \mathbb{C}$.

Example 18. Take a look at the series

$$L(z) := \sum_{n=0}^{\infty} n!(-z)^{-n} = 1 - \frac{1!}{z} + \frac{2!}{z^2} - \frac{3!}{z^3} + \cdots\,.$$

It diverges for all $z \in \mathbb{C}$. A function closely associated with this series is the function F, given as an integral

$$\begin{aligned} F(z) &:= \int_0^\infty \frac{ze^{-t}}{z+t}dt = \int_0^\infty e^{-t} \cdot \frac{1}{1-(-t/z)}\,dt \\ &= \int_0^\infty e^{-t}\left(1 - \frac{t}{z} + \frac{t^2}{z^2} - \cdots + \frac{(-t)^n}{z^n} + \frac{(-t)^{n+1}}{z^n(z+t)}\right)dt \\ &= 1 - \frac{1!}{z} + \frac{2!}{z^2} - \frac{3!}{z^3} + \cdots + (-1)^n\frac{n!}{z^n} + \int_0^\infty e^{-t}\frac{(-t)^{n+1}}{z^n(z+t)}\,dt. \end{aligned}$$

This function is holomorphic in the cut plane $|\arg(z)| < \pi$, and

$$F(z) - \sigma_n(z) = \int_0^\infty e^{-t}\frac{(-t)^{n+1}}{z^n(z+t)}dt \quad \text{for } \sigma_n(z) = \sum_{m=0}^{n}(-1)^m\frac{m!}{z^m}.$$

For an arbitrary, but fixed value of $\alpha \in [0, \frac{\pi}{2})$ we assume in the following that z is in a part of the plane where $|\arg(z)| \le \alpha$. Then

$$\begin{aligned} |F(z) - \sigma_n(z)| &\le \frac{1}{|z|^n}\int_0^\infty \frac{e^{-t}t^{n+1}}{\mathrm{Re}(z+t)}\,dt \\ &\le \frac{1}{|z|^n}\int_0^\infty \frac{e^{-t}t^{n+1}dt}{|z|\cos\alpha} = \frac{(n+1)!}{|z|^{n+1}\cos\alpha}. \end{aligned}$$

Now, keep n fixed, and let $z \to \infty$. Then the right hand side (and hence the left hand side) tends to 0. This property is expressed as follows: The series $L(z)$ is an *asymptotic expansion* of $F(z)$ as $z \to \infty$ in the angular opening where $|\arg(z)| \le \alpha$, and we write $F(z) \sim L(z)$ as $z \to \infty$.

This is one bridge between the divergent series $L(z)$ and the function F. Another bridge is by way of continued fractions. It can be shown that the continued fraction

$$\frac{1}{1+}\,\frac{1/z}{1}\,+\,\frac{1/z}{1}\,+\,\frac{2/z}{1}\,+\,\frac{2/z}{1}\,+\,\frac{3/z}{1}\,+\,\frac{3/z}{1}\,+\,\frac{4/z}{1}\,+\,\frac{4/z}{1}\,+\cdots$$

corresponds at ∞ to the series $L(z)$. It converges in the cut plane $|\arg(z)| < \pi$ to an analytic function f (Theorem 3.21 on page 124). Indeed, in volume 2 we shall see that $f = F$. So again we have given the divergent series a meaning by converting it to a convergent continued fraction. $\diamond$

What we have seen here is closely connected to the *moment problem*, or more precisely *the Stieltjes moment problem*. A real non-decreasing function Ψ on $\mathbb{R}^+$ with infinitely many points of increase is called a *distribution function*, and

$$c_n := \int_0^\infty t^n d\Psi(t) \qquad (5.2.1)$$

is called the nth *moment* with respect to Ψ. With

$$F(z) := \int_0^\infty \frac{z\,d\Psi(t)}{z+t} = \int_0^\infty \left(\sum_{n=0}^\infty (-1)^n \left(\frac{t}{z}\right)^n\right) d\Psi(t)\,, \tag{5.2.2}$$

termwise integration leads to the series

$$L(z) := \sum_{n=0}^\infty (-1)^n \frac{c_n}{z^n}\,. \tag{5.2.3}$$

The Stieltjes moment problem is as follows: *For a given sequence* $\{c_n\}_{n=0}^\infty$ *of real numbers, find a distribution function* Ψ *on* $(0,\infty)$ *such that*

$$c_n = \int_0^\infty t^n\,d\Psi(t) \quad \textit{for } n = 0,1,2,\ldots\,.$$

Any such function Ψ is called a *solution of the moment problem.* Related tasks are now:

1. Find necessary and sufficient conditions on $\{c_n\}$ for *existence* of a solution Ψ.
2. Find necessary and sufficient conditions for a solution Ψ to be *unique* up to an additive constant.

It is the beauty of this theory that the Stieltjes moment problem for $\{c_n\}$ has a solution if and only if the series (5.2.3) corresponds at $z = \infty$ to a continued fraction of the form

$$\frac{a_1}{1} \mathop{+}\limits \mathop{\mathrm{K}}_{n=2}^{\infty} \frac{a_n/z}{1} \quad \text{where all } a_n > 0$$

called an *S-fraction.* The solution is unique if and only if this continued fraction converges for $z > 0$. The value of the continued fraction is then the function $F(z)$ in (5.2.2), from which the distribution function can be derived. This was proved by Stieltjes in his famous 1894-paper ([Stie94]).

Example 19. As a consequence of Example 18, the moment problem with the moments $c_n := n!$ has a unique solution $\Psi(t)$ with $d\Psi(t) = e^{-t}\,dt = \Psi'(t)dt$. The solution is usually written $\Psi(t) := 1 - e^{-t}$, since it then increases from 0 to 1 when t increases from 0 to ∞ . $\diamond$

Terminology from measure theory is often used in this connection. In Example 19 the *measure* Ψ is *absolutely continuous* with derivative e^{-t}. Since

$$\int_0^\infty d\Psi(t) = \int_0^\infty e^{-t}dt = 1,$$

it is an example of a *probability measure.*

1.5.3 Orthogonal polynomials

Example 20. Let us define the inner product

$$\langle f, g\rangle := \int_{-1}^{1} f(x)g(x)\,dx$$

in the space of real functions continuous on $[-1,1]$. The *Legendre polynomials* $\{P_n\}_{n=0}^{\infty}$ are the real polynomials P_n of degree n with the property that

$$\langle P_m, P_n\rangle = \int_{-1}^{1} P_m(x)P_n(x)dx = \frac{2}{2n+1}\delta_{m,n} \quad \text{for all } m,n\geq 0,$$

where $\delta_{m,n}$ is the *Kronecker delta*; i.e., $\delta_{m,n}=1$ if $m=n$ and 0 otherwise. We say that the Legendre polynomials are *orthogonal* on $[-1,1]$ with respect to this inner product.

The Legendre polynomials $\{P_n\}$ are known to be the solution of the recurrence relation

$$(n+1)P_{n+1}(x) = (2n+1)xP_n(x) - nP_{n-1}(x) \quad \text{for } n = 1,2,3,\ldots$$

with initial expressions

$$P_0(x) := 1, \quad P_1(x) := x,$$

so for instance

$$P_2(x) = \frac{3}{2}x^2 - \frac{1}{2}, \quad P_3(x) = \frac{5}{2}x^3 - \frac{3}{2}x, \quad P_4(x) = \frac{35}{8}x^4 - \frac{15}{4}x^2 + \frac{3}{8}.$$

An alternative way of writing this recurrence relation is

$$P_{n+1}(x) = \frac{2n+1}{n+1}xP_n(x) - \frac{n}{n+1}P_{n-1}(x)\,.$$

Therefore $P_n(x)$ are exactly the canonical denominators $B_n(x)$ for the continued fraction

$$\frac{1}{x+}\,\frac{-1/2}{3x/2+}\,\frac{-2/3}{5x/3+}\,\frac{-3/4}{7x/4+}\cdots\,.$$

◇

Example 21. The *Chebyshev polynomials* $U_n(x)$ of the second kind are the solutions of the recurrence relation

$$U_{n+1}(x) = 2xU_n(x) - U_{n-1}(x)$$

with initial values $U_0(x) := 1$ and $U_1(x) := 2x$. Therefore $\{U_n(x)\}$ are the canonical denominators of the continued fraction

$$\frac{-1}{2x+}\,\frac{-1}{2x+}\cdots\frac{-1}{2x+}\cdots\,.$$

One can prove that

$$\int_{-1}^{1} U_m(x)U_n(x)(1-x^2)^{1/2}dx = \frac{\pi}{2}\delta_{m,n}.$$

We therefore say that $\{U_n(x)\}$ are orthogonal on $[-1,1]$ with respect to the weight function $W(x) := (1-x^2)^{1/2}$. ◇

These are merely two simple cases of a general theory. The connection between orthogonal sequences and continued fractions is the three term recurrence relation: a famous theorem on orthogonal polynomials, commonly but incorrectly called *Favard's theorem*, says that $\{P_n\}$ is a sequence of orthogonal polynomials with respect to some real measure $d\Psi(t)$ if and only if $\{P_n\}$ is a solution of a three term linear recurrence relation

$$P_n(x) = (x-c_n)P_{n-1}(x) - \lambda_n P_{n-2}(x) \quad \text{for } n = 1, 2, 3, \ldots$$

with initial values $P_{-1}(x) := 0$, $P_0(x) := 1$ (or some other positive constant), where $\lambda_n > 0$ and $c_n \in \mathbb{R}$. On the other hand, a three term recurrence relation of this form gives rise to a continued fraction. The continued fractions in question are the *Jacobi continued fractions*, briefly called *J-fractions*

$$\frac{\lambda_1}{x-c_1} - \frac{\lambda_2}{x-c_2} - \cdots - \frac{\lambda_n}{x-c_n} - \cdots.$$

This connection can be exploited to find asymptotic properties and zero-free regions for $\{P_n(x)\}$.

1.5.4 Thiele interpolation

Let f be an unknown function with known values $f(z_n)$ at given distinct points $z_0, z_1, z_2 \ldots$ in $\mathbb{C}$. We want to find f. What we do is successively to find constants $\phi_m \in \mathbb{C}$ such that the functions

$$F_n(z) := \phi_0 + \mathop{\mathrm{K}}_{m=1}^{n} \frac{z - z_{m-1}}{\phi_m} \quad \text{for } n = 0, 1, 2, \ldots$$

satisfy

$$F_n(z_k) = f(z_k) \quad \text{for } k = 0, 1, \ldots n-1. \tag{5.4.1}$$

The idea is then: if the resulting *Thiele continued fraction*

$$\phi_0 + \mathop{\mathrm{K}}_{n=1}^{\infty} \frac{z - z_{n-1}}{\phi_n}$$

converges, then its value is such a function $f(z)$. Normally we do not get the whole continued fraction — we have to stop after a finite number of terms, say at $F_n(z)$.

Then we have a rational function $F_n(z)$ which interpolates $f(z)$ at the first n points z_k.

There exists an algorithm for computing the numbers ϕ_m (inverse differencies). It fails if not all operations are meaningful, but otherwise it works as follows:

$$\phi_0 := \phi_0[z_0] := f(z_0), \quad \phi_1 := \phi_1[z_0, z_1] := \frac{z_1 - z_0}{f(z_1) - f(z_0)},$$

$$\phi_k := \phi_k[z_0, z_1, \ldots, z_k] := \frac{z_k - z_{k-1}}{\phi_{k-1}[z_0, \ldots, z_{k-2}, z_k] - \phi_{k-1}[z_0, \ldots, z_{k-1}]}$$

for $k = 2, 3, 4, \ldots$. Then (5.4.1) holds if

$$\begin{aligned} &F_n(z_k) = \phi_0 + \frac{z_k - z_0}{\phi_1} \underset{+}{} \frac{z_k - z_1}{\phi_2} \underset{+ \cdots +}{} \frac{z_k - z_{k-1}}{\phi_k} \\ &\text{and } \phi_{k+1} + \frac{z_k - z_{k+1}}{\phi_{k+2}} \underset{+ \cdots +}{} \frac{z_k - z_{n-1}}{\phi_n} \neq 0 \quad \text{for } k < n. \end{aligned} \tag{5.4.2}$$

Example 22. Let

$$z_0 := 0, \quad z_1 := 4, \quad z_2 := 9, \quad z_3 := 25,$$

$$f(z_0) := 0, \quad f(z_1) := 2, \quad f(z_2) := 3, \quad f(z_3) := 5.$$

In this case we find $\phi_0[z_0] = 0$ and

$$\phi_1[z_0, z_1] = \frac{4-0}{2-0} = 2, \quad \phi_1[z_0, z_2] = \frac{9-0}{3-0} = 3, \quad \phi_1[z_0, z_3] = \frac{25-0}{5-0} = 5,$$

$$\phi_2[z_0, z_1, z_2] = \frac{9-4}{3-2} = 5, \quad \phi_2[z_0, z_1, z_3] = \frac{25-4}{5-2} = 7.$$

$$\phi_3[z_0, z_1, z_2, z_3] = \frac{25-9}{7-5} = 8.$$

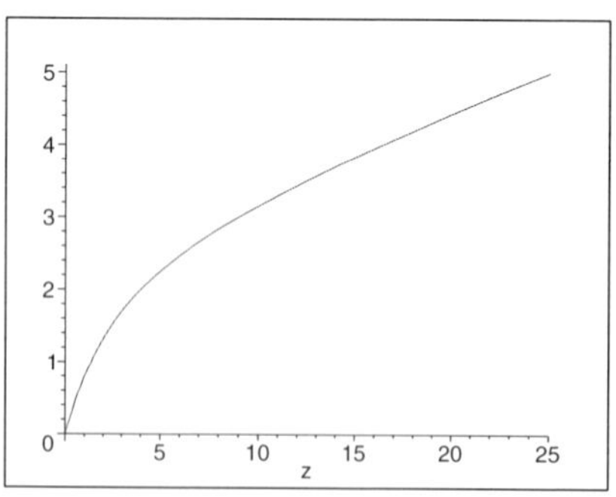

Therefore

$$\begin{aligned} F_3(z) &= \frac{z}{2+} \frac{z-4}{5} \underset{+}{} \frac{z-9}{8} \\ &= \frac{z(31+z)}{10(3+z)}. \end{aligned}$$

We verify that $F_3(0) = 0$, $F_3(4) = 2$, $F_3(9) = 3$, $F_3(25) = 5$. This is therefore one function (of many possibilities) with the required values at 0, 2, 9 and 25. (Another possibility is $\sqrt{z}$.) The figure shows the graph of $F_3(z)$ for real $z > 0$. ◇

1.5.5 Stable polynomials

A polynomial

$$Q_n(r) := r^n + a_1 r^{n-1} + \cdots + a_n \tag{5.5.1}$$

is called *stable* if all zeros lie in the left half plane $\mathrm{Re}(r) < 0$. The concept is important in the theory of linear differential equations with constant coefficients, as illustrated in the example below.

Example 23. To solve the differential equation

$$\ddot{y} + 3\dot{y} + 2y = 0$$

where $\dot{y}$ denotes the derivative with respect to the real variable t, one solves the corresponding characteristic equation

$$Q_2(r) := r^2 + 3r + 2 = 0.$$

It has the roots $r_1 = -1$ and $r_2 = -2$. Hence the differential equation has the general solution

$$y(t) = C_1 e^{-t} + C_2 e^{-2t},$$

which means that $y(t) \to 0$ as $t \to +\infty$. This corresponds to the fact that Q_2 is a stable polynomial.

Similarly, the characteristic equation for the differential equation

$$\dddot{y} + 4\ddot{y} + 6\dot{y} + 4y = 0$$

is

$$Q_3(r) := r^3 + 4r^2 + 6r + 4 = 0$$

which has the solutions $r_1 = -2$, $r_{2,3} = -1 \pm i$, so Q_3 is stable. Every solution $y(t)$ of the differential equation is therefore damped, i.e., $y(t) \to 0$ as $t \to \infty$. The general solution is of course

$$y(t) = C_1 e^{-2t} + e^{-t}(C_2 \cos t + C_3 \sin t).$$

◇

We want to decide whether a polynomial is stable or not, without having to compute its zeros. This has been done for the case of real coefficients by Hurwitz ([Hurw95]). For a given polynomial (5.5.1) he introduced the auxiliary polynomial

$$P_n(r) := a_1 r^{n-1} + a_3 r^{n-3} + a_5 r^{n-5} \ldots,$$

and proved that Q_n is stable if and only if $P_n(r)/Q_n(r)$ can be written as a terminating continued fraction of the form

$$\frac{1}{1 + d_1 r +} \frac{1}{d_2 r +} \cdots \frac{1}{+ d_n r} \quad \text{where all } d_k > 0.$$

The proof can be found in [LoWa92], p 470, along with a result for the complex case. Here we only check the result for $Q_2(r)$ and $Q_3(r)$ in Example 23.

$$Q_2(r) = r^2 + 3r + 2, \quad P_2(r) := 3r, \quad \frac{P_2(r)}{Q_2(r)} = \frac{1}{1 + \frac{1}{3}r} \, \frac{1}{+ \frac{3}{2}r} ,$$

$$Q_3(r) = r^3 + 4r^2 + 6r + 4, \quad P_3(r) := 4r^2 + 4$$

$$\frac{P_3(r)}{Q_3(r)} = \frac{1}{1 + \frac{1}{4}r} \, \frac{1}{+ \frac{4}{5}r} \, \frac{1}{+ \frac{5}{4}r} .$$

1.6 Remarks

1. **The birth of continued fractions.** The euclidean algorithm, the basis for the regular continued fraction, goes back to around 300 B.C., but in a slightly different form, as a subtraction algorithm rather than a division algorithm ([Eucl56]). But at that time it did not lead to a continued fraction.

   The birth of continued fractions, like many other things in the culture of mankind, took place in Italy in the renaissance, by Bombelli in 1572 ([Bomb72]) and Cataldi in 1613 ([Cata13]), in both cases as approximate values for a square root. But already Fibonacci, around 1200, i.e., long before the renaissance, touched upon the idea of a continued fraction, but an ascending one ([Fibo02]). Of the further development we mention, as we did in Section 1.2.4 of the present chapter, Huygens' use of continued fractions for designing cogwheels ([Huyg95]). In 1965 Lord Brouncker derived a nonterminating continued fraction for $\pi/4$ on request by Wallis who was working on his book ([Wallis85]) at that time. Lord Brouncker neither published his result nor his method, but Wallis writes enough about the ideas for Khrushchev to reconstruct them in his very interesting paper ([Khru06a]). In the same book Wallis also gives recurrence relations for the approximants of a continued fraction.

   With Euler ([Euler67]) a general theory for continued fractions began to develop, a theory where Gauss' famous continued fractions for ratios of hypergeometric functions ([Gauss13]) was an early high point, still of great value in modern theory for computing special functions.

2. **Additional literature.** For those who want to go deeper into the analytic theory of continued fractions we refer to the three standard monographs in the field: the classical text-book by Perron ([Perr54], [Perr57] (in German)), Wall's book ([Wall48]) and the exposition by Jones and Thron ([JoTh80]). In Henrici's 3 volume work on Applied and Computational Complex Analysis ([Henr77]) a large portion of volume 2 is devoted to analytic theory of continued fraction. Khovanskii's book ([Khov63]) contains some interesting applications.

   The history of continued fractions up to 1939 is described in Brezinski's book ([Brez91]). Also the texts mentioned above, in particular the book by Jones and Thron, contain interesting comments on the historic development of concepts, methods and applications. For the computational aspects of continued fractions we recommend the handbook ([CJPVW7]).

   To most people (meaning mathematicians) continued fractions are the regular continued fractions in number theory. More information on this side of the continued fraction theory can be found in [Perr54], [Ries85] and the recent book by Hensley ([Hens06]).

3. **The definition of a continued fraction.** Our definition of a continued fraction $b_0 + \mathbf{K}(a_n/b_n)$ on page 5 is inspired by the classical definition by Henrici and Pfluger ([HePf66]) who defined $b_0 + \mathbf{K}(a_n/b_n)$ as the ordered pair $((\{a_n\}, \{b_n\}), \{f_n\})$ where $\{f_n\}$ is the sequence of classical approximants.
Neither of these two definitions is an absolutely obvious choice. Approximants

$$\tilde{f}_n := b_0 + \frac{a_1}{b_1} + \frac{a_2}{b_2} + \cdots + \frac{a_{n-1}}{b_{n-1} + a_n}$$

are for instance also known from the classical literature. And the sequence $\{S_n\}$ in our definition could be replace by

$$\tilde{s}_n := \varphi_{n-1}^{-1} \circ s_n \circ \varphi_n, \quad \tilde{S}_n := \tilde{s}_0 \circ \tilde{s}_1 \circ \cdots \tilde{s}_n = S_n \circ \varphi_n$$

for any sequence $\{\varphi_n\}$ from $\mathcal{M}$ with $\varphi_{-1}(w) \equiv w$.
Or we could follow Beardon ([Bear04]) and define a continued fraction as a sequence $\{S_n^*\}$ from $\mathcal{M}$ with the properties that

$$S_n^* := s_1^* \circ s_2^* \circ \cdots \circ s_n^* \quad \text{where } s_k \in \mathcal{M} \text{ with } s_k^*(a) = b$$

for two fixed (possibly coinciding) constants a, $b \in \widehat{\mathbb{C}}$.

4. **G-continued fractions.** A continued fraction $\mathbf{K}(a_n/b_n)$ is closely connected to the recurrence relation

$$X_n = b_n X_{n-1} + a_n X_{n-2} \quad \text{for } n = 1, 2, 3, \ldots$$

where $a_n \neq 0$. For one thing, $\{A_n\}$ and $\{B_n\}$ are solutions of this recurrence. Moreover,

$$-\frac{X_{n-1}}{X_{n-2}} = \frac{a_n}{b_n - X_n/X_{n-1}} = s_n(-X_n/X_{n-1})$$

for every non-trivial solution. Hence $\{-X_n/X_{n-1}\}$ is a tail sequence for $\mathbf{K}(a_n/b_n)$.
A G-continued fraction of dimension N is connected to a longer recurrence relation

$$X_n + a_n^{(N)} X_{n-1} + a_n^{(N-1)} X_{n-2} + \cdots + a_n^{(1)} X_{n-N} = 0$$

where $a_n^{(1)} \neq 0$. Let $\{B_n\}_{n=1-N}^{\infty}$ and $\{A_n^{(k)}\}_{n=1-N}^{\infty}$ for $k = 1, 2, \ldots, N-1$ be the solutions of this recurrence with initial values

$$\begin{pmatrix} A_{1-N}^{(1)} & A_{2-N}^{(1)} & \cdots & A_0^{(1)} \\ \vdots & \vdots & & \vdots \\ A_{1-N}^{(N-1)} & A_{2-N}^{(N-1)} & \cdots & A_0^{(N-1)} \\ B_{1-N} & B_{2-N} & \cdots & B_0 \end{pmatrix} = \begin{pmatrix} 1 & 0 & \cdots & 0 \\ 0 & 1 & \cdots & 0 \\ \vdots & \vdots & & \vdots \\ 0 & 0 & \cdots & 1 \end{pmatrix}$$

Then $(\frac{A_n^{(1)}}{B_n}, \frac{A_n^{(2)}}{B_n}, \ldots, \frac{A_n^{(N-1)}}{B_n})$ are the approximants of the G-continued fraction. These continued fractions are also called vector-valued continued fractions. For more information and further references we refer to ([LoWa92], p 225).

1.7 Problems

1. **Continued fractions with given $\mathbf{A_n}$, $\mathbf{B_n}$.** Construct the continued fraction with

   (a) $A_n = 2n$, $B_n = 3n + 1$ for $n \geq 0$,

   (b) $A_n = \sin \frac{n\pi}{2}$, $B_n = \cos \frac{n\pi}{2}$ for $n \geq 0$,

   (c) $A_{2n-1} = n^2$, $B_{2n-1} = n^2$, $A_{2n} = 2n^2 + 1$ and $B_{2n} = 2n^2$ for $n \geq 1$.

2. **Continued fraction identities.** Construct a continued fraction $\mathbf{K}(1/b_n)$ which converges to

   (a) 1. (b) -1. (c) ∞.

3. **Periodic continued fraction.** Assume that the continued fraction

$$\frac{1}{1} + \frac{1}{1} + \frac{1}{1} + \cdots$$

   converges. Prove that it its value is $(\sqrt{5} - 1)/2$.

4. **Denominators = 1.** Prove that $B_n = 1$ for all $n \geq 0$ for $\mathbf{K}(a_n/b_n)$ if and only if $b_1 = 1$ and $a_n + b_n = 1$ for $n \geq 2$.

5. **Reversed terminating continued fraction.** Let $\mathbf{K}_{n=1}^{N}(a_n/b_n)$ for some $N \in \mathbb{N}$ be a terminating continued fraction (with all $a_n \neq 0$), and let

$$b_N + \frac{a_N}{b_{N-1}} + \frac{a_{N-1}}{b_{N-2}} + \cdots + \frac{a_2}{b_1}$$

   be its *reversed continued fraction.* Prove that the $(N-1)$th denominator B_{N-1} is the same for the two terminating continued fractions.

6. **Periodic continued fraction.** The periodic continued fraction

$$\operatorname*{K}_{n=1}^{\infty} \frac{z}{2} = \frac{z}{2} + \frac{z}{2} + \frac{z}{2} + \cdots$$

   converges for all $z \in D := \{z \in \mathbb{C};\ |\arg(z+1)| < \pi\}$.

   (a) What possible values can it have?

   (b) What is its value for $z > 0$? Use this to find a continued fraction expansion for $\sqrt{5}$.

   (c) Let $z := 1 - 2i$. Compute the first classical approximants for $\mathbf{K}(z/2)$ and guess its value.

7. **Approximants.** Compute the first three classical approximants and the first three approximants $S_n(1)$ of the continued fraction $\mathbf{K}(n/1)$ by means of

   (a) the forward recurrence algorithm,

   (b) the backwards recurrence algorithm,

(c) the Euler-Minding formula and its generalization.

8. **Euclidean algorithm.** Use the Euclidean algorithm to find the greatest common divisor

   (a) $\gcd(119, 221)$, (b) $\gcd(3839, 1711)$, (c) $\gcd(49907, 22243)$.

9. **Terminating regular continued fraction.** Find the regular continued fraction expansion and its first few approximants for the following numbers:

   (a) $47/99$, (b) $3839/1711$, (c) $15015/7429$.

10. **Periodic regular continued fractions.** For each of the given numbers, find its regular continued fraction expansion and show that it is periodic (possibly after a few fraction terms). Further compute some of its first approximants.

    (a) $\sqrt{82}$, (b) $\sqrt{51}$, (c) $\sqrt{53}$.

11. ♠ **Periodic regular continued fractions.** Prove that the value f of a (non-terminating) regular continued fraction $\mathbf{K}(1/b_n)$ has the form $f = \dfrac{M+\sqrt{Q}}{N}$ with M, N, Q are integers with $N \neq 0$, $Q > 0$ and $\sqrt{Q} \notin \mathbb{N}$, if and only if $\{b_n\}$ is periodic (from some n on). (Lagrange [Lagr70], [Perr54], p 66.)

12. ♠ **The family $\mathcal{M}$ and matrices.** Show that if we identify transformations $\tau_n(w) := \dfrac{a_n w + b_n}{c_n w + d_n}$ from $\mathcal{M}$ with the matrix $T_n := \begin{pmatrix} a_n & b_n \\ c_n & d_n \end{pmatrix}$, then $\tau_1 \circ \tau_2$ corresponds to the matrix product $T_1 T_2$.

13. ♠ **Periodic continued fractions.** Prove that

$$f_n = -xy \frac{x^n - y^n}{x^{n+1} - y^{n+1}}$$

for the continued fraction

$$-\frac{xy}{x+y} - \frac{xy}{x+y} - \frac{xy}{x+y} - \cdots; \quad x, y \neq 0, \quad x \neq y.$$

14. **Best rational approximation.** Find the first five classical approximants of the regular continued fraction expansion of $e = 2.718281828459\ldots$ and compare them to other rational approximations with no larger denominators.

15. **Diophantine equation.** Find all positive integer solutions (x, y) with $y < 20$ of the diophantine equation $19x - 13y = 5$.

16. **Diophantine equations.** Find the general solution of the linear diophantine equation

    (a) $11x - 3y = 5$, (b) $99x - 47y = 3$,

    (c) $3839x - 1711y = 1$, (d) $3839x + 1711y = 1$.

17. **Fibonacci sequence.**

    (a) Use the identity
$$\frac{\sqrt{5}-1}{2}=\cfrac{1}{1+\cfrac{\sqrt{5}-1}{2}}$$
    to produce a continued fraction by the procedure of Example 8. Compute the first 7 classical approximants f_n and compare the values to $(\sqrt{5}-1)/2$.

    (b) Prove that $f_n = F_{n-1}/F_n$ where $F_0 = 1$, $F_1 = 1$, $F_2 = 2$, $F_3 = 3$, $F_4 = 5$, and generally
$$F_{n+1} = F_n + F_{n-1} \quad \text{for } n \geq 1\,.$$

    (c) Prove that $f_n \to (\sqrt{5}-1)/2$. (The sequence $\{F_n\}$ is the sequence of *Fibonacci numbers*, and $(\sqrt{5}-1)/2 \approx 0.61803\ldots$ is the *golden ratio.*)

18. ♠ **Regular continued fractions.** Let $\{A_n\}$ and $\{B_n\}$ be the canonical numerators and denominators for the regular continued fraction $\mathbf{K}(1/b_n)$ as in Theorem 1.7 on page 17.

    (a) Show that $B_n f - A_n = B_n(f - f_n)$.

    (b) Show that $|B_{n+1}f - A_{n+1}| = -b_{n+1}|B_n f - A_n| + |B_{n-1}f - A_{n-1}| > 0$. (Hint: Use the recurrence relations for $\{A_n\}$ and $\{B_n\}$ and the fact that $f - f_n$ alternates in sign.)

    (c) Show that $|B_{n-1}f - A_{n-1}| > |B_n f - A_n|$.

19. ♠ **Divergence of periodic continued fraction.** Prove that if the continued fraction
$$\frac{a}{1}\,\frac{}{+}\,\frac{a}{1}\,\frac{}{+}\,\frac{a}{1}\,\frac{}{+}\cdots$$
converges, then it converges to one of the roots of the equation $x = a/(1+x)$. Use this to prove that the continued fraction diverges for all $a < -1/4$.

20. ♠ **The binomial series.** Let α be real and not a positive integer, and let ${}_2F_1(a,b;c;z)$ be the hypergeometric series (3.2.2) on page 28. Prove that
$${}_2F_1(-\alpha,1;1;-z) = (1+z)^{\alpha} \quad \text{for } |z|<1.$$
Use (3.2.5) - (3.2.6) on page 29 to find a continued fraction expansion of the form $\mathbf{K}(a_n z/1)$ for $z(1+z)^{\alpha}$.

21. **From power series to continued fraction.** Use the procedure of Example 13 on page 30 to find the first terms
$$\frac{a_1 z}{1} + \frac{a_2 z}{1} + \frac{a_3 z}{1} + \frac{a_4 z}{1} + \cdots$$
of a continued fraction expansion of $f(z) := e^z - 1$.

22. **From continued fraction to power series.** Use the procedure of Example 14 on page 33 to find the first terms of the power series expansion at 0 corresponding to the continued fraction

$$\frac{z}{1+}\frac{-z/2}{1}+\frac{z/6}{1}+\frac{-z/6}{1}+\cdots$$

23. ♠ **From approximants to continued fraction.** Let $\{f_n\}_{n=0}^{\infty}$ with $f_0 := 0$ and $f_n \neq f_{n-1}$ for all n be a given sequence of complex numbers. Prove that the continued fraction $\mathbf{K}(a_n/b_n)$ with $b_1 := 1$, $a_1 := f_1$ and

$$a_n := -\frac{f_n - f_{n-1}}{f_{n-1} - f_{n-2}}, \qquad b_n := \frac{f_n - f_{n-2}}{f_{n-1} - f_{n-2}} \quad \text{for } n = 2, 3, 4, \dots$$

has classical approximants $\{f_n\}$. (Bernoulli [Berno75].) (Hint: see Problem 4.)

24. **From approximants to continued fraction.** Prove that a given sequence $\{f_n\}$ of complex numbers is a sequence of classical approximants for a continued fraction $b_0 + \mathbf{K}(a_n/1)$ if and only if $b_0 = f_0$, $f_n \neq f_{n-1}$, $f_n \neq f_{n-2}$ and

$$a_1 = f_1 - f_0, \quad a_2 = -\frac{f_1 - f_2}{f_0 - f_2} \text{ and } a_n = -\frac{(f_{n-3} - f_{n-2})(f_{n-1} - f_n)}{(f_{n-3} - f_{n-1})(f_{n-2} - f_n)}$$

for $n > 2$. (Bernoulli [Berno75].)

25. ♠ **From series to continued fraction.** Let $\sigma_n := \sum_{k=1}^{n} c_k$ be the partial sums of the series $\sum_{k=1}^{\infty} c_n$ where all $c_n \neq 0$. Show that the continued fraction $\mathbf{K}(a_n/b_n)$ with

$$a_1 := c_1, \quad a_n := -\frac{c_n}{c_{n-1}}, \quad b_1 := 1, \quad b_n := \frac{c_{n-1} + c_n}{c_{n-1}}$$

has canonical numerators $A_n = \sigma_n$ and canonical denominators $B_n = 1$ and thus classical approximants $f_n = \sigma_n$. (Euler ([Euler48]), Stern ([Stern32]), Glaisher ([Glai74]).)

26. ♠ **From infinite product to power series.** Let $p_n := \prod_{k=0}^{n} a_k$ be the partial products of the infinite product $\prod_{k=0}^{\infty} a_k$ where all $a_k \neq 0, 1$. Show that the continued fraction

$$a_0 + \frac{a_0(a_1 - 1)}{1} - \frac{a_1(a_2 - 1)/(a_1 - 1)}{(a_1 a_2 - 1)/(a_1 - 1)} - \frac{a_2(a_3 - 1)/(a_2 - 1)}{(a_2 a_3 - 1)/(a_2 - 1)} -$$

$$\cdots - \frac{a_{n-1}(a_n - 1)/(a_{n-1} - 1)}{(a_{n-1} a_n - 1)/(a_{n-1} - 1)} - \cdots$$

has canonical numerators $A_n = p_n$ and canonical denominators $B_n = 1$ and thus classical approximants $f_n = p_n$. (Stern [Stern32], Glaisher [Glai74].)

27. **From continued fraction to continued fraction.** Let $\{A_n\}_{n=0}^{\infty}$ and $\{B_n\}_{n=0}^{\infty}$ be the canonical numerators and denominators of $1 + \mathbf{K}(a_n/1)$.

(a) Show that

$$A_1, A_0, A_3, A_2, A_5, A_4, \ldots \quad \text{and} \quad B_1, B_0, B_3, B_2, B_5, B_4, \ldots$$

are the sequences of canonical numerators and denominators for the continued fraction

$$1 + a_1 - \frac{a_1}{1+} \frac{1+a_3}{a_2} + \frac{\frac{a_2 a_3}{1+a_3}}{\frac{1}{1+a_3}} + \frac{1+a_4}{a_4} + \frac{\frac{a_4 a_5}{1+a_5}}{\frac{1}{1+a_5}} + \cdots .$$

(Perron [Perr57], p 7.)

(b) Show that the sequence of classical approximants for

$$1 + a_1 - \frac{a_1}{1+} \frac{1+a_3}{a_2} + \frac{a_2 a_3}{1} + \frac{(1+a_3)(1+a_5)}{a_4} + \frac{a_4 a_5}{1} + \frac{(1+a_5)(1+a_7)}{a_6} + \cdots$$

is the same as for the continued fraction in (a).

(c) Find the continued fraction with canonical numerators and denominators

$$A_0, A_2, A_1, A_4, A_3, A_6, \ldots \quad \text{and} \quad B_0, B_2, B_1, B_4, B_3, B_6, \ldots .$$

28. **Padé table.** Given the power series $\sum_{n=0}^{\infty} n! z^n$. Find the 3×3 upper left entries in the Padé table for the series.

29. **Differential equation.** Solve the differential equation $y = 2y' + 3y''$ by using the "method" in Subsection 1.5.1.

30. **Moments.** Let a be an arbitrary positive number, and let Ψ be the distribution function given by $\Psi(t) = 0$ for $0 \le t < a$ and $\Psi(t) = 1$ for $a \le t < \infty$.

   (a) Find the moments c_n with respect to $\Psi(t)$.

   (b) Find the S-fraction corresponding to the series $\sum_{m=0}^{\infty} (-1)^n c_n z^{-n}$. (The series can be written in closed form, and the S-fraction is terminating.)

31. **Moments.** Let Ψ be the distribution function given by $\Psi(t) = 0$ for $0 \le t < 1$ and $\Psi(t) = 1/2$ for $1 \le t < 2$ and $\Psi(t) = 1$ for $2 \le t < \infty$. Try to answer questions (a) and (b) in Problem 30. Do you recognize anything from your everyday life?

32. **Thiele fractions.** Find the shortest Thiele fraction which has the values 1, 2 and 3 at the points 0, 1 and 2 respectively.

33. ♠ **Thiele fractions.** Find the shortest Thiele fraction $F(z)$ which has the values 1, 2, 3, b at the points 0, 1, 2 and 3 respectively. What can be said about $F(z)$ for different values of b?

34. ♠ **Stable polynomials.**

   (a) For which values of k is the polynomial $Q_3(x) := x^3 + 3x^2 + 3x + 1 + k$ stable?

   (b) For which values of p and q is the polynomial $Q_2(x) := x^2 + px + q$ stable?

   (c) For which values of p q and r is the polynomial $Q_3(x) := x^3 + px^2 + qx + r$ stable?

Chapter 2

Basics

The transformations $s_n(w) := a_n/(b_n + w)$ are linear fractional transformations with $s_n(\infty) = 0$. Hence $S_n := s_1 \circ s_2 \circ \cdots \circ s_n$ are linear fractional transformations with $S_n(\infty) = S_{n-1}(0)$, and a continued fraction is essentially a sequence of linear fractional transformations $\{S_n\}$ with $S_n(\infty) = S_{n-1}(0)$ for all n. It is therefore natural to define convergence of continued fractions as convergence of $\{S_n\}$ in some sense. Traditionally what is required is that $\{S_n(0)\}$ converges in $\widehat{\mathbb{C}}$. This definition is easy to grasp, but not quite what we are after. In this chapter we shall define an alternative concept, *general convergence*, which essentially requires that $\{S_n\}$ converges to a constant function in $\widehat{\mathbb{C}}$. The only snag is that we have to accept *exceptional sequences* to be defined.

Two important tools in the convergence theory for continued fractions are *tail sequences* and *value sets* which both will be defined in this chapter. The interplay between these two gives a deeper understanding of a continued fraction, but also useful methods for proving convergence and deriving truncation error bounds. Indeed, they are two of the main pillars on which the theory in this book is built.

The last part of this chapter is devoted to transformations of one continued fraction to another. The idea is that the new one shall have some desired properties lacking in the old one.

2.1 Convergence

2.1.1 Properties of linear fractional transformations

As always, $\mathcal{M}$ denotes the family of linear fractional transformations

$$\tau(w) := \frac{aw+b}{cw+d}; \quad ad-bc \neq 0. \tag{1.1.1}$$

These transformations have a number of nice properties. First of all, τ is a bijective meromorphic mapping of $\widehat{\mathbb{C}}$ onto $\widehat{\mathbb{C}}$; that is, $\tau(\widehat{\mathbb{C}}) = \widehat{\mathbb{C}}$ and τ is univalent (one-to-one). Some additional properties are given below. Others will be described later.

Properties:

1. The cross ratio is invariant under linear fractional transformations; that is, if u, v, w and z are four distinct points in $\widehat{\mathbb{C}}$, then

   $$\frac{\tau(u)-\tau(z)}{\tau(u)-\tau(w)} \cdot \frac{\tau(v)-\tau(w)}{\tau(v)-\tau(z)} = \frac{u-z}{u-w} \cdot \frac{v-w}{v-z} \tag{1.1.2}$$

   with the natural limiting forms if one or two of the quantities involved are infinite. This property follows by straight forward checking.

2. From (1.1.2) it follows that whenever w_1, w_2, w_3 are three distinct points in $\widehat{\mathbb{C}}$ and γ_1, γ_2, γ_3 are three distinct points in $\widehat{\mathbb{C}}$, then there exists a unique τ from $\mathcal{M}$ such that

   $$\tau(w_1) = \gamma_1, \quad \tau(w_2) = \gamma_2 \quad \text{and} \quad \tau(w_3) = \gamma_3.$$

   This does not mean that the coefficients of τ are unique, but the mapping τ is unique in the sense described on page 5.

3. From 2. it follows immediately that if a sequence $\{\tau_n\}$ from $\mathcal{M}$ converges (pointwise) at three distinct points to distinct values, then it converges to some $\tau \in \mathcal{M}$ in all of $\widehat{\mathbb{C}}$.

4. If $\{\tau_n\}$ converges at $n \geq 3$ distinct points, but not to a $\tau \in \mathcal{M}$, then the limit is the same at at least two of the three points.

The point $\infty \in \widehat{\mathbb{C}}$ is not a special point for a linear fractional transformation. Indeed, for τ in (1.1.1)

$$\tau(\infty) = a/c \qquad \text{and} \qquad \tau(-d/c) = \infty$$

where $a/c = \infty$ and $d/c = \infty$ if $c = 0$, and finite numbers otherwise. This calls for a representation of $\widehat{\mathbb{C}}$ which makes no distinction between ∞ and the complex numbers. The Riemann sphere is a perfect choice.

In the complex plane, $|w_1 - w_2|$ is the natural way to define the distance between two points (the euclidean metric). On the Riemann sphere $\widehat{\mathbb{C}}$ one way to do it is to measure the euclidean length of the chord connecting the two points. This gives us the *chordal distance* (or *chordal metric*) introduced by Ahlfors ([Ahlf53]):

$$\mathfrak{m}(w_1, w_2) := \begin{cases} \dfrac{2|w_1 - w_2|}{\sqrt{1+|w_1|^2}\sqrt{1+|w_2|^2}} & \text{for } w_1, w_2 \in \mathbb{C} \\ \dfrac{2}{\sqrt{1+|w_1|^2}} & \text{for } w_1 \in \mathbb{C},\ \ w_2 = \infty \\ 0 & \text{for } w_1 = w_2 = \infty\,. \end{cases} \tag{1.1.3}$$

For our purpose the main advantages of this metric are:

(i) it is bounded, since $\mathfrak{m}(w_1, w_2) \leq 2$ for all $w_1, w_2 \in \widehat{\mathbb{C}}$. (The radius of the Riemann sphere is taken to be 1.)

(ii) $\widehat{\mathbb{C}}$ equipped with this metric is compact; that is, every sequence $\{w_n\}$ on $\widehat{\mathbb{C}}$ has a convergent subsequence.

(iii) $w_n \to \widehat{w} \in \widehat{\mathbb{C}}$ if and only if $\mathfrak{m}(w_n, \widehat{w}) \to 0$.

(iv) if $\tau_n \to \tau \in \mathcal{M}$, then the convergence is uniform on $\widehat{\mathbb{C}}$ with respect to the chordal metric.

(v) the cross ratio (1.1.2) implies that

$$\frac{\mathfrak{m}(\tau(u),\, \tau(z))}{\mathfrak{m}(\tau(u),\, \tau(w))} \cdot \frac{\mathfrak{m}(\tau(v),\, \tau(w))}{\mathfrak{m}(\tau(v),\, \tau(z))} = \frac{\mathfrak{m}(u,\, z)}{\mathfrak{m}(u,\, w)} \cdot \frac{\mathfrak{m}(v,\, w)}{\mathfrak{m}(v,\, z)} \tag{1.1.4}$$

when u, v, w, z are four distinct points on $\widehat{\mathbb{C}}$.

The euclidean distance is not bounded, and one has to distinguish between the cases $\widehat{w} \neq \infty$ and $\widehat{w} = \infty$ to define the convergence $w_n \to \widehat{w}$ in this metric. If $\widehat{w} \neq \infty$, then $w_n \to \widehat{w}$ if and only if $|w_n - \widehat{w}| \to 0$. If $\widehat{w} = \infty$, this characterization fails.

We use the chordal distance to define convergence in $\mathcal{M}$, or rather the metric

$$\sigma(\tau_1, \tau_2) := \sup_{w \in \widehat{\mathbb{C}}} \mathfrak{m}(\tau_1(w), \tau_2(w)). \tag{1.1.5}$$

Definition 2.1. *A sequence $\{\tau_n\}$ from $\mathcal{M}$ converges to some $\tau \in \mathcal{M}$ if and only if* $\lim \sigma(\tau_n, \tau) = 0$.

That is, $\tau_n(w) \to \tau(w)$ uniformly in $\widehat{\mathbb{C}}$ with respect to the chordal metric. The type of convergence described in Property 4 requires some closer attention:

Example 1. We shall see three examples of convergence of a sequence $\{\tau_n\}$ from $\mathcal{M}$ given by

$$\tau_n(w) := \frac{a_n w + b_n}{c_n w + d_n}, \qquad \Delta_n := a_n d_n - b_n c_n \neq 0 \quad \text{for } n = 1, 2, 3, \ldots .$$

Example 1A: Let the coefficients a_n, b_n, c_n and d_n converge to finite values a, b, c and d respectively, where $ad - bc \neq 0$. Then $\tau_n(w) \to \tau(w) := (aw+b)/(cw+d)$ for all $w \in \widehat{\mathbb{C}}$. In the euclidean metric this convergence is uniform on compact subsets of $\mathbb{C} \setminus \{-d/c\}$. In the chordal metric it is uniform on $\widehat{\mathbb{C}}$. In short, $\tau_n \to \tau$.

Example 1B: The coefficients converge to finite values a, b, c and d respectively, where $c \neq 0$ and $ad - bc = 0$. This time $\{\tau_n\}$ approaches a constant function

$$\tau(w) := \frac{aw+b}{cw+d} = \frac{a}{c} - \frac{ad-bc}{c(cw+d)} = \frac{a}{c} \quad \text{for } w \neq -\frac{d}{c}.$$

So $\tau_n(w) \to a/c$ for every $w \in \widehat{\mathbb{C}}$ except possibly at the point $w^\dagger := -d/c$. The convergence is locally uniform in $\widehat{\mathbb{C}} \setminus \{-d/c\}$ with respect to the chordal metric, and $\tau_n(w_n) \to a/c$ whenever $\liminf \mathfrak{m}(w_n, -d/c) > 0$. For $w_n := -d_n/c_n$ we always have $\tau_n(w_n) = \infty \neq a/c$ for all n, so there exist sequences $\{w_n\}$ approaching $-d/c$ for which $\tau_n(w_n) \not\to a/c$.

Example 1C: Let $a_n \to a \neq \infty$, $c_n \to c \neq 0, \infty$, and $(a_n d_n - b_n c_n) \to 0$. $\tau_n(w)$ still seems to converge to a/c, but we have a problem. This time $\{d_n\}$ may have a large number of limit points d, and w ought to stay away from $-d/c$ for all of them. This can be very restrictive and even impossible since the limit points for $\{d_n\}$ may be dense in $\widehat{\mathbb{C}}$. On the other hand, for every limit point d there exists a subsequence $\{d_{n_k}\}$ converging to d, and for this subsequence we are back to the former case: $\{\tau_{n_k}\}$ converges (locally uniformly with respect to the chordal metric) to a/c in $\widehat{\mathbb{C}} \setminus \{-d/c\}$. So there is some extensive converging to a/c going on. To get hold of this convergence we look at

$$\tau_n(w_n) = \frac{a_n w_n + b_n}{c_n w_n + d_n} = \frac{a_n}{c_n} - \frac{a_n d_n - b_n c_n}{c_n^2 (w_n + d_n/c_n)} .$$

Let $w_n^* := -d_n/c_n$ from some n on. Then $\tau_n(w_n) \to a/c$ whenever w_n stays a uniform distance away from w_n^*; i.e., whenever

$$\liminf_{n\to\infty} \mathfrak{m}(w_n, w_n^*) > 0.$$

◇

Definition 2.2. *A sequence $\{\tau_n\}$ from $\mathcal{M}$ converges generally to a constant $\gamma \in \widehat{\mathbb{C}}$ if and only if there exists a sequence $\{w_n^\dagger\}$ from $\widehat{\mathbb{C}}$ such that*

$$\lim_{n\to\infty} \tau_n(w_n) = \gamma \quad \textit{whenever} \quad \liminf_{n\to\infty} \mathfrak{m}(w_n, w_n^\dagger) > 0 . \qquad (1.1.6)$$

Remarks.

1. We write $\tau_n \to \gamma$ to denote that $\{\tau_n\}$ converges generally to the constant $\gamma \in \widehat{\mathbb{C}}$, and $\{w_n^\dagger\}$ is called an *exceptional sequence* for$\{\tau_n\}$ in this case.

2. The exceptional sequence is not unique. If the sequence $\{\tau_n\}$ of transformations from $\mathcal{M}$ converges generally to γ with exceptional sequence $\{w_n^\dagger\}$, then every sequence $\{w_n^*\}$ with $\lim \mathfrak{m}(w_n^\dagger, w_n^*) = 0$ is exceptional. It is only its asymptotic behavior that matters.

3. We say that two sequences $\{u_n\}$ and $\{v_n\}$ from $\widehat{\mathbb{C}}$ are *equivalent* if and only if $\mathfrak{m}(u_n, v_n) \to 0$. Thereby the exceptional sequences for $\{\tau_n\}$ form an equivalence class, and $\{w_n^*\}$ or $\{w_n^\dagger\}$ are just representatives for this equivalence class.

4. If $\tau_n \to \gamma$ with exceptional sequence $\{w_n^\dagger\}$, it is not always clear what happens to $\tau_n(w_n^\dagger)$. It may converge to γ or to some other value, or it may diverge. Its behavior depends on how $\{w_n^\dagger\}$ is chosen compared to $\{\tau_n\}$

5. A sequence $\{\tau_n\}$ from $\mathcal{M}$ can not converge uniformly in $\widehat{\mathbb{C}}$ to a constant γ, not even in the chordal metric. One always has to accept exceptional sequences. This follows since $\tau_n(w_n) = \mu \neq \gamma$ for all n whenever $w_n := \tau_n^{-1}(\mu)$. Indeed, it is more of a mystery that we can get away with only one equivalence class of exceptional sequences. We shall return to this question in Section 2.1.3 on page 60.

Condition (1.1.6) is not always so easy to check since it requires that we know an exceptional sequence $\{w_n^\dagger\}$. The following equivalent definition makes checking easier:

Definition 2.3. *A sequence $\{\tau_n\}$ from $\mathcal{M}$ converges generally to a constant $\gamma \in \widehat{\mathbb{C}}$ if and only if there exist two sequences $\{v_n\}$ and $\{w_n\}$ from $\widehat{\mathbb{C}}$ such that*

$$\liminf_{n\to\infty} \mathfrak{m}(v_n, w_n) > 0 \tag{1.1.7}$$

and

$$\lim_{n\to\infty} \tau_n(v_n) = \lim_{n\to\infty} \tau_n(w_n) = \gamma\,. \tag{1.1.8}$$

In other words, we just require that $\tau_n(v_n) \to \gamma$ and $\tau_n(w_n) \to \gamma$ for two sufficiently distinct sequences $\{v_n\}$ and $\{w_n\}$. We shall prove that this definition is equivalent to the former one, and thus that the limit γ in Definition 2.3 is unique.

Proof of the equivalence of Definition 2.2 and 2.3: Let first $\{\tau_n\}$ satisfy (1.1.6). Then there clearly exist two sequences $\{v_n\}$ and $\{w_n\}$ bounded away from the exceptional sequence $\{w_n^\dagger\}$ so that (1.1.7) - (1.1.8) holds.

Next, let there exist two sequences $\{v_n\}$ and $\{w_n\}$ such that (1.1.7) - (1.1.8) holds. Let $\gamma^{(n)} := \tau_n^{-1}(\gamma)$ and

$$p_n := \begin{cases} v_n & \text{if } \mathfrak{m}(v_n, \gamma^{(n)}) > \mathfrak{m}(w_n, \gamma^{(n)}), \\ w_n & \text{otherwise}. \end{cases}$$

Then $\liminf \mathfrak{m}(p_n, \gamma^{(n)}) > 0$ and $\gamma \neq \tau_n(p_n) \to \gamma$. Let $w_n^\dagger := \tau_n^{-1}(\mu)$ for some $\mu \in \widehat{\mathbb{C}}$, $\mu \neq \gamma$, and let $\{u_n\}$ be any sequence from $\widehat{\mathbb{C}}$ with $\liminf \mathfrak{m}(u_n, w_n^\dagger) > 0$. We shall prove that $\lim \tau_n(u_n) = \gamma$. Since we already know that $\tau_n(p_n) \to \gamma$ and $\tau_n(\gamma^{(n)}) = \gamma$, we only have to consider indices n for which

$$u_n \neq \gamma^{(n)}, p_n. \tag{1.1.9}$$

For these indices we can apply the invariance of the cross ratio (1.1.4) with $\tau := \tau_n$, $z := u_n$, $u := \gamma^{(n)}$, $w := p_n$ and $v := w_n^\dagger = \tau_n^{-1}(\mu)$:

$$\frac{\mathfrak{m}(\gamma, \tau_n(u_n))}{\mathfrak{m}(\gamma, \tau_n(p_n))} \cdot \frac{\mathfrak{m}(\mu, \tau_n(p_n))}{\mathfrak{m}(\mu, \tau_n(u_n))} = \frac{\mathfrak{m}(\gamma^{(n)}, u_n)}{\mathfrak{m}(\gamma^{(n)}, p_n)} \cdot \frac{\mathfrak{m}(w_n^\dagger, p_n)}{\mathfrak{m}(w_n^\dagger, u_n)}.$$

The right hand side of this equality stays bounded as $n \to \infty$. Since $\tau_n(p_n) \to \gamma$, this means that $\mathfrak{m}(\gamma, \tau_n(u_n)) \to 0$; i.e. $\lim \tau_n(u_n) = \gamma$. □

Summary. If a sequence $\{F_n\}$ of univalent functions converges in a domain D, then it either converges to a univalent function or to a constant. So also for linear fractional transformations, but of course, for $\tau_n \in \mathcal{M}$ we know more:

(i) if $\{\tau_n(w)\}$ converges at three distinct points to distinct values, then $\{\tau_n\}$ converges to a $\tau \in \mathcal{M}$ on $\widehat{\mathbb{C}}$, uniformly with respect to the chordal metric.

(ii) if $\{\tau_n(w)\}$ converges at two distinct points to the same value, then $\{\tau_n(w)\}$ converges to this constant whenever $\{\tau_n(w)\}$ converges, except possibly at one single point.

Here (ii) is fairly trivial, but it may also be very restrictive. General convergence generalizes (ii) by allowing w to vary with n. Thereby we only have to stay away from one single point for each n, and we get convergence to the constant in "all of $\widehat{\mathbb{C}}$ except at one point for each n". Therefore there are only two possibilities for a convergent sequence $\{\tau_n\}$ from $\mathcal{M}$: either

(i) $\{\tau_n\}$ converges to some (non-singular) $\tau \in \mathcal{M}$, and we write $\tau_n \to \tau$ as $n \to \infty$,

(ii) or $\{\tau_n\}$ converges generally to some constant $\gamma \in \widehat{\mathbb{C}}$ with exceptional sequence $\{w_n^\dagger\}$. Then $\tau_n(w_n) \to \gamma$ whenever $\liminf \mathfrak{m}(w_n, w_n^\dagger) > 0$, and we write $\tau_n \to \gamma$ as $n \to \infty$.

In all other cases we say that $\{\tau_n\}$ diverges.

2.1.2 Convergence of continued fractions

So far we have used the classical convergence concept for continued fractions; i.e., $\mathbf{K}(a_n/b_n)$ converges to f if $\{S_n(0)\}$ converges to f. Since

$$S_{n+1}(\infty) = S_n(0), \tag{1.2.1}$$

classical convergence implies that $\{S_n\}$ converges generally to the value f. Incidentally, this is the reason why the classical definition of convergence works so well. Still, it has some drawbacks, as illustrated by the following example:

Example 2. Let $a_{3n+1} := 2$, $a_{3n+2} := 1$ and $a_{3n+3} := -1$ for all $n \geq 0$ in the continued fraction $\mathbf{K}(a_n/1)$; that is,

$$\mathop{\mathrm{K}}_{n=1}^{\infty} \frac{a_n}{1} := \frac{2}{1} + \frac{1}{1} - \frac{1}{1} + \frac{2}{1} + \frac{1}{1} - \frac{1}{1} + \cdots .$$

By the recurrence relations (1.2.7) on page 6 and induction, we find that

$$\begin{aligned} A_{3n-2} &= 2^n, & A_{3n-1} &= 2^n, & A_{3n} &= 0, \\ B_{3n-2} &= 2^{n+1} - 3, & B_{3n-1} &= 2^{n+1} - 2, & B_{3n} &= 1, \end{aligned}$$

and so $\lim S_{3n-2}(0) = \lim S_{3n-1}(0) = \frac{1}{2}$ and $\lim S_{3n}(0) = 0$, and thus $\{S_n(0)\}$ diverges. On the other hand

$$S_n(w_n) = \frac{A_{n-1}w_n + A_n}{B_{n-1}w_n + B_n},$$

so

$$\begin{aligned} \lim\nolimits_{n\to\infty} S_{3n-2}(w_{3n-2}) &= \tfrac{1}{2} && \text{whenever } \{w_{3n-2}\} \text{ is bounded} \\ \lim\nolimits_{n\to\infty} S_{3n-1}(w_{3n-1}) &= \tfrac{1}{2} && \text{whenever } \{w_{3n-1}\} \text{ is bounded} \\ & && \text{away from } -1 \\ \lim\nolimits_{n\to\infty} S_{3n}(w_{3n}) &= \tfrac{1}{2} && \text{whenever } \{w_{3n}\} \text{ is bounded} \\ & && \text{away from } 0 \end{aligned}$$

Indeed, with $w^\dagger_{3n-2} := \infty$, $w^\dagger_{3n-1} := -1$ and $w^\dagger_{3n} := 0$ for all $n \geq 1$, we have

$$\lim_{n\to\infty} S_n(w_n) = \frac{1}{2} \quad \text{whenever} \quad \liminf \mathfrak{m}(w_n, w^\dagger_n) > 0. \tag{1.2.2}$$

Therefore, $\{S_n\}$ converges generally to $\frac{1}{2}$ with exceptional sequence $\{w^\dagger_n\}$. ◇

We therefore define general convergence of continued fractions ([Jaco86]):

Definition 2.4. *The continued fraction* $\mathbf{K}(a_n/b_n)$ *converges generally to* $f \in \widehat{\mathbb{C}}$ *with exceptional sequence* $\{w_n^\dagger\}$ *if and only if* $\{S_n\}$ *converges generally to* f *with exceptional sequence* $\{w_n^\dagger\}$.

Properties of general convergence:

1. The classical convergence to f implies general convergence to f. Hence the new concept includes the classical one, and picks up additional cases, for instance the one in Example 2.

2. The definitions of general convergence add some insight to the classical concept of convergence. In particular it makes sense to talk about exceptional sequences $\{w_n^\dagger\}$ for a (classically) convergent continued fraction.

3. If $\mathbf{K}(a_n/b_n)$ fails to converge generally, we say that $\mathbf{K}(a_n/b_n)$ *diverges generally.*

General convergence has also another important advantage: it is sometimes much easier to prove! (This will be evident in Chapter 4.) We still use the classical definition of convergence alongside with general convergence, mainly because the $S_n(0)$–definition is so well established, but also because it has some merits of its own. For instance: it is simple and uniquely defined, and it is important in many applications. To distinguish between the two, the $S_n(0)$–convergence will be referred to as just *convergence* or *classical convergence*, whereas the convergence in Definition 2.4 always will be called *general convergence.*

2.1.3 Restrained sequences

Let $\mathbf{K}(a_n/b_n)$ converge generally to f, and let $q \neq f$. Then $S_n(w_n) = q$ for all n for the choice $w_n := S_n^{-1}(q)$. That is, $\{S_n^{-1}(q)\}$ is an exceptional sequence for $\mathbf{K}(a_n/b_n)$. This is true for every $q \neq f$. Hence all these sequences $\{S_n^{-1}(q)\}$ belong to the same equivalence class. (See Remark 3 on page 57.)

Now, $\{S_n^{-1}\}$ is also a sequence of linear fractional transformations. Since $\mathfrak{m}(S_n^{-1}(q_1), S_n^{-1}(q_2)) \to 0$ whenever $q_1 \neq f$ and $q_2 \neq f$, no subsequence of $\{S_n^{-1}\}$ can converge to a transformation from $\mathcal{M}$. Following Jacobsen and Thron ([JaTh87]) we say that $\{S_n^{-1}\}$ is a *restrained sequence.* Note that $\{S_n^{-1}\}$ need not be generally convergent since $\{S_n^{-1}(q)\}$ with $q \neq f$ may have more than one limit point. But every subsequence of $\{S_n^{-1}\}$ must have a generally convergent subsequence.

Definition 2.5. *A sequence* $\{\tau_n\}$ *from* $\mathcal{M}$ *is restrained if and only if there exists a sequence* $\{w_n^\dagger\}$ *from* $\widehat{\mathbb{C}}$ *such that whenever*

$$\liminf \mathfrak{m}(v_n, w_n^\dagger) > 0 \quad \textit{and} \quad \liminf \mathfrak{m}(w_n, w_n^\dagger) > 0, \tag{1.3.1}$$

$$\lim_{n\to\infty} \mathfrak{m}(\tau_n(v_n), \tau_n(w_n)) = 0. \tag{1.3.2}$$

Definition 2.6. *A sequence* $\{\tau_n\}$ *from* $\mathcal{M}$ *is restrained if and only if there exist two sequences* $\{v_n\}$ *and* $\{w_n\}$ *from* $\widehat{\mathbb{C}}$ *with* $\liminf \mathfrak{m}(v_n, w_n) > 0$ *such that (1.3.2) holds.*

Definition 2.7. *A sequence* $\{\tau_n\}$ *from* $\mathcal{M}$ *is restrained if and only if no subsequence of* $\{\tau_n\}$ *converges to some* $\tau \in \mathcal{M}$.

You are asked to prove that these three definitions are equivalent in Problem 3 on page 91. Definition 2.5 shows that $\{\tau_n\}$ is restrained if and only if the asymptotic behavior of $\{\tau_n(w_n)\}$ is independent of $\{w_n\}$ in the (1.3.2)-sense whenever $\liminf \mathfrak{m}(w_n, w_n^\dagger) > 0$. That is, all sequences $\{\tau_n(w_n)\}$ with $\liminf \mathfrak{m}(w_n, w_n^\dagger) > 0$ are equivalent (in the sense of Remark 3 on page 57). A sequence from this equivalence class is called a *generic sequence* for $\{\tau_n\}$. We evidently have:

Theorem 2.1. *Let the sequence* $\{\tau_n\} \subseteq \mathcal{M}$ *be restrained with generic sequence* $\{z_n\}$ *and exceptional sequence* $\{w_n^\dagger\}$. *Then* $\{\tau_n^{-1}\}$ *is restrained with generic sequence* $\{w_n^\dagger\}$ *and exceptional sequence* $\{z_n\}$.

Remarks.

1. If $\{\tau_n\}$ converges generally to some constant, then $\{\tau_n\}$ is in particular restrained.

2. If $\{\tau_n\}$ converges generally to f with exceptional sequence $\{w_n^\dagger\}$, then the constant sequence $\{f\}$ is a generic sequence for $\{\tau_n\}$, and $\{\tau_n^{-1}(q)\}$ is an exceptional sequence for every $q \neq f$. Therefore $\{\tau_n^{-1}(q)\}$ is generic for $\{\tau_n^{-1}\}$ when $q \neq f$ and $\{f\}$ is exceptional for $\{\tau_n^{-1}\}$.

Definition 2.8. *A continued fraction* $\mathbf{K}(a_n/b_n)$ *is restrained if and only if* $\{S_n\}$ *is restrained.*

Example 3. The continued fraction in Example 2 on page 59 has the approximants

$$S_n(w_n) = \frac{A_{n-1}w_n + A_n}{B_{n-1}w_n + B_n}$$

where $A_{3n-2} = A_{3n-1} = 2^n$, $A_{3n} = 0$, $B_{3n-2} = 2^{n+1} - 3$, $B_{3n-1} = 2^{n+1} - 2$ and $B_{3n} = 1$ for all n. We showed that $\{S_n\}$ converges generally to $\frac{1}{2}$ with exceptional sequence $\{w_n^\dagger\}$ given by

$$w_{3n-2}^\dagger = \infty, \quad w_{3n-1}^\dagger = -1 \quad \text{and } w_{3n}^\dagger = 0 \quad \text{for all } n.$$

Every sequence $\{z_n\}$ given by $z_n := S_n(w_n)$ with $\liminf \mathfrak{m}(w_n, w_n^\dagger) > 0$ is a generic sequence for this continued fraction. Indeed, every sequence $\{\mu_n\}$ converging to $\frac{1}{2}$ is generic for $\mathbf{K}(a_n/1)$, including the constant sequence $\{\frac{1}{2}\}$. By Theorem 2.1, $\{S_n^{-1}\}$ should therefore be restrained with exceptional sequence $\{\frac{1}{2}\}$ and generic sequence $\{w_n^\dagger\}$. Let us check this out. Simple computation shows that

$$S_n^{-1}(w) = -\frac{B_n w - A_n}{B_{n-1}w - A_{n-1}},$$

so

$$S_{3n-1}^{-1}(w_n) = -\frac{(2^{n+1}-2)w_n - 2^n}{(2^{n+1}-3)w_n - 2^n} \to -1,$$

$$S_{3n}^{-1}(w_n) = -\frac{w_n - 0}{(2^{n+1}-2)w_n - 2^n} \to 0,$$

$$S_{3n+1}^{-1}(w_n) = -\frac{(2^{n+2}-3)w_n - 2^{n+1}}{w_n - 0} \to \infty$$

whenever $\liminf d(w_n, \frac{1}{2}) > 0$. ◇

A sequence $\{\tau_n\}$ from $\mathcal{M}$ is either restrained, or it has a subsequence $\{\tau_{n_k}\}$ which converges to a non-singular transformation. These are the only two possibilities. If $\{\tau_n\}$ has no restrained subsequence, we say that $\{\tau_n\}$ is *totally non-restrained.* To determine whether $\{\tau_n\}$ is restrained or not, the following observation may be of help.

Theorem 2.2. *Let the sequence* $\{\tau_n\}$ *from* $\mathcal{M}$ *converge to a transformation* $\tau \in \mathcal{M}$. *Then* $\sigma_n := \tau_{n-1}^{-1} \circ \tau_n$ *converges to the identity transformation* $I(w) \equiv w$ *in* $\widehat{\mathbb{C}}$.

Proof : $\sigma_n = \tau_{n-1}^{-1} \circ \tau_n$ where $\tau_n \longrightarrow \tau$ and thus $\tau_n^{-1} \longrightarrow \tau^{-1}$. □

2.1.4 Tail sequences

Let us return to continued fractions $\mathbf{K}(a_n/b_n)$; i.e., to sequences $\{S_n\}$ of linear fractional transformations characterized by $S_{n+1}(\infty) = S_n(0)$ (although we might as well work with general sequences $\{\tau_n\}$; $\tau_n \in \mathcal{M}$).

Definition 2.9. *For every $t \in \widehat{\mathbb{C}}$, the sequence*

$$t_n := S_n^{-1}(t) \quad \text{for } n = 0, 1, 2, \dots \tag{1.4.1}$$

is called a tail sequence for $\mathbf{K}(a_n/b_n)$ or for $\{S_n\}$.

Properties:

1. On page 6 we defined the sequence $\{f^{(n)}\}$ of tail values for a convergent continued fraction $\mathbf{K}(a_n/b_n)$. Evidently $\{f^{(n)}\}$ is an example of a tail sequence.

2. If $\mathbf{K}(a_n/b_n)$ converges generally to f and $t \neq f$, we recognize $\{t_n\}$ as a representative for the equivalence class of exceptional sequences for $\mathbf{K}(a_n/b_n)$.

3. Since $S_n = s_1 \circ s_2 \circ \cdots \circ s_n$, we have

$$t_{n-1} = s_n(t_n) \quad \text{for } n = 1, 2, 3, \dots . \tag{1.4.2}$$

4. If $\{t_n\}_{n=0}^{\infty}$ is a tail sequence for $\mathbf{K}(a_n/b_n)$, then $\{t_n\}_{n=N}^{\infty}$ is a tail sequence for its Nth tail. More generally, if $\{t_n\}$ is a tail sequence for $\{S_n\}$, then the subsequence $\{t_{n_k}\}$ is a tail sequence for $\{S_{n_k}\}$.

5. If $\{t_n\}$ and $\{\tilde{t}_n\}$ are two tail sequences for $\mathbf{K}(a_n/b_n)$ with $t_k = \tilde{t}_k$ for one index k, then $t_n = \tilde{t}_n$ for all n by (1.4.2) since s_n, $s_n^{-1} \in \mathcal{M}$ are univalent functions.

Theorem 2.3. *Let $\{t_n\}$ be a tail sequence for $\mathbf{K}(a_n/b_n)$. Then*

$$\begin{aligned} t_n &= S_n^{-1}(t_0) = s_n^{-1} \circ s_{n-1}^{-1} \circ \cdots \circ s_1^{-1}(t_0) \\ &= -\left\{ b_n + \frac{a_n}{b_{n-1}} \underset{+}{} \frac{a_{n-1}}{b_{n-2}} \underset{+\cdots+}{} \frac{a_2}{b_1} \underset{+}{} \frac{a_1}{(-t_0)} \right\} \quad \text{for } n \geq 1 . \end{aligned} \tag{1.4.3}$$

Proof : The result follows since

$$s_k^{-1}(w) = -b_k + \frac{a_k}{w} = -\left\{b_k + \frac{a_k}{(-w)}\right\} \quad \text{for } k \geq 1\,. \tag{1.4.4}$$

□

Theorem 2.4. *Let* $\mathbf{K}(a_n/b_n)$ *have three distinct tail sequences* $\{t_n\}$, $\{\tilde{t}_n\}$ *and* $\{t_n^*\}$ *where*

$$\lim \mathfrak{m}(\tilde{t}_n, t_n^*) = 0 \quad \textit{and} \quad \liminf \mathfrak{m}(t_n, t_n^*) > 0.$$

Then $\mathbf{K}(a_n/b_n)$ *converges generally to* t_0 *with exceptional sequence* $\{\tilde{t}_n\}$.

Proof : Since $\lim \mathfrak{m}(\tilde{t}_n, t_n^*) = \lim \mathfrak{m}(S_n^{-1}(\tilde{t}_0),\, S_n^{-1}(t_0^*)) = 0$ where $\tilde{t}_0 \neq t_0^*$, it follows that $\{S_n^{-1}\}$ is restrained with generic sequence $\{\tilde{t}_n\}$. On the other hand we know that $\liminf \mathfrak{m}(t_n, t_n^*) = \liminf \mathfrak{m}(S_n^{-1}(t_0),\, S_n^{-1}(t_0^*)) > 0$. Therefore the constant sequence $\{t_0\}$ is exceptional for $\{S_n^{-1}\}$. Hence $\{S_n\}$ is restrained with generic sequence $\{t_0\}$ and exceptional sequence $\{\tilde{t}_n\}$ (Theorem 2.1 on page 61). That is, $\mathbf{K}(a_n/b_n)$ converges generally to t_0. □

The tail sequence $\{\zeta_n\}$ defined by

$$\zeta_n := S_n^{-1}(\infty) \quad \text{for } n = 0, 1, 2, \dots\,, \tag{1.4.5}$$

plays a special role in our theory. Here $\zeta_0 = \infty$, $\zeta_1 = -b_1$, and by (1.4.3)

$$\zeta_n = -\frac{B_n}{B_{n-1}} = -b_n - \frac{a_n}{b_{n-1}} \genfrac{}{}{0pt}{}{}{+} \frac{a_{n-1}}{b_{n-2}} \genfrac{}{}{0pt}{}{}{+} \cdots \genfrac{}{}{0pt}{}{}{+} \frac{a_2}{b_1} \tag{1.4.6}$$

for $n \geq 2$. In earlier work one has rather used the notation

$$h_n := -S_n^{-1}(\infty), \tag{1.4.7}$$

and the sequence $\{h_n\}$ is called the *critical tail sequence* for $\mathbf{K}(a_n/b_n)$ (although, strictly speaking, it is $\{-h_n\}$ which is a tail sequence). The importance of $\{\zeta_n\}$ or $\{h_n\}$ is that if $\mathbf{K}(a_n/b_n)$ converges generally to $f \neq \infty$, then $\{-h_n\}$ is an exceptional sequence which often shows up in useful formulas.

Example 4. In Example 2 on page 59 we saw that the 3-periodic continued fraction

$$\mathop{\mathrm{K}}_{n=1}^{\infty} \frac{a_n}{1} = \frac{2}{1} \genfrac{}{}{0pt}{}{}{+} \frac{1}{1} \genfrac{}{}{0pt}{}{}{-} \frac{1}{1} \genfrac{}{}{0pt}{}{}{+} \frac{2}{1} \genfrac{}{}{0pt}{}{}{+} \frac{1}{1} \genfrac{}{}{0pt}{}{}{-} \frac{1}{1} \genfrac{}{}{0pt}{}{}{+} \frac{2}{1} \genfrac{}{}{0pt}{}{}{+} \dots$$

converges generally to $f = 1/2$ with exceptional sequence

$$w_n^\dagger: \quad 0,\ \infty,\ -1,\ 0,\ \infty,\ -1,\ 0,\ \ldots$$

(starting with $w_0^\dagger = 0$). Every tail sequence $\{t_n\}$ with $t_0 \neq \frac{1}{2}$ must have the same asymptotic behavior; i.e.,

$$\lim t_{3n} = 0, \quad \lim t_{3n+1} = \infty, \quad \lim t_{3n+2} = -1.$$

The tail values $\{f^{(n)}\}$ is the tail sequence starting with $t_0 = \frac{1}{2}$:

$$\begin{aligned} f^{(0)} &= \frac{1}{2}, \\ f^{(1)} &= s_1^{-1}(f^{(0)}) = -1 + \frac{2}{1/2} = 3, \\ f^{(2)} &= s_2^{-1}(f^{(1)}) = -1 + \frac{1}{3} = -\frac{2}{3}, \\ f^{(3)} &= s_3^{-1}(f^{(2)}) = -1 + \frac{-1}{-2/3} = \frac{1}{2}, \end{aligned}$$

and so on. That is, $\{f^{(n)}\}$ is periodic with period 3, as it just has to be. In Problem 4 on page 92 you are asked to prove that $\{w_n^\dagger\}$ as given above and $\{f^{(n)}\}$ are the only periodic tail sequences for $\mathbf{K}(a_n/1)$. $\diamond$

2.1.5 Tail sequences and three term recurrence relations

There are strong connections between tail sequences for a continued fraction $\mathbf{K}(a_n/b_n)$ and solutions of the corresponding recurrence relation

$$X_n = b_n X_{n-1} + a_n X_{n-2} \quad \text{for } n = 1, 2, 3, \ldots . \tag{1.5.1}$$

We have already seen that $\zeta_n = -B_n/B_{n-1}$ forms a tail sequence, where B_n is a solution of (1.5.1). The following is therefore not so surprising:

Theorem 2.5. *Let $\{X_n\}_{n=-1}^\infty$ be a non-trivial solution of (1.5.1) (i.e., not all $X_n = 0$). Then $t_n := -X_n/X_{n-1}$ is well defined in $\widehat{\mathbb{C}}$ and $\{t_n\}_{n=0}^\infty$ is a tail sequence for $\mathbf{K}(a_n/b_n)$.*

Proof : The sequence $\{t_n\}$ is well defined since $X_n = X_{n-1} = 0$ for a fixed n implies that all $X_n = 0$, something we have excluded. From (1.5.1) we find that

$$-\frac{X_n}{X_{n-1}} = -b_n + \frac{a_n}{-X_{n-1}/X_{n-2}},$$

that is, $t_n = -b_n + a_n/t_{n-1} = s_n^{-1}(t_{n-1})$ for all n. The result follows therefore from (1.4.2). □

There is also an interesting connection the other way around: we can express solutions of (1.5.1) in terms of tail sequences. We shall only do so for the particular solutions $\{A_n\}$ and $\{B_n\}$, which are the solutions of (1.5.1) starting with

$$A_{-1} = 1, \quad A_0 = 0, \quad A_1 = a_1, \quad B_{-1} = 0, \quad B_0 = 1, \quad B_1 = b_1, \tag{1.5.2}$$

in other words: the canonical numerators and denominators of $\mathbf{K}(a_n/b_n)$. But it is not difficult to extend the results to general solutions $\{X_n\}$ of (1.5.1), since it turns out that every solution $\{X_n\}$ is a linear combination of $\{A_n\}$ and $\{B_n\}$. The formulas may seem rather complicated, but they are indeed quite useful. As always: an empty sum is equal to 0 and an empty product is equal to 1. For some history on this subject we refer to the remark section on page 89.

Theorem 2.6. *Let $\{t_n\}_{n=0}^{\infty}$ be a tail sequence for $\mathbf{K}(a_n/b_n)$ with all $t_n \neq \infty$. Then all $t_n \neq 0, -b_n$ and*

$$B_n + B_{n-1}t_n = \prod_{k=1}^{n}(b_k + t_k), \tag{1.5.3}$$

$$A_n - B_n t_0 = \prod_{k=0}^{n}(-t_k), \tag{1.5.4}$$

$$A_n = t_0 \sum_{k=1}^{n}\prod_{j=1}^{k}(b_j + t_j) \prod_{j=k+1}^{n}(-t_j), \tag{1.5.5}$$

$$B_n = \sum_{k=0}^{n}\prod_{j=1}^{k}(b_j + t_j) \prod_{j=k+1}^{n}(-t_j), \tag{1.5.6}$$

$$t_0 - \frac{A_n}{B_n} = t_0/\Sigma_n \tag{1.5.7}$$

$$\textit{where } \Sigma_n := \sum_{k=0}^{n} P_k \ \textit{ and } P_k := \prod_{j=1}^{k}\frac{b_j + t_j}{-t_j}, \tag{1.5.8}$$

$$t_0 - S_n(w) = \frac{t_0(w - t_n)}{w\Sigma_{n-1} - t_n\Sigma_n}. \tag{1.5.9}$$

Proof : That $t_n \neq 0, -b_n$ follows from (1.4.2) which says that $t_{n-1} = a_n/(b_n + t_n)$ for all n. Now, $B_0 + B_{-1}t_0 = 1$ and $B_1 + B_0 t_1 = b_1 + t_1$. Since $a_n = t_{n-1}(b_n + t_n)$

by (1.4.2), and thus, by the recurrence relation,

$$\begin{aligned}B_n + B_{n-1}t_n &= b_nB_{n-1} + a_nB_{n-2} + B_{n-1}t_n \\ &= (b_n + t_n)B_{n-1} + t_{n-1}(b_n + t_n)B_{n-2} = (b_n + t_n)(B_{n-1} + B_{n-2}t_{n-1}),\end{aligned}$$

equality (1.5.3) follows by induction.

To prove (1.5.4) we again use induction. Since $A_{-1} - B_{-1}t_0 = 1$, $A_0 - B_0t_0 = -t_0$ and $a_m = t_{m-1}(b_m + t_m)$, we have

$$\begin{aligned}A_m - B_mt_0 &= b_mA_{m-1} + a_mA_{m-2} - t_0(b_mB_{m-1} + a_mB_{m-2}) \\ &= b_m(A_{m-1} - B_{m-1}t_0) + t_{m-1}(b_m + t_m)(A_{m-2} - B_{m-2}t_0).\end{aligned}$$

Hence, if (1.5.4) holds for $n := m - 1$ and $n := m - 2$, then

$$\begin{aligned}A_m - B_mt_0 &= b_m\prod_{k=0}^{m-1}(-t_k) + t_{m-1}(b_m + t_m)\prod_{k=0}^{m-2}(-t_k) \\ &= b_m\prod_{k=0}^{m-1}(-t_k) - (b_m + t_m)\prod_{k=0}^{m-1}(-t_k) = \prod_{k=0}^{m}(-t_k).\end{aligned}$$

The expression (1.5.6) for B_n follows from (1.5.3):

$$\begin{aligned}B_n &= \prod_{k=1}^{n}(b_k + t_k) - t_nB_{n-1} \\ &= \prod_{k=1}^{n}(b_k + t_k) - t_n\prod_{k=1}^{n-1}(b_k + t_k) + t_nt_{n-1}B_{n-2} = \cdots = \\ &= \prod_{k=1}^{n}(b_k + t_k) - t_n\prod_{k=1}^{n-1}(b_k + t_k) + t_nt_{n-1}\prod_{k=1}^{n-2}(b_k + t_k) - \cdots + \prod_{k=1}^{n}(-t_k)\end{aligned}$$

which is exactly the expression for B_n. The expression (1.5.5) follows since by Lemma 1.4 on page 10, $A_n = a_1B_{n-1}^{(1)}$ where $B_k^{(1)}$ denotes the canonical denominators of the first tail of $\mathbf{K}(a_n/b_n)$. Finally, (1.5.7) and (1.5.9) follow from (1.5.4) and (1.5.6) (after some work). □

Formulas (1.5.7) and (1.5.9) are in particular useful for proving classical convergence and estimate the speed of this convergence. For instance, the following corollary follows easily:

Corollary 2.7. (Waadeland's Tail Theorem.) *Let $\{t_n\}_{n=0}^{\infty}$ be a tail sequence for $\mathbf{K}(a_n/b_n)$ with all $t_n \neq \infty$. Then $\mathbf{K}(a_n/b_n)$ converges in the classical sense if and only if $\{\Sigma_n\}$ given by (1.5.8) converges in $\widehat{\mathbb{C}}$.*
If $\{\Sigma_n\}$ converges to $\Sigma_\infty \in \widehat{\mathbb{C}}$, then $\mathbf{K}(a_n/b_n)$ converges to $f := t_0\Big(1 - \frac{1}{\Sigma_\infty}\Big)$, and $f - f_n = t_0\Big(\frac{1}{\Sigma_n} - \frac{1}{\Sigma_\infty}\Big)$.

Example 5. For given $b, t \in \mathbb{C}$ with $t \neq 0, -b$ and $b \neq 0$, the periodic continued fraction

$$\frac{t(b+t)}{b} \, + \, \frac{t(b+t)}{b} \, + \, \frac{t(b+t)}{b} \, + \cdots$$

has a tail sequence $\{t_n\}$ with all $t_n := t$. Therefore, by (1.5.7)

$$t - f_n = t - \frac{A_n}{B_n} = \frac{t}{\sum_{k=0}^{n} \left(\frac{b+t}{-t}\right)^k} = \frac{t\left(1 + \frac{b+t}{t}\right)}{1 - \left(\frac{b+t}{-t}\right)^{n+1}} \tag{1.5.10}$$

which shows that $f_n \to t$ if $|t| < |b+t|$ and $f_n \to -(b+t)$ if $|t| > |b+t|$.
If $|t| = |b+t|$, then still $t \neq b+t$ under our conditions, and $\{t - f_n\}$ oscillates. If in particular $(b+t)/t$ is an Nth root of unity; i.e., $((b+t)/(-t))^N = 1$ for some $N \in \mathbb{N}$, then $\{t - f_n\}$ is periodic. Otherwise $\{((b+t)/(-t))^{n+1}\}$ is a dense sequence of distinct points on the unit circle. ◇

Formulas (1.5.3) and (1.5.6) imply that

$$\begin{aligned}\zeta_n &= -\frac{B_n}{B_{n-1}} = -\frac{B_n + B_{n-1}t_n}{B_{n-1}} + t_n = t_n - \frac{\prod_{k=1}^{n}(b_k + t_k)}{B_{n-1}} \\ &= t_n - \frac{\prod_{k=1}^{n}(b_k + t_k)}{\sum_{k=0}^{n-1}\prod_{j=1}^{k}(b_j + t_j)\prod_{j=k+1}^{n-1}(-t_j)} = t_n + \frac{t_n P_n}{\Sigma_{n-1}}\end{aligned} \tag{1.5.11}$$

where we have used the notation (1.5.8). That is, the critical tail sequence is given in terms of $\{t_n\}$. More generally, we can express any tail sequence $\{T_n\}$ in terms of $\{t_n\}$ when all $t_n \neq \infty$:

Corollary 2.8. *Let $\{t_n\}$ be a tail sequence for $\mathbf{K}(a_n/b_n)$ with all $t_n \neq \infty$. Then the tail sequence $\{T_n\}$ for $\mathbf{K}(a_n/b_n)$ beginning with T_0 is given by*

$$T_n = t_n \frac{(t_0 - T_0)\Sigma_n - t_0}{(t_0 - T_0)\Sigma_{n-1} - t_0} \quad \text{where } \Sigma_n := \sum_{k=0}^{n} \prod_{j=1}^{k} \frac{b_j + t_j}{-t_j}.$$

Proof : Since

$$T_n = S_n^{-1}(T_0) = -\frac{A_n - B_n T_0}{A_{n-1} - B_{n-1} T_0}$$
$$= -\frac{A_n - B_n t_0 + B_n(t_0 - T_0)}{A_{n-1} - B_{n-1} t_0 + B_{n-1}(t_0 - T_0)},$$

the result follows by use of (1.5.4) and (1.5.6). □

To any given sequence $\{t_n\}$ from $\mathbb{C}\setminus\{0\}$ we can always construct continued fractions $\mathbf{K}(a_n/b_n)$ for which $\{t_n\}$ is a tail sequence. The only requirement is that

$$t_{n-1}(b_n + t_n) = a_n.$$

If $\{b_n\}$ is chosen such that $\{\Sigma_n\}$ given by (1.5.8) converges in $\widehat{\mathbb{C}}$, then this continued fraction converges. Its value is t_0 if $\Sigma_\infty = \infty$ and $t_0(1 - 1/\Sigma_\infty)$ otherwise (Corollary 2.7). Thereby we have derived a continued fraction identity

$$f = \mathbf{K}(a_n/b_n).$$

Example 6. Let $\{t_n\}$ be given by

$$t_{2n} := n + 1, \quad t_{2n+1} := 1 \quad \text{for } n \geq 0,$$

and choose $b_n := 1$ for all n. Then $\{t_n\}$ is a tail sequence for $\mathbf{K}(a_n/1)$ given by

$$a_{2n-1} := t_{2n-2}(1 + t_{2n-1}) = 2n, \quad a_{2n} := t_{2n-1}(1 + t_{2n}) = n + 2.$$

Since

$$P_{2k} = \prod_{j=1}^{k} \frac{1 + t_{2j-1}}{-t_{2j-1}} \cdot \frac{1 + t_{2j}}{-t_{2j}} = \prod_{j=1}^{k} 2 \cdot \frac{j+2}{j+1} = 2^k \frac{k+2}{2} = 2^{k-1}(k+2)$$
$$P_{2k-1} = P_{2k-2} \cdot \frac{1 + t_{2k-1}}{-t_{2k-1}} = -2\,P_{2k-2} = -2^{k-1}(k+1),$$

it follows that

$$\Sigma_{2n} = 1 + \sum_{k=1}^{n} (P_{2k-1} + P_{2k}) = 1 + \sum_{k=1}^{n} 2^{k-1} = 1 + \frac{1 - 2^n}{1 - 2} = 2^n \to \infty,$$
$$\Sigma_{2n+1} = \Sigma_{2n} + P_{2n+1} = 2^n - 2^n(n+2) \to -\infty.$$

Therefore

$$\underset{n=1}{\overset{\infty}{\mathrm{K}}} \frac{a_n}{1} := \frac{2}{1} \, {}_{+} \, \frac{3}{1} \, {}_{+} \, \frac{4}{1} \, {}_{+} \, \frac{4}{1} \, {}_{+} \, \frac{6}{1} \, {}_{+} \, \frac{5}{1} \, {}_{+} \, \frac{8}{1} \, {}_{+} \, \frac{6}{1} \, {}_{+} \, \frac{10}{1} \, {}_{+} \, \frac{7}{1} \, {}_{+} \cdots = 1\,.$$

◇

2.1.6 Value sets

Tail sequences are useful to determine convergence properties of certain continued fractions. We shall return to this later. We also have another potent tool to determine such properties: *value sets*!

Definition 2.10. *A sequence $\{V_n\}_{n=0}^{\infty}$ of sets $V_n \subset \widehat{\mathbb{C}}$ is a sequence of value sets for $\mathbf{K}(a_n/b_n)$ if and only if*

$$s_n(V_n) = \frac{a_n}{b_n + V_n} \subseteq V_{n-1} \quad \textit{for} \quad n = 1, 2, 3, \ldots\,. \tag{1.6.1}$$

If $\{V_n\}$ is 1-periodic, so that all $V_n = V$, we say that V is a *simple value set* for $\mathbf{K}(a_n/b_n)$. If $\{V_n\}$ is 2-periodic, so that all $V_{2n} = V_0$ and all $V_{2n+1} = V_1$, we say that (V_0, V_1) are *twin value sets* for $\mathbf{K}(a_n/b_n)$.

In the literature $\{V_n\}$ is often referred to as *prevalue sets*, and the term *value sets* is reserved for prevalue sets with the property $a_n/b_n \in V_{n-1}$ for all $n \geq 1$. This is a natural choice when convergence is based on the asymptotic behavior of classical approximants. Our wider view of approximants $S_n(w_n)$ and convergence calls for the wider concept of value sets in Definition 2.10.

The importance of value sets follows from the nestedness

$$K_n := S_n(V_n) = S_{n-1}(s_n(V_n)) \subseteq S_{n-1}(V_{n-1}) = K_{n-1} \tag{1.6.2}$$

which means that

$$S_n(w_n) \in K_n \subseteq K_{n-1} \subseteq V_0 \quad \text{for } w_n \in V_n. \tag{1.6.3}$$

Let us assume that V_n are closed, non-empty sets. Then K_n are closed, non-empty sets, and thus the nestedness (1.6.2) implies that the limit set

$$K := \lim K_n := \bigcap_{n=1}^{\infty} K_n \tag{1.6.4}$$

exists and is closed and non-empty. If $\operatorname{diam}(K) = 0$, the *limit point case*, then we have hit the jackpot, since then:

- K consists of just one point, say $f \in \widehat{\mathbb{C}}$, and $S_n(w_n) \to f$ whenever $w_n \in V_n$ for all n.
- if $0 \in V_n$ for all n, then $\mathbf{K}(a_n/b_n)$ converges to f in the classical sense.
- if $\liminf \operatorname{diam}_{\mathfrak{m}}(V_n) > 0$ for the chordal diameter

$$\operatorname{diam}_{\mathfrak{m}}(V_n) := \sup\{\mathfrak{m}(v, w);\ v, w \in V_n\} \tag{1.6.5}$$

  of V_n, then $\mathbf{K}(a_n/b_n)$ converges generally to f.
- the truncation error bound

$$|f - S_n(w_n)| \leq \operatorname{diam}(K_n) \quad \text{for } w_n \in V_n \tag{1.6.6}$$

  approaches 0 as $n \to \infty$. (Clearly, $S_n(w_n) \in K_n$ and $f \in K \subseteq K_n$.)

But also if the limit point case fails to occur, the existence of a sequence of value sets for a continued fraction is useful. With the notation (1.6.4) we have:

Theorem 2.9. *Let $\{V_n\}$ be a sequence of closed value sets for $\mathbf{K}(a_n/b_n)$. Then $t_n \in S_n^{-1}(K) \subseteq V_n$ for every tail sequence $\{t_n\}$ starting with a $t_0 \in K$, and every tail sequence $\{t_n\}$ with $t_k \in \widehat{\mathbb{C}} \setminus V_k$ for some $k \in \mathbb{N} \cup \{0\}$ satisfies $t_n \in \widehat{\mathbb{C}} \setminus V_n$ for all $n \geq k$.*

Proof : Let first $t_0 \in K$. Then $t_n = S_n^{-1}(t_0) \in S_n^{-1}(K) \subseteq S_n^{-1}(K_n) = V_n$. Next, let $t_k \in \widehat{\mathbb{C}} \setminus V_k$. Since $t_k = s_{k+1}(t_{k+1}) \notin V_k$ whereas $s_{k+1}(V_{k+1}) \subseteq V_k$, it follows that $t_{k+1} \in \widehat{\mathbb{C}} \setminus V_{k+1}$. By induction it follows therefore that $t_n \in \widehat{\mathbb{C}} \setminus V_n$ for all $n \geq k$. □

Corollary 2.10. *Let $\{V_n\}$ be a sequence of closed value sets for the generally convergent continued fraction $\mathbf{K}(a_n/b_n)$ with $\limsup_{n\to\infty} \operatorname{diam}_{\mathfrak{m}}(V_n) > 0$. Then $f^{(n)} \in V_n$ for all n for the tail values $\{f^{(n)}\}$ of $\mathbf{K}(a_n/b_n)$. If moreover $V_k \neq \widehat{\mathbb{C}}$ for some $k \in \mathbb{N} \cup \{0\}$, then $\mathbf{K}(a_n/b_n)$ has an exceptional sequence $\{w_n^\dagger\}$ with $w_n^\dagger \in \widehat{\mathbb{C}} \setminus V_n$ for all $n \geq k$.*

Proof : We first prove that the value f of $\mathbf{K}(a_n/b_n)$ belongs to K. Let $\{w_n^\dagger\}$ be an exceptional sequence for $\mathbf{K}(a_n/b_n)$. Under our conditions there exists a subsequence $\{V_{n_k}\}$ of $\{V_n\}$ with $\operatorname{diam}_{\mathfrak{m}}(V_{n_k}) \geq d$ for some $d > 0$. Therefore there exists a $w_{n_k} \in V_{n_k}$ with $\mathfrak{m}(w_{n_k}, w_{n_k}^\dagger) \geq d/2$ for each $k \in \mathbb{N}$. Hence $S_{n_k}(w_{n_k}) \to f$, and the result follows since $S_{n_k}(w_{n_k}) \in K_{n_k} \to K$ Hence $f^{(n)} \in V_n$ for all n by Theorem 2.9.

Let $V_k \neq \widehat{\mathbb{C}}$ and $t_k \in \widehat{\mathbb{C}} \setminus V_k$. Then $t_k \neq f^{(k)}$, and the corresponding tail sequence $\{t_n\}$ is an exceptional sequence with $t_n \in \widehat{\mathbb{C}} \setminus V_n$ for all $n \geq k$ by Theorem 2.9. □

Remark. In the particular case where all $b_n = b \neq 0$ and all $V_n = V$, then

$$s_n(V) := \frac{a_n}{b+V} \subseteq V \iff s_n^{-1}(-b-V) = -b + \frac{a_n}{-b-V} \subseteq -b-V,$$

so $\{w_n^\dagger\}$ has all its limit points in $-b-V$. A similar result holds if all $b_{2n-1} = b_1$, $b_{2n} = b_2$ and $V_{2n} = V_0$, $V_{2n+1} = V_1$.

Corollary 2.10 generalizes a result from [Jaco86b]. Of course, it also holds if we replace the condition on $\mathrm{diam}_{\mathrm{m}}(V_n)$ with the condition $0 \in V_n$ for infinitely many n. Another question is: how can we find value sets for a given continued fraction?

Example 7. Let the real interval $V := [-\frac{1}{2}, \frac{1}{2}]$ be a simple value set for $\mathbf{K}(a_n/1)$; that is,

$$\frac{a_n}{1+V} \subseteq V \quad \text{for all } n.$$

Since $1 + V = [\frac{1}{2}, \frac{3}{2}]$, we have $1/(1+V) = [\frac{2}{3}, 2]$, and thus $a_n/(1+V) \subseteq V$ if and only if either $a_n > 0$ and $[\frac{2a_n}{3}, 2a_n] \subseteq [-\frac{1}{2}, \frac{1}{2}]$ or $a_n < 0$ and $[2a_n, \frac{2a_n}{3}] \subseteq [-\frac{1}{2}, \frac{1}{2}]$; i.e., if and only if $a_n \in [-\frac{1}{4}, \frac{1}{4}]$. ◇

In this example we started with the value set, and found sufficient conditions on a continued fraction for V to be its simple value set. This is the easy way. To find value sets for a given continued fraction is harder:

Example 8. By Example 4 on page 64 the 3-periodic continued fraction

$$\mathop{\mathrm{K}}_{n=1}^{\infty} \frac{a_n}{1} := \frac{2}{1} + \frac{1}{1} - \frac{1}{1} + \frac{2}{1} + \frac{1}{1} - \frac{1}{1} + \frac{2}{1} + \dots$$

converges generally and has tail values

$$f^{(3n)} = \frac{1}{2}, \quad f^{(3n+1)} = 3, \quad f^{(3n+2)} = -\frac{2}{3}.$$

We want to find useful value sets for $\mathbf{K}(a_n/1)$. Therefore we want V_n to contain $f^{(n)}$ (Corollary 2.10). The easiest procedure seems to be to look for circular disks

V_n centered at $f^{(n)}$. This leads for instance to the three-periodic sequence

$$V_{3n} = \left\{w \in \mathbb{C};\ \left|w - \frac{1}{2}\right| < \frac{1}{6}\right\}$$
$$V_{3n+1} = \left\{w \in \mathbb{C};\ \left|w - 3\right| < 1\right\}$$
$$V_{3n+2} = \left\{w \in \mathbb{C};\ \left|w + \frac{2}{3}\right| < \frac{1}{12}\right\}.$$

We shall return to this idea in Section 5.2 on page 238. ◇

Now, $\mathbf{K}(a_n/b_n)$ does not necessarily converge generally, even if it is *ε-contractive* with respect to V; i.e., if there exists an $\varepsilon > 0$ such that $s_n(V) \subseteq V$, but omits a chordal ε-disk

$$\mathfrak{B}_{\mathfrak{m}}(\gamma_n, \varepsilon) := \{w \in \widehat{\mathbb{C}};\ \mathfrak{m}(w, \gamma_n) \le \varepsilon\} \tag{1.6.7}$$

with some center $\gamma_n \in V$ for all $n \in \mathbb{N}$. (The parabola theorem on page 151 constitutes a counterexample.) But we can conclude that $\mathbf{K}(a_n/b_n)$ is restrained:

Theorem 2.11. *Let V be a simple value set for $\mathbf{K}(a_n/b_n)$. If $\mathbf{K}(a_n/b_n)$ is ε-contractive with respect to V, then $\mathbf{K}(a_n/b_n)$ is restrained.*

Proof : Assume that $\mathbf{K}(a_n/b_n)$ is not restrained. Then there exists a subsequence $\{S_{n_k}\}$ of $\{S_n\}$ converging to some $S \in \mathcal{M}$. Therefore $\sigma_k := S_{n_k}^{-1} \circ S_{n_{k+1}} = s_{n_k+1} \circ \cdots \circ s_{n_{k+1}}$ converges uniformly to $I(w) \equiv w$ (as in Theorem 2.2 on page 62). This is impossible when $\mathbf{K}(a_n/b_n)$ is ε-contractive with respect to V. □

Remark. A similar result holds if $\{V_n\}$ is a periodic sequence of value sets for $\mathbf{K}(a_n/b_n)$.

2.1.7 Element sets

Families of continued fractions can be described by means of *element sets* $\{\Omega_n\}_{n=1}^{\infty}$ where $\Omega_n \subseteq \mathbb{C}^2$. A continued fraction $\mathbf{K}(a_n/b_n)$ belongs to the family if $(a_n, b_n) \in \Omega_n$ for all $n \in \mathbb{N}$. For short, we say that $\mathbf{K}(a_n/b_n)$ is a continued fraction from $\{\Omega_n\}$.

Similarly we say that $\mathbf{K}(a_n/1)$ is a continued fraction from the element sets $\{E_n\}_{n=1}^{\infty}$ where $E_n \subseteq \mathbb{C}$ if $a_n \in E_n$ for all $n \in \mathbb{N}$. And $\mathbf{K}(1/b_n)$ is a continued fraction from the element sets $\{G_n\}_{n=1}^{\infty}$ where $G_n \subseteq \mathbb{C}$ if $b_n \in G_n$ for all $n \in \mathbb{N}$.

If $\{V_n\}_{n=0}^{\infty}$; $V_n \subseteq \widehat{\mathbb{C}}$ is a sequence of value sets for every continued fraction from a sequence $\{\Omega_n\}$ (or $\{E_n\}$ or $\{G_n\}$) of element sets, we say that $\{V_n\}$ is a sequence of value sets for $\{\Omega_n\}$ (or $\{E_n\}$ or $\{G_n\}$). To a given sequence of value sets, there exists an extreme sequence of element sets:

Definition 2.11. *For a given sequence* $\{V_n\}_{n=0}^{\infty}$ *with* $V_n \subseteq \widehat{\mathbb{C}}$, *the sequence* $\{\Omega_n\}$ *given by*

$$\Omega_n := \{(a,b) \in \mathbb{C}^2;\ a/(b+V_n) \subseteq V_{n-1}\} \tag{1.7.1}$$

is called the element sets (for continued fractions $\mathbf{K}(a_n/b_n)$ *) corresponding to* $\{V_n\}$.

These element sets characterize all continued fractions for which $\{V_n\}$ is a sequence of value sets (which may be an empty family). Similarly

$$E_n := \{a \in \mathbb{C};\ a/(1+V_n) \subseteq V_{n-1}\} \quad \text{for } n = 1,2,3,\dots \tag{1.7.2}$$

and

$$G_n := \{b \in \mathbb{C};\ 1/(b+V_n) \subseteq V_{n-1}\} \quad \text{for } n = 1,2,3,\dots \tag{1.7.3}$$

are called element sets for continued fractions $\mathbf{K}(a_n/1)$ and $\mathbf{K}(1/b_n)$ respectively, corresponding to $\{V_n\}$.

If all $\Omega_n = \Omega$, we say that Ω is a *simple element set.* Similarly, if $\{\Omega_n\}$ is 2-periodic, we say that (Ω_1, Ω_2) are *twin element sets* for continued fractions $\mathbf{K}(a_n/b_n)$. Of course we use the same vocabulary to describe the element sets for $\mathbf{K}(a_n/1)$ and $\mathbf{K}(1/b_n)$. For instance, the set $E := [-\frac{1}{4}, \frac{1}{4}]$ in Example 7 on page 72 is a simple element set corresponding to the simple value set $V := [-\frac{1}{2}, \frac{1}{2}]$.

Example 9. Let $V := \overline{\mathbb{D}}$, the closure of the unit disk $\mathbb{D} := \{w \in \mathbb{C};\ |w| < 1\}$. Then $a/(b+V) \subseteq V$ if and only if $|b| \geq |a| + 1$. Therefore

$$\Omega := \{(a,b) \in \mathbb{C}^2;\ |b| \geq |a| + 1\}$$

is a simple element set corresponding to V, and V is a simple value set for every continued fraction $\mathbf{K}(a_n/b_n)$ from Ω. On page 129 we shall prove that every continued fraction from Ω converges. $\diamond$

Example 10. Let

$$V_0 := \mathfrak{B}(1,3) \quad \text{and} \quad V_1 := \widehat{\mathbb{C}} \setminus \mathfrak{B}(0,2)^{\circ}$$

where $\mathfrak{B}(a,r) := \{w \in \mathbb{C};\ |w-a| \le r\}$ for $a \in \mathbb{C}$ and $r > 0$, and $\mathfrak{B}(a,r)^\circ$ is its interior. Then $1+V_0 = \mathfrak{B}(2,3)$, $1/(1+V_0) = \widehat{\mathbb{C}} \setminus \mathfrak{B}(-\frac{2}{5}, \frac{3}{5})^\circ$, and thus $a/(1+V_0) = \widehat{\mathbb{C}} \setminus \mathfrak{B}(-\frac{2a}{5}, \frac{3|a|}{5})^\circ$. Therefore $a/(1+V_0) \subseteq V_1$ if and only if

$$\left|-\frac{2a}{5} - 0\right| + 2 \le \frac{3|a|}{5}; \quad \text{i.e., } |a| \ge 10.$$

Similarly, $1+V_1 = \widehat{\mathbb{C}} \setminus \mathfrak{B}(1,2)^\circ$, $1/(1+V_1) = \mathfrak{B}(-\frac{1}{3}, \frac{2}{3})$, and $a/(1+V_1) = \mathcal{B}(-\frac{a}{3}, \frac{2|a|}{3})$ which is contained in V_0 if and only if

$$\left|-\frac{a}{3} - 1\right| + \frac{2|a|}{3} \le 3; \quad \text{i.e., } |a+3| + 2|a| \le 9.$$

Therefore

$$E_1 := \{a \in \mathbb{C};\ |a+3| + 2|a| \le 9\}, \quad E_2 := \{a \in \mathbb{C};\ |a| \ge 10\}$$

are the twin element sets for continued fractions $\mathbf{K}(a_n/1)$ corresponding to (V_0, V_1). ◇

To a given sequence of element sets there also exists an extreme sequence of value sets:

Theorem 2.12. *For a given sequence $\{\Omega_n\}_{n=1}^\infty$ with $\Omega_n \subseteq \widehat{\mathbb{C}}^2$, let for each $n \in \mathbb{N} \cup \{0\}$, $V_n \in \widehat{\mathbb{C}}$ be the set of nth tail values for every generally convergent continued fraction from $\{\Omega_n\}$. Then $\{V_n\}$ is a sequence of value sets for $\{\Omega_n\}$.*

Proof : Let $(a,b) \in \Omega_n$ with $a \ne 0$ and $f^{(n)} \in V_n$ be arbitrarily chosen. Then

$$f^{(n)} = \frac{a_{n+1}}{b_{n+1}} + \frac{a_{n+2}}{b_{n+2}} + \frac{a_{n+3}}{b_{n+3}} + \cdots$$

for some generally convergent continued fraction $\mathbf{K}(a_n/b_n)$ from $\{\Omega_n\}$, and thus also

$$f^{(n-1)} := \frac{a}{b+f^{(n)}} = \frac{a}{b} + \frac{a_{n+1}}{b_{n+1}} + \frac{a_{n+2}}{b_{n+2}} + \frac{a_{n+3}}{b_{n+3}} + \cdots$$

is the $(n-1)$th tail value for some generally convergent continued fraction from $\{\Omega_n\}$. That is, $f^{(n-1)} \in V_{n-1}$. □

This particular sequence of value sets for a sequence $\{\Omega_n\}$ of element sets is called the sequence of *limit sets* for $\{\Omega_n\}$, and we use the terms *simple limit set* and *twin limit sets* with obvious interpretation.

More importantly: when does $\mathbf{K}(a_n/b_n)$ from $\{\Omega_n\}$ converge? This will be the topic of Chapter 3. The following slight generalization of a result from [Jaco86b] gives a simple answer:

Theorem 2.13. *Let (E_1, E_2) be twin element sets corresponding to the twin value sets (V_0, V_1), and let E_1^*, E_2^* be closed, bounded subsets of E_1° and E_2° respectively. Then every continued fraction $\mathbf{K}(a_n/1)$ from (E_1^*, E_2^*) converges.*

The proof is postponed to Chapter 4. The last example in this section indicates how the knowledge of corresponding element and value sets can be used to estimate continued fraction values when we already know convergence.

Example 11. We shall see later (Worpitzky's theorem on page 135) that the continued fraction

$$\operatorname*{K}_{n=1}^{\infty} \frac{a_n}{1} := \frac{-1/4}{1} + \frac{1/8}{1} + \frac{a_3}{1} + \frac{a_4}{1} + \cdots \tag{1.7.4}$$

with $-\frac{1}{4} \leq a_n \leq \frac{1}{4}$ for all n converges. What can be said about the value f of $\mathbf{K}(a_n/1)$?

Since the interval $V := [-\frac{1}{2}, \frac{1}{2}]$ is a simple value set for $\mathbf{K}(a_n/1)$ (Example 7), it follows that the value $f^{(2)}$ of the second tail

$$\frac{a_3}{1} + \frac{a_4}{1} + \cdots + \frac{a_n}{1} + \cdots$$

lies in this interval. Hence $f \in S_2(V)$ where

$$S_2(w) = \cfrac{-1/4}{1 + \cfrac{1/8}{1+w}} = -\frac{1}{4} \cdot \frac{1+w}{\frac{9}{8}+w} = -\frac{1}{4}\left(1 - \frac{1/8}{w+\frac{9}{8}}\right).$$

Now, $V + \frac{9}{8} = [\frac{5}{8}, \frac{13}{8}]$, so $1/(V + \frac{9}{8}) = [\frac{8}{13}, \frac{8}{5}]$ and $-\frac{1}{8}/(V + \frac{9}{8}) = [-\frac{1}{5}, -\frac{1}{13}]$, and thus

$$S_2(V) = -\frac{1}{4}\left[1 - \frac{1}{5}, 1 - \frac{1}{13}\right] = \left[-\frac{3}{13}, -\frac{1}{5}\right].$$

That is,

$$-0.23\cdots < -\frac{3}{13} \leq f \leq -\frac{1}{5} = -0.20\ldots.$$

◇

2.2 Transformations of continued fractions

2.2.1 Introduction

We have already seen a number of transformations:

- from power series to continued fractions (Section 1.4.1 on page 30)
- from continued fraction to power series (Section 1.4.2 on page 33)
- from A_n and B_n to $b_0 + \mathbf{K}(a_n/b_n)$ (Section 1.1.2 on page 5)
- from $\{f_n\}$ to $b_0 + \mathbf{K}(a_n/b_n)$ (Problem 23 on page 51)
- from an infinite product to a continued fraction (Problem 26 on page 51)

In this section we shall introduce some transformations between continued fractions. The idea is that the transformed continued fraction may be easier to work with than the original one.

2.2.2 Equivalence transformations

Definition 2.12. *We say that two continued fractions are equivalent if they have the same sequence of classical approximants.*

We write $\mathbf{K}(a_n/b_n) \sim \mathbf{K}(c_n/d_n)$ to express that $\mathbf{K}(a_n/b_n)$ and $\mathbf{K}(c_n/d_n)$ are equivalent. Let the canonical numerators and denominators be denoted by A_n and B_n for $\mathbf{K}(a_n/b_n)$ and by C_n and D_n for $\mathbf{K}(c_n/d_n)$. If we require that all $C_n = A_n$ and $D_n = B_n$, then it follows from Theorem 1.2 on page 8 that the two continued fractions are identical; that is, $c_n = a_n$ and $d_n = b_n$ for all n. So that has no point. What we have done is to require that $A_n/B_n = C_n/D_n$ for all n.

The idea of equivalent continued fractions is due to Seidel ([Seid55]) who also proved:

Theorem 2.14. $\mathbf{K}(a_n/b_n) \sim \mathbf{K}(c_n/d_n)$ *if and only if there exists a sequence* $\{r_n\}$ *of complex numbers with* $r_0 = 1, r_n \neq 0$ *for all* $n \in \mathbb{N}$, *such that*

$$c_n = r_n r_{n-1} a_n\,, \quad d_n = r_n b_n \quad \textit{for all } n \in \mathbb{N}\,. \tag{2.2.1}$$

Proof : Let A_n, B_n be the canonical numerators and denominators of $\mathbf{K}(a_n/b_n)$. Then $\mathbf{K}(a_n/b_n) \sim \mathbf{K}(c_n/d_n)$ if and only if there exist numbers $r_n \neq 0$ such that the canonical numerators C_n and denominators D_n of $\mathbf{K}(c_n/d_n)$ can be written

$$C_{-1} = 1, \quad D_{-1} = 0, \quad C_n = A_n \prod_{k=0}^{n} r_k, \quad D_n = B_n \prod_{k=0}^{n} r_k \tag{2.2.2}$$

for all n. Since $D_0 = B_0 = 1$ we need $r_0 = 1$. From Theorem 1.2 on page 8 it follows then that $\mathbf{K}(c_n/d_n)$ is given by (2.2.1). □

Properties:

1. If (2.2.1) holds and $\mathbf{K}(a_n/b_n)$ has canonical numerators A_n and denominators B_n, then $\mathbf{K}(c_n/d_n)$ has canonical numerators C_n and denominators D_n given by (2.2.2).

2. The concept of equivalence is tied to the classical approximants. If $\mathbf{K}(a_n/b_n) \sim \mathbf{K}(c_n/d_n)$ by the relations (2.2.1), then

$$S_n(w) = T_n(r_n w) \quad \text{for } n = 0, 1, 2, \ldots, \tag{2.2.3}$$

where $S_n(w)$ are approximants of $\mathbf{K}(a_n/b_n)$, and $T_n(w)$ are approximants of $\mathbf{K}(c_n/d_n)$.

3. If $\{t_n\}$ is a tail sequence for $\mathbf{K}(a_n/b_n)$, then $\{t_n r_n\}$ is a tail sequence for $\mathbf{K}(c_n/d_n)$.

Example 12. The continued fraction

$$\mathop{\mathrm{K}}_{n=1}^{\infty} \frac{a_n z}{1} := \frac{z}{1+} \frac{z/2}{1} + \frac{z/6}{1} + \frac{2z/6}{1} + \frac{2z/10}{1} + \frac{3z/10}{1} + \frac{3z/14}{1} + \frac{4z/14}{1} + \cdots,$$

where

$$a_1 := 1, \quad a_{2k} := \frac{k}{2(2k-1)}, \quad a_{2k+1} := \frac{k}{2(2k+1)} \quad \text{for } k \geq 1,$$

converges to $\mathrm{Ln}(1+z)$ for $|\arg(1+z)| < \pi$. An equivalence transformation with

$$r_1 := 1, \ r_2 := 2, \ r_3 := 3, \ r_4 := 2, \ r_5 := 5, \ r_6 := 2, \ r_7 := 7, \ldots$$

brings it over to the form

$$\frac{z}{1+}\frac{z}{2+}\frac{z}{3+}\frac{2z}{2}+\frac{2z}{5}+\frac{3z}{2}+\frac{3z}{7}+\frac{4z}{2}+\frac{4z}{9}+\frac{5z}{2}+\frac{5z}{11+}\frac{6z}{2}+\cdots$$

which therefore also converges to $\mathrm{Ln}(1+z)$. ◇

Example 13. In Section 1.3.2 on page 27 we found that

$$1+\mathop{\mathrm{K}}_{n=1}^{\infty}\frac{a_n z}{1} := 1 - \cfrac{\frac{a(c-b)}{c(c+1)}z}{1} - \cfrac{\frac{(b+1)(c-a+1)}{(c+1)(c+2)}z}{1} - \cfrac{\frac{(a+1)(c-b+1)}{(c+2)(c+3)}z}{1} - \cfrac{\frac{(b+2)(c-a+2)}{(c+3)(c+4)}z}{1} - \cdots$$

where $a, b, c \notin -\mathbb{N}\cup\{0\}$, converges to the ratio $F(a,b;c;z)/F(a,b+1;c+1;z)$ of hypergeometric functions in the cut plane $D := \{z \in \mathbb{C};\ 0 < \arg(1+z) < 2\pi\}$. An equivalence transformation with $r_n := c+n$ for $n \geq 1$, and multiplication with c, gives the equivalent identity

$$c\frac{F(a,b;c;z)}{F(a,b+1;c+1;z)} = c - \frac{a(c-b)z}{c+1} - \frac{(b+1)(c-a+1)z}{c+2} - \frac{(a+1)(c-b+1)z}{c+3} - \cdots$$

for $z \in D$. ◇

Example 14. A regular C-fraction is a continued fraction of the form

$$\mathop{\mathrm{K}}_{n=1}^{\infty}\frac{a_n z}{1} = \frac{a_1 z}{1} + \frac{a_2 z}{1} + \frac{a_3 z}{1} + \cdots,$$

and if all $a_n > 0$, it is called an S-fraction. The equivalence transformation with $r_0 := 1$, $r_{2n-1} := w := 1/z$ and $r_{2n} := 1$ for all $n \in \mathbb{N}$ brings it over to the form

$$\frac{a_1}{w} + \frac{a_2}{1} + \frac{a_3}{w} + \frac{a_4}{1} + \cdots; \quad w := \frac{1}{z}.$$

If we instead use $r_0 := 1$, $r_n := w := 1/\sqrt{z}$ for all $n \in \mathbb{N}$, we find that $\mathbf{K}(a_n z/1)$ is equivalent to

$$\frac{a_1/w}{w} + \frac{a_2}{w} + \frac{a_3}{w} + \frac{a_4}{w} + \frac{a_5}{w} + \cdots; \quad w := \frac{1}{\sqrt{z}}.$$

◇

Example 15. The continued fraction in Problem 26 on page 51 has the form

$$a_0 + \frac{a_0(a_1-1)}{1} - \frac{a_1(a_2-1)/(a_1-1)}{(a_1a_2-1)/(a_1-1)} - \frac{a_2(a_3-1)/(a_2-1)}{(a_2a_3-1)/(a_2-1)} - \cdots - \frac{a_{n-1}(a_n-1)/(a_{n-1}-1)}{(a_{n-1}a_n-1)/(a_{n-1}-1)} - \cdots.$$

An equivalence transformation with $r_0 := 1$, $r_1 := 1$, $r_n := a_{n-1} - 1$ for $n \geq 2$ leads to

$$a_0 + \frac{a_0(a_1-1)}{1} \underset{-}{} \frac{a_1(a_2-1)}{a_1a_2-1} \underset{-}{} \frac{a_2(a_3-1)(a_1-1)}{a_2a_3-1} \underset{-}{} \frac{a_3(a_4-1)(a_2-1)}{a_3a_4-1} \underset{-}{} \frac{a_4(a_5-1)(a_3-1)}{a_4a_5-1} \underset{-}{} \cdots$$

which therefore also has classical approximants $f_n = \prod_{k=0}^{n} a_k$. $\diamond$

Two equivalence transformations are of particular interest, namely the ones we get by using

$$r_n := \prod_{k=1}^{n} a_k^{(-1)^{n-k+1}} \quad \text{and} \quad r_n := \frac{1}{b_n} \quad \text{for all } n \geq 1 :$$

Corollary 2.15.

A. $\mathbf{K}(a_n/b_n) \sim \mathbf{K}(1/d_n)$ *where*

$$d_n := b_n \prod_{k=1}^{n} a_k^{(-1)^{n+1-k}} \quad \textit{for } n = 1, 2, 3, \ldots . \tag{2.2.4}$$

B. If $b_n \neq 0$ *for all* $n \geq 1$*, then* $\mathbf{K}(a_n/b_n) \sim \mathbf{K}(c_n/1)$*, where*

$$c_1 := \frac{a_1}{b_1}, \qquad c_n := \frac{a_n}{b_n b_{n-1}} \quad \textit{for } n = 2, 3, 4, \ldots . \tag{2.2.5}$$

Remark: The transformation in A can always be performed. That is, every continued fraction $\mathbf{K}(a_n/b_n)$ can be brought to the equivalent form $\mathbf{K}(1/d_n)$. The elements d_n have the structure $d_n = b_n/a_n d_{n-1}$, so

$$d_1 = b_1 \cdot \frac{1}{a_1}, \quad d_2 = b_2 \frac{a_1}{a_2}, \quad d_3 = b_3 \frac{a_2}{a_1 a_3}, \quad d_4 = b_4 \frac{a_1 a_3}{a_2 a_4}, \ldots .$$

The transformation in B can only be applied if all $b_n \neq 0$, since otherwise c_n is not a well defined complex number.

The equivalence transformation is extremely useful in the continued fraction theory. Of course, if $\mathbf{K}(a_n/b_n) \sim \mathbf{K}(c_n/d_n)$, then $\mathbf{K}(a_n/b_n)$ converges if and only if $\mathbf{K}(c_n/d_n)$ converges. This is no longer true for general convergence. If $\mathbf{K}(a_n/b_n)$ converges generally, but not in the classical sense, then some of its equivalent forms $\mathbf{K}(c_n/d_n)$ may still diverge generally.

Example 16. Let $\mathbf{K}(a_n/1)$ be as in Example 2 on page 59, and let $r_n := -B_{n-1}/B_n$ for all $n \in \mathbb{N}$. Then $\mathbf{K}(a_n/1) \sim \mathbf{K}(c_n/d_n)$ where $c_n := r_{n-1}r_n a_n$ and $d_n := r_n b_n$ where $r_0 := 1$, so by (2.2.3) the approximants of $\mathbf{K}(c_n/d_n)$ are $T_n(w) = S_n(w/r_n)$. In particular

$$T_{3n}(w) = \frac{A_{3n-1}w/r_{3n} + A_{3n}}{B_{3n-1}w/r_{3n} + B_{3n}} = \frac{A_{3n} - A_{3n-1}wB_{3n}/B_{3n-1}}{B_{3n}(1-w)},$$

where we know that $A_{3n} = 0$, $A_{3n-1} = 2^n$, $B_{3n} = 1$ and $B_{3n-1} = 2^{n+1} - 2$. Therefore

$$T_{3n}(w) = -\frac{2^n}{2^{n+1}-2} \cdot \frac{w}{1-w} \to \frac{1}{2}\frac{w}{w-1},$$

a non-singular transformation from $\mathcal{M}$. Hence $\{T_n\}$ is not restrained, and $\mathbf{K}(c_n/d_n)$ diverges generally. ◇

Indeed, this is an example of a general phenomenon:

Theorem 2.16. *Let $\mathbf{K}(a_n/b_n)$ diverge in the classical sense, and let $\{t_n\}$ be a tail sequence for $\mathbf{K}(a_n/b_n)$ with t_0 not a limit point for $\{S_n(0)\}$. Then $\{T_n\}$ given by $T_n(w) := S_n(t_n w)$ is not a restrained sequence.*

Proof : Since $\{S_n(0)\}$ diverges, there exists a subsequence $\{n_k\}$ of the natural numbers such that

$$f_{n_k-1} = S_{n_k-1}(0) \to g_1 \quad \text{and} \quad f_{n_k} = S_{n_k}(0) \to g_2 \neq g_1.$$

Assume first that $t_0 = \infty$ and $g_1, g_2 \neq 0, \infty$. Then $t_n = \zeta_n = -B_n/B_{n-1}$, and f_{n_k-1} and f_{n_k} are $\neq 0, \infty$ from some k on. So, for $n := n_k$ with k sufficiently large,

$$S_n(t_n w) = \frac{A_{n-1}t_n w + A_n}{B_{n-1}t_n w + B_n} = \frac{-f_{n-1}B_n w + A_n}{-B_n w + B_n} = \frac{-f_{n-1}w + f_n}{-w+1}$$

which converges to the non-singular transformation

$$\lim_{k\to\infty} S_{n_k}(t_{n_k} w) = \frac{g_2 - g_1 w}{1-w}.$$

If $t_0 \neq \infty$ and/or $g_k \in \{0, \infty\}$ for $k = 1$ or 2, there always exist complex numbers c_1, c_2, d_1, d_2 such that $c_1 c_2 \neq 0$ and

$$\frac{c_1}{d_1 +} \frac{c_2}{d_2 + t_0} = \infty, \qquad \frac{c_1}{d_1 +} \frac{c_2}{d_2 + g_k} \neq 0, \infty \quad \text{for } k = 1, 2.$$

Since therefore

$$\frac{c_1}{d_1 +} \frac{c_2}{d_2 + S_n(t_n w)}$$

is non-restrained, so is $\{S_n(t_n w)\}$. □

This means in particular that if $\mathbf{K}(a_n/b_n)$ converges generally with exceptional sequence $\{w_n^\dagger\}$, but not in the classical sense, then one should stay away from equivalence transformations (2.2.1) with $\liminf \mathfrak{m}(r_n, w_n^\dagger) = 0$.

Typical in this situation is that $\{w_n^\dagger\}$ has limit points at 0 and ∞. Indeed, if $\{r_n\}$ is bounded away from 0 and ∞, the equivalence transformation preserves general convergence:

Theorem 2.17. *Let $\mathbf{K}(a_n/b_n)$ converge generally to f. If the sequence $\{r_n\}$ of complex numbers is bounded and bounded away from 0, then $\mathbf{K}(r_{n-1}r_n a_n/r_n b_n)$ converges generally to $r_0 f$.*

Proof : This follows from (2.2.3) since

$$\liminf_{n\to\infty} \mathfrak{m}(u_n, v_n) > 0 \quad \Longleftrightarrow \quad \liminf_{n\to\infty} \mathfrak{m}(u_n/r_n, v_n/r_n) > 0.$$

□

2.2.3 The Bauer-Muir transformation

For given continued fraction $b_0 + \mathbf{K}(a_n/b_n)$ and given sequence $\{w_n\}$ from $\mathbb{C}$, the idea is to construct a new continued fraction $d_0 + \mathbf{K}(c_n/d_n)$ whose sequence of classical approximants $\{T_n(0)\}$ is exactly $\{S_n(w_n)\}$. The new continued fraction is only unique up to an equivalence transformation. The Bauer-Muir transform, the way we define it, is the canonical one in this equivalence class:

Definition 2.13. *The Bauer-Muir transform of a continued fraction $b_0 + \mathbf{K}(a_n/b_n)$ with respect to a sequence $\{w_n\}$ from $\mathbb{C}$ is the continued fraction $d_0 + \mathbf{K}(c_n/d_n)$ whose canonical numerators C_n and denominators D_n are given by*

$$\begin{aligned} &C_{-1} = 1\,, && D_{-1} = 0\,, \\ &C_n = A_{n-1}w_n + A_n\,, && D_n = B_{n-1}w_n + B_n;\quad n \geq 0. \end{aligned} \tag{2.3.1}$$

This transformation dates back to the 1870's when Bauer ([Bauer72]) and Muir ([Muir77]), independently of each other, proved what can be formulated as follows:

Theorem 2.18. *The Bauer-Muir transform of* $b_0 + \mathbf{K}(a_n/b_n)$ *with respect to* $\{w_n\}$ *from* $\mathbb{C}$ *exists if and only if*

$$\lambda_n := a_n - w_{n-1}(b_n + w_n) \neq 0 \quad \text{for } n = 1, 2, 3, \dots . \tag{2.3.2}$$

If it exists, then it is given by

$$\begin{aligned} &b_0 + w_0 + \frac{\lambda_1}{b_1 + w_1} + \frac{c_2}{d_2} + \frac{c_3}{d_3} + \cdots \quad \text{where} \\ &c_n := a_{n-1}\frac{\lambda_n}{\lambda_{n-1}} \quad \text{and} \quad d_n := b_n + w_n - w_{n-2}\frac{\lambda_n}{\lambda_{n-1}} . \end{aligned} \tag{2.3.3}$$

Proof : Let $\{C_n\}$ and $\{D_n\}$ be given by (2.3.1). Then $\{C_n\}$ and $\{D_n\}$ are canonical numerators and denominators of a continued fraction $d_0 + \mathbf{K}(c_n/d_n)$ if and only if $C_{-1} = D_0 = 1, D_{-1} = 0$ and $\widehat{\Delta}_n := C_{n-1}D_n - D_{n-1}C_n \neq 0$ for all $n \geq 1$. (See Theorem 1.2 on page 8.) The initial conditions for C_n and D_n are satisfied. Moreover

$$\begin{aligned} \widehat{\Delta}_n &:= C_{n-1}D_n - D_{n-1}C_n \\ &= (A_{n-2}w_{n-1} + A_{n-1})(B_{n-1}w_n + B_n) \\ &\quad - (B_{n-2}w_{n-1} + B_{n-1})(A_{n-1}w_n + A_n) \\ &= (A_{n-2}w_{n-1} + A_{n-1})(B_{n-1}w_n + b_nB_{n-1} + a_nB_{n-2}) \\ &\quad -(B_{n-2}w_{n-1} + B_{n-1})(A_{n-1}w_n + b_nA_{n-1} + a_nA_{n-2}) \\ &= -(A_{n-2}B_{n-1} - A_{n-1}B_{n-2})(a_n - w_{n-1}b_n - w_{n-1}w_n) \end{aligned}$$

where the first factor is different from 0 by the determinant formula on page 7, and the second factor is equal to λ_n in (2.3.2). This proves the existence part of Theorem 2.18. The expressions for c_n and d_n follow from Theorem 1.2 on page 8. □

This transformation is in particular useful if $b_0+\mathbf{K}(a_n/b_n)$ and its Bauer-Muir transform converge to the same value. This was proved to be the case for positive continued fractions and positive w_n by Perron ([Perr57], p 27), but of course it holds whenever $\mathbf{K}(a_n/b_n)$ converges with exceptional sequence $\{w_n^\dagger\}$ and $\liminf \mathfrak{m}(w_n, w_n^\dagger) > 0$, ([Lore94c]).

Example 17. Stable computation. We want to compute the approximants $S_n(-6)$ of the continued fraction

$$\mathrm{K}\frac{a_n}{1} := \frac{30 + 0.5}{1} + \frac{30 + (0.5)^2}{1} + \frac{30 + (0.5)^3}{1} + \cdots .$$

It will be evident later that $\mathbf{K}(a_n/1)$ converges to a finite number $f > 0$ with exceptional sequence $\{-6\}$. Computation shows that $f = 5.05859$ correctly rounded to 6 digits. But what happens to $S_n(-6)$? As noted in Remark 4 on page 57 we do not know off-hand whether $\{S_n(-6)\}$ diverges or converges, and if it converges, we do not know its limit. If we try to compute $S_n(-6)$ from the continued fraction, we have a problem. Small inaccuracies in the input or computation will have the effect that $\{-6\}_{n=0}^{\infty}$ is not recognized as an exceptional sequence. Our numerically computed value for $S_n(-6)$ will approach f as $n \to \infty$.

The Bauer-Muir transformation gives a more stable method to compute $S_n(-6)$. Since

$$\lambda_n = 30 + (0.5)^n - (-6)(1-6) = (0.5)^n ,$$

the Bauer-Muir transform of $\mathbf{K}((30+(0.5)^n)/1)$ with respect to $w_n = -6$ exists. It is given by

n	$S_n(-6)$	$T_n(0)$
1	−6.09999	−6.03960
2	−6.03962	−6.07582
3	−6.07580	−6.05404
12	−6.06215	−6.06228
13	−6.06218	−6.06215
14	−6.06229	−6.06223
15	−6.06208	−6.06218
18	−6.06247	−6.06221
19	−6.06189	−6.06220
20	−6.06257	−6.06220
161	5.05858	−6.06220
162	5.05859	−6.06220
163	5.05859	−6.06220

$$-6 + \frac{0.5}{-5} \underset{+}{} \frac{(30+0.5)\cdot 0.5}{-5-(-6)\cdot 0.5} \underset{+}{} \frac{(30+(0.5)^2)\cdot 0.5}{-5-(-6)\cdot 0.5} \underset{+}{} \cdots$$

$$= -6 + \frac{0.5}{-5} \underset{+}{} \frac{15+(0.5)^2}{-2} \underset{+}{} \frac{15+(0.5)^3}{-2} \underset{+}{} \cdots ,$$

and its classical approximants $T_n(0)$ (which can be computed stably) are exactly $S_n(-6)$. In the table we have computed $S_n(-6)$ and $T_n(0)$. The two columns should have been identical. The computation is done with only 6 digits to illustrate the instability in the computation of $S_n(-6)$. ◇

The Bauer-Muir transformation has many applications. Let us look at one more:

Example 18. Functional equation. The continued fraction

$$\mathop{\mathrm{K}}_{n=1}^{\infty} \frac{z+n}{n} := \frac{z+1}{1} \underset{+}{} \frac{z+2}{2} \underset{+}{} \frac{z+3}{3} \underset{+}{} \cdots$$

converges to some $f(z)$ with exceptional sequence $\{w_n^\dagger\}$ where $(w_n^\dagger/n) \to -1$ for all $z \in \mathbb{C}$. (This is a consequence of Theorem 4.13 on page 188 after an equivalence transformation.) Hence also $S_n(1)$ converges to $f(z)$. Therefore its Bauer-Muir transform $d_0 + \mathbf{K}(c_n/d_n)$ with respect to $w_n := 1$ converges to $f(z)$. Since

$$\lambda_n = z + n - 1(n+1) = z - 1 \quad \text{for all } n$$

with this choice for w_n, we have for $z \neq 1$

$$f(z) = d_0 + \mathop{\mathrm{K}}_{n=1}^{\infty} \frac{c_n}{d_n} = 1 + \frac{z-1}{2} \underset{+}{} \frac{z+1}{2} \underset{+}{} \frac{z+2}{3} \underset{+}{} \frac{z+3}{4} \underset{+}{} \cdots ,$$

which looks similar to $\mathbf{K}((z+n)/n)$. Indeed, the first tail $g^{(1)}(z)$ of $d_0+\mathbf{K}(c_n/d_n)$ is such that

$$\frac{z}{1+g^{(1)}(z)} = \frac{z}{1+}\,\frac{z+1}{2}\,{}_{+}\,\frac{z+2}{3}\,{}_{+\cdots} = f(z-1)\,,$$

that is, $g^{(1)}(z) = -1 + z/f(z-1)$. This means that

$$f(z) = 1 + \frac{z-1}{2+g^{(1)}(z)} = 1 + \frac{z-1}{1+\dfrac{z}{f(z-1)}} = z\cdot\frac{f(z-1)+1}{f(z-1)+z}$$

for $z \neq 1$. This is a functional equation for $f(z)$. For $z=1$ the original continued fraction is

$$f(1) = \frac{2}{1}\,{}_{+}\,\frac{3}{2}\,{}_{+}\,\frac{4}{3}\,{}_{+}\,\frac{5}{4}\,{}_{+\cdots} = 1$$

since the constant sequence $\{1\}$ is a tail sequence for this continued fraction, and

$$\sum_{n=0}^{\infty}\prod_{k=1}^{n}\frac{b_k+t_k}{-t_k} = \sum_{n=0}^{\infty}\prod_{k=1}^{n}\frac{k+1}{-1} = \sum_{n=0}^{\infty}(-1)^n(n+1)! = \infty$$

(corollary 2.7 on page 68). Hence,

$$f(2) = 2\cdot\frac{1+1}{1+2} = \frac{4}{3}\,, \qquad f(3) = 3\cdot\frac{\frac{4}{3}+1}{\frac{4}{3}+3} = \frac{21}{13}\,,$$

$$f(4) = 4\cdot\frac{\frac{21}{13}+1}{\frac{21}{13}+4} = \frac{136}{73}\,, \qquad f(5) = 5\cdot\frac{\frac{136}{73}+1}{\frac{136}{73}+5} = \frac{1045}{501}\,,$$

and so on. ◇

2.2.4 Contractions and extensions

We say that $d_0+\mathbf{K}(c_n/d_n)$ is a *contraction* of $b_0+\mathbf{K}(a_n/b_n)$ if its classical approximants $\{g_n\}$ is a subsequence of the classical approximants $\{f_n\}$ of $b_0+\mathbf{K}(a_n/b_n)$. We call $b_0+\mathbf{K}(a_n/b_n)$ an *extension* of $d_0+\mathbf{K}(c_n/d_n)$ in this case. We say in particular that $d_0+\mathbf{K}(c_n/d_n)$ is a *canonical contraction* of $b_0+\mathbf{K}(a_n/b_n)$ if

$$C_k - A_{n_k}\,, \qquad D_k = B_{n_k} \quad \text{for } k=0,1,2,\ldots\,, \tag{2.4.1}$$

where C_n, D_n, and A_n, B_n are the canonical numerators and denominators of $d_0+\mathbf{K}(c_n/d_n)$ and $b_0+\mathbf{K}(a_n/b_n)$ respectively. To derive a general expression for a canonical contraction we can use Theorem 1.2 on page 8. This idea is due to Seidel ([Seid55]). Rather than considering the general case we shall look at two important special cases:

- an *even part* of $b_0 + \mathbf{K}(a_n/b_n)$ is a contraction with classical approximants $g_k := f_{2k}$. (That is, an even part is a continued fraction!) It is the *canonical even part* if $C_k = A_{2k}$ and $D_k = B_{2k}$ for all k.

- an *odd part* of $b_0 + \mathbf{K}(a_n/b_n)$ is a contraction with classical approximants $g_k := f_{2k+1}$. It is the *canonical odd part* if $C_k = A_{2k+1}$ and $D_k = B_{2k+1}$ for all k.

Theorem 2.19. *$\mathbf{K}(a_n/b_n)$ has an even part if and only if all $b_{2n} \neq 0$. Its canonical even part is then given by*

$$\frac{b_2a_1}{b_2b_1+a_2} \frac{a_2a_3b_4/b_2}{-a_4+b_3b_4+a_3b_4/b_2} \frac{a_4a_5b_6/b_4}{-a_6+b_5b_6+a_5b_6/b_4} - \cdots \tag{2.4.2}$$

If $\{t_n\}_{n=0}^{\infty}$ is a tail sequence for $\mathbf{K}(a_n/b_n)$ with all $t_n \neq \infty$, then $t_0, -t_1t_2, -t_3t_4, -t_5t_6, \ldots$ is a tail sequence for (2.4.2).

Proof : From Theorem 1.2 on page 8 it follows that $\mathbf{K}(a_n/b_n)$ has an even part if and only if $f_{2n} \neq f_{2n-2}$ for all n, that is, if and only if all $b_{2n} \neq 0$ (Theorem 1.3 on page 9). From Theorem 1.2 on page 8 we find that the canonical even part $d_0 + \mathbf{K}(c_n/d_n)$ has elements $d_0 := 0$,

$$\begin{aligned} d_1 &:= D_1 = B_2 = b_2b_1 + a_2\,, \\ c_1 &:= C_1 - C_0D_1 = A_2 - A_0B_2 = b_2a_1\,, \\ c_n &:= -\widehat{\Delta}_n/\widehat{\Delta}_{n-1}\,, \qquad d_n := \widehat{\Delta}_n^{(1)}/\widehat{\Delta}_{n-1}\,, \end{aligned}$$

where

$$\begin{aligned} \widehat{\Delta}_n &:= C_{n-1}D_n - D_{n-1}C_n = A_{2n-2}B_{2n} - B_{2n-2}A_{2n} \\ \widehat{\Delta}_n^{(1)} &:= C_{n-2}D_n - D_{n-2}C_n = A_{2n-4}B_{2n} - B_{2n-4}A_{2n}\,. \end{aligned}$$

The recurrence relation for $\{A_n\}$ and $\{B_n\}$ (page 8) and the determinant formula (page 7) lead to

$$\begin{aligned} \widehat{\Delta}_n &= b_{2n} \prod_{j=1}^{2n-1} (-a_j)\,, \\ \widehat{\Delta}_n^{(1)} &= \big(b_{2n}(b_{2n-1}b_{2n-2} + a_{2n-1}) + a_{2n}b_{2n-2}\big) \prod_{j=1}^{2n-3} (-a_j)\,. \end{aligned}$$

This proves (2.4.2). Let $\{t_n\}$ be a tail sequence for $\mathbf{K}(a_n/b_n)$ with all $t_n \neq \infty$. To see that $t_0, -t_1t_2, -t_3t_4, \ldots$ is a tail sequence for (2.4.2) it suffices to prove that $-t_{2n-3}t_{2n-2} = c_n/(d_n - t_{2n-1}t_{2n})$ for $n \geq 2$; i.e.,

$$c_n = -t_{2n-3}t_{2n-2}(d_n - t_{2n-1}t_{2n}) \quad \text{and } c_1 = t_0(d_1 - t_1t_2)\,.$$

This follows by straight forward computation using that $a_n = t_{n-1}(b_n + t_n)$. □

The even part (2.4.2) is equivalent to

$$b_0 + \frac{b_2 a_1}{b_2 b_1 + a_2} - \frac{a_2 a_3 b_4}{b_2(a_4 + b_3 b_4) + a_3 b_4} - \frac{a_4 a_5 b_6 b_2}{b_4(a_6 + b_5 b_6) + a_5 b_6} - \frac{a_6 a_7 b_8 b_4}{b_6(a_8 + b_7 b_8) + a_7 b_8} - \cdots, \tag{2.4.3}$$

which is more widely used (but which is no longer canonical). This even part was established by Seidel ([Seid55]) although Lagrange had some special cases already in 1774–76 ([Lagr76]).

Theorem 2.20. $\mathbf{K}(a_n/b_n)$ *has an odd part if and only if all* $b_{2k+1} \neq 0$. *Its canonical odd part is then given by*

$$\frac{a_1}{b_1} - \frac{a_1 a_2 b_3/b_1}{b_1(a_3 + b_2 b_3) + a_2 b_3} - \frac{a_3 a_4 b_5 b_1/b_3}{a_5 + b_4 b_5 + a_4 b_5/b_3} - \frac{a_5 a_6 b_7/b_5}{a_7 + b_6 b_7 + a_6 b_7/b_5} - \frac{a_7 a_8 b_9/b_7}{a_9 + b_8 b_9 + a_8 b_9/b_7} - \cdots \tag{2.4.4}$$

If $\{t_n\}$ *is a tail sequence for* $b_0 + \mathbf{K}(a_n/b_n)$ *with all* $t_n \neq \infty$, *then* $-t_0 t_1/b_1, -b_1 t_2 t_3, -t_4 t_5, -t_6 t_7, \ldots$ *is a tail sequence for (2.4.4).*

The proof follows the same lines as the proof of Theorem 2.19 and is omitted. The odd part (2.4.4) is equivalent to

$$\frac{a_1}{b_1} - \frac{a_1 a_2 b_3/b_1}{b_1(a_3 + b_2 b_3) + a_2 b_3} - \frac{a_3 a_4 b_5 b_1}{b_3(a_5 + b_4 b_5) + a_4 b_5} - \frac{a_5 a_6 b_7 b_3}{b_5(a_7 + b_6 b_7) + a_6 b_7} - \cdots. \tag{2.4.5}$$

2.2.5 Contractions and convergence

Example 19. For given $0 < q < 1$ we study the continued fraction

$$\operatorname*{K}_{n=1}^{\infty}(1/q^n) = \frac{1}{q} + \frac{1}{q^2} + \frac{1}{q^3} + \cdots. \tag{2.5.1}$$

Now, $A_0 = 0$, $A_1 = 1$, $B_0 = 1$ and $B_1 = q$, and the recurrence relations

$$A_n = q^n A_{n-1} + A_{n-2}, \qquad B_n = q^n B_{n-1} + B_{n-2}$$

show that $\{A_{2n}\}$, $\{A_{2n+1}\}$, $\{B_{2n}\}$ and $\{B_{2n+1}\}$ are strictly increasing, positive sequences, and that by induction

$$A_n < \prod_{k=1}^{n}(1+q^k), \qquad B_n \le \prod_{k=1}^{n}(1+q^k).$$

Hence the four sequences are bounded, and thus convergent, and the finite, positive limits

$$\begin{aligned} \mathcal{A}_0 &:= \lim A_{2n}, & \mathcal{A}_1 &:= \lim A_{2n+1}, \\ \mathcal{B}_0 &:= \lim B_{2n}, & \mathcal{B}_1 &:= \lim B_{2n+1} \end{aligned}$$

all exist. Moreover, the determinant formula shows that

$$|A_{n-1}B_n - B_{n-1}A_n| = 1; \quad \text{i.e.,} \ |\mathcal{A}_0\mathcal{B}_1 - \mathcal{B}_0\mathcal{A}_1| = 1,$$

and thus $\{S_{2n}\}$ and $\{S_{2n+1}\}$ converge to non-singular linear fractional transformations

$$S(w) =: \frac{\mathcal{A}_1 w + \mathcal{A}_0}{\mathcal{B}_1 w + \mathcal{B}_0}, \quad S^*(w) =: \frac{\mathcal{A}_0 w + \mathcal{A}_1}{\mathcal{B}_0 w + \mathcal{B}_1}$$

respectively. In particular this means that both $\{S_{2n}\}$ and $\{S_{2n+1}\}$ are totally non-restrained. Still

$$\lim S_{2n}(0) = \mathcal{A}_0/\mathcal{B}_0 \quad \text{and} \quad \lim S_{2n+1}(0) = \mathcal{A}_1/\mathcal{B}_1,$$

so the even and odd parts of $\mathbf{K}(1/q^n)$ converge to separate values. The canonical even part of $\mathbf{K}(1/q^n)$ is

$$\frac{q^2}{1+q^{1+2}} \underset{-}{} \frac{q^2}{1+q^2+q^{3+4}} \underset{-}{} \frac{q^2}{1+q^2+q^{5+6}} \underset{-}{} \cdots \tag{2.5.2}$$

and the canonical odd part is

$$\begin{aligned} \frac{1}{q} - \frac{q^2}{q(1+q^2+q^{2+3})} \underset{-}{} \frac{q^3}{1+q^2+q^{4+5}} \underset{-}{} \frac{q^2}{1+q^2+q^{6+7}} \underset{-}{} \\ \frac{q^2}{1+q^2+q^{8+9}} \underset{-}{} \cdots \underset{-}{} \frac{q^2}{1+q^2+q^{4n+1}} \underset{-}{} \cdots . \end{aligned} \tag{2.5.3}$$

The sequences

$$\begin{aligned} S_n(w) &:= \frac{A_{n-1}w + A_n}{B_{n-1}w + B_n}, \quad T_n(w) := \frac{A_{2n-2}w + A_{2n}}{B_{2n-2}w + B_{2n}}, \\ U_n(w) &:= \frac{A_{2n-1}w + A_{2n+1}}{B_{2n-1}w + B_{2n+1}} \end{aligned}$$

belonging to (2.5.1), (2.5.2) and (2.5.3) respectively, have different properties: $\{S_{2n}\}$ and $\{S_{2n+1}\}$ converge to distinct non-singular linear fractional transformations, so $\{S_n\}$ is totally non-restrained. On the other hand, both (2.5.2) and (2.5.3) converge in the classical sense, so $\{T_n\}$ and $\{U_n\}$ are restrained, even generally convergent. ◇

There is an important lesson to be learned from this example: contractions are continued fractions, so if they converge, they converge generally. But $\mathbf{K}(a_n/b_n)$ itself does not even have to be restrained. This is no longer so if $\liminf |b_n| > 0$, which in particular includes the continued fractions $\mathbf{K}(a_n/1)$:

Theorem 2.21. *Let* $\liminf |b_n| > 0$ *for* $\mathbf{K}(a_n/b_n)$. *If* $\{S_{2n+m}(0)\}$ *converges for a fixed* $m \in \{0, 1\}$, *then* $\{S_{2n+m}\}$ *converges generally.*

Proof : Let $S_{2n+m}(0) \to f$ as $n \to \infty$. The result follows since

$$S_{2n+m}(-b_{2n+m}) = S_{2n+m-1}(\infty) = S_{2n+m-2}(0),$$

so $\lim S_{2n+m}(0) = \lim S_{2n+m}(-b_{2n+m}) = f$ where $\liminf \mathfrak{m}(0, -b_{2n+m}) > 0$. □

Contractions are defined in terms of classical approximants. It is not much to be gained from extending the concept to more general approximants, but it is interesting to study the connection between the linear fractional transformations

$$S_n(w) = \frac{A_{n-1}w + A_n}{B_{n-1}w + B_n}$$

for $\mathbf{K}(a_n/b_n)$ and

$$S_n^e(w) = \frac{A_{2n-2}w + A_{2n}}{B_{2n-2}w + B_{2n}}, \quad S_n^o(w) = \frac{A_{2n-1}w + A_{2n+1}}{B_{2n-1}w + B_{2n+1}}$$

for its canonical even and odd parts when these continued fractions exist. The recurrence relation (1.2.7) for $\{A_n\}$ and $\{B_n\}$ on page 6 then shows that

$$\begin{aligned} S_n^e(w) &= \frac{A_{2n-2}w + b_{2n}A_{2n-1} + a_{2n}A_{2n-2}}{B_{2n-2}w + b_{2n}B_{2n-1} + a_{2n}B_{2n-2}} \\ &= \frac{A_{2n-2}\frac{w+a_{2n}}{b_{2n}} + A_{2n-1}}{B_{2n-2}\frac{w+a_{2n}}{b_{2n}} + B_{2n-1}} = S_{2n-1}\left(\frac{w + a_{2n}}{b_{2n}}\right), \end{aligned} \tag{2.5.4}$$

and similarly

$$S_n^o(w) = S_{2n}\left(\frac{w + a_{2n+1}}{b_{2n+1}}\right). \tag{2.5.5}$$

2.3 Remarks

1. **The convergence concept.** The natural way to define convergence of a continued fraction $\mathbf{K}(a_n/b_n)$ is to truncate it after n terms to get either

$$f_n := \frac{a_1}{b_1} \genfrac{}{}{0pt}{}{}{+} \frac{a_2}{b_2} \genfrac{}{}{0pt}{}{}{+\cdots+} \frac{a_n}{b_n} \quad \text{or} \quad \widehat{f}_n := \frac{a_1}{b_1} \genfrac{}{}{0pt}{}{}{+} \frac{a_2}{b_2} \genfrac{}{}{0pt}{}{}{+\cdots+} \frac{a_n}{b_n + a_{n+1}}$$

and to require convergence of $\{f_n\}$ or $\{\widehat{f}_n\}$. This was also the idea for a long time. The first proper definition is due to Seidel ([Seid46]). However, as time passed on, one got a little uneasy with this definition, in particular since also approximants of the type $S_n(w)$ were in use.

In 1918 Hamel ([Hamel18]) suggested that convergence should be defined in terms of $S_n(w)$ instead of $S_n(0)$. His rather vague ideas was made into a proper definition of *strong convergence* by Thron and Waadeland in 1982 ([ThWa82] p 42).

But even this was not quite right. The definition of general convergence dates back to 1986 ([Jaco86]). It is really a convergence concept for sequences of functions from $\mathcal{M}$ (or rather functions from quasi-normal families [Lore03a]).

Now, there exist continued fractions for which $\{f_n\}$ converges and $\{\widehat{f}_n\}$ diverges and vice versa. More generally, for any sequence $\{\varphi_n\}$ from $\mathcal{M}$ with $\varphi_0(w) \equiv w$

$$\tilde{s}_n := \varphi_{n-1}^{-1} \circ s_n \circ \varphi_n \quad \text{and} \quad \tilde{S}_n := \tilde{s}_1 \circ \cdots \circ \tilde{s}_n = S_n \circ \varphi_n$$

could in principle have been used to define $\mathbf{K}(a_n/b_n)$, and the convergence properties of $\{\tilde{S}_n\}$ may differ from the ones of $\{S_n\}$.

2. **Tail sequences.** In early works on continued fractions the critical tail sequence $h_n := -S_n^{-1}(\infty)$ pops up in formulas of different types, mainly because $h_n = B_n/B_{n-1}$. Otherwise tail sequences have not been recognized as objects of its own right until Waadeland in a series of papers starting in 1966 ([Waad66]) used periodic tail sequences of periodic continued fractions. His aim was to continue analytic functions defined by limit periodic continued fractions to larger domains. Later on, Thron and Waadeland ([ThWa80a]) applied sequences of tail values for periodic continued fractions to accelerate the convergence of limit periodic ones. These applications were the inspiration for the names of these tail sequences: *right tail sequence* for the tail values, and *wrong tail sequence* for the other tail sequences of a convergent continued fraction.

   In this book the tail sequences are used for a number of purposes, such as proving convergence, estimating the value and finding truncation error bounds for a given continued fraction. The vital point is the asymptotic behavior of the sequence of tail values on the one hand and the remaining tail sequences which all work as exceptional sequences for a restrained continued fraction on the other hand. This use of tail sequences has been advocated by the authors for a long time.

   Chihara ([Chih78]) has derived a number of results by use of what he called *chain sequences* $\{a_n\}$; i.e., sequences of positive numbers which can be written $a_n = g_n(1-g_{n-1})$ with $0 < g_n < 1$. The sequence $\{g_n\}$ was called a *parameter sequence* for $\{a_n\}$. The reason why this works so well is that $\{g_n - 1\}$ actually is a tail sequence for $\mathbf{K}(-a_n/1)$.

   The first formulas in Theorem 2.6 on page 66 was published in [JaWa82], and the last one by Waadeland ([Waad84]) for continued fractions $\mathbf{K}(a_n/1)$ and extended by Jacobsen ([Jaco86a]).

3. **Value sets.** The early convergence results were obtained by manipulating the recurrence relation $X_n = b_nX_{n-1} + a_nX_{n-2}$. The idea of value sets possibly started with Weyl ([Weyl10]) and Hamel ([Hamel18]). The concept was applied with great success by a number of authors, such as Jones, Leighton, Paydon, Thron, Scott,

Wall. They required that either $0 \in V_n$ or $a_n/b_n \subset V_n$ for all n so that $f_n = S_n(0) \in S_{n-1}(V_{n-1}) \subseteq V_0$. The shift to our definition started with Jacobsen ([Jaco82]) and Thron and Waadeland ([ThWa82], p 43). Indeed, the idea of general convergence came as a result of struggling with these more general value sets.

4. **Limit sets.** The (unique) sequence of *best value sets* $\{W_n\}$ for a given sequence $\{\Omega_n\}$ of element sets was defined by Jones and Thron ([JoTh80], p 65) as

$$W_n := \{s_{n+1} \circ s_{n+2} \circ \cdots \circ s_{n+m}(0);\ m \in \mathbb{N},\ (a_k, b_k) \in \Omega_k \text{ for all } k\}$$

for $n = 0, 1, 2, \ldots$. That is, W_n is the set of all classical approximants $f_m^{(n)}$ for the nth tail of every continued fraction from $\{\Omega_k\}$. It was best in the sense that $W_n \subseteq V_n$ for all n for every sequence of value sets $\{V_n\}$ of the traditional type for $\{\Omega_k\}$.

Since we are no longer limited to look at classical approximants, these best value sets are too large in most cases; we can do better. We want our value sets to be "small" closed sets. The concept of *limit sets* $\{V_n\}$ was introduced in [Jaco86b]. With the present definition, they are also "best" in the sense that $V_n \subseteq \tilde{V}_n$ for all n for every sequence $\{\tilde{V}_n\}$ of closed value sets for $\{\Omega_n\}$.

5. **Transformations of continued fractions.** One can construct a number of different types of transformations for continued fractions, for instance based on approximants, classical or non-classical. But one must always keep in mind what the transformation actually does, because it may change certain convergence properties.

Equivalence transformations are totally based on classical approximants. Hence $\mathbf{K}(a_n/b_n)$ converges if and only if it is equivalent to a convergent continued fraction, whereas general convergence does not have this same property.

One can define some kind of *general equivalence* with respect to *two* sequences $\{u_n\}$ and $\{v_n\}$ by demanding

$$S_n(u_n) = T_n(u_n), \quad S_n(v_n) = T_n(v_n) \quad \text{and} \quad \liminf \mathfrak{m}(u_n, v_n) > 0.$$

This would be the counterpart to traditional equivalence, there we actually have $S_n(0) = T_n(0)$, and thus also $S_n(\infty) = T_n(\infty)$, but it would not be half as nice ....

2.4 Problems

1. **Linear fractional transformations.** Determine the linear fractional transformation τ with $\tau(0) = -1$, $\tau(1) = 0$ and $\tau(\infty) = \frac{1}{2}$.

2. **Chordal metric.** Prove that $\lim a_n = \gamma \in \widehat{\mathbb{C}}$ if and only if $\lim \mathfrak{m}(a_n, \gamma) = 0$.

3. **Restrained sequences.** Prove that the three definitions 2.5, 2.6 and 2.7 of a restrained sequence beginning on page 61 are equivalent.

4. **Periodic tail sequences.** Prove that $\{w_n^\dagger\}$ in Example 4 on page 64 is a tail sequence for $\mathbf{K}(a_n/1)$. Prove that $\{w_n^\dagger\}$ and $\{f^{(n)}\}$ are the only periodic tail sequences for $\mathbf{K}(a_n/1)$.

5. **Computation of tail sequences.** Given the periodic continued fraction

$$\mathop{\mathrm{K}}_{n=1}^{\infty}\frac{2}{1}=\frac{2}{1}{+}\frac{2}{1}{+}\frac{2}{1}{+}\cdots.$$

(a) Prove that

$$S_n(w)=\frac{(1+2\cdot(-\frac{1}{2})^n)w+2(1-(-\frac{1}{2})^n)}{(1-(-\frac{1}{2})^n)w+2+(-\frac{1}{2})^n}.$$

(b) Prove that $\{S_n\}$ converges generally to 1.

(c) For $t_0:=1+h\in\widehat{\mathbb{C}}$, find the tail sequence $\{t_n\}$ of $\mathbf{K}(2/1)$. Distinguish between the three cases $h=0$, $h=-3$ and $h\in\widehat{\mathbb{C}}\setminus\{0,-3\}$. Which ones of these tail sequences are exceptional sequences for $\mathbf{K}(2/1)$? What is the critical tail sequence for $\mathbf{K}(2/1)$?

(d) Find an explicit expression for $S_n^{-1}(w)$, and prove that $\{S_n^{-1}\}$ converges generally to -2 with exceptional sequence $\{1\}$.

(e) Use the results above to illustrate Theorem 2.1 on page 61.

6. **Computation of tail sequences.** Given the continued fraction

$$\mathop{\mathrm{K}}_{n=1}^{\infty}\frac{a_n}{1}:=\mathop{\mathrm{K}}_{n=1}^{\infty}\frac{n(n+2)}{1}=\frac{1\cdot 3}{1}{+}\frac{2\cdot 4}{1}{+}\frac{3\cdot 5}{1}{+}\cdots.$$

(a) Prove that $\{n+1\}$ is a tail sequence for $\mathbf{K}(a_n/1)$.

(b) Prove that the continued fraction converges to 1.

(c) Use Corollary 2.8 on page 69 to find an expression for the terms in the tail sequence $\{t_n\}$ for $\mathbf{K}(a_n/1)$ starting with $t_0:=1+h\neq 1$.

7. ♠ **Critical tail sequence.** Let $h_n:=-S_n^{-1}(\infty)$ for $\mathbf{K}(a_n/b_n)$. Show that

$$h_n=b_n+a_n/h_{n-1}\quad\text{for } n=1,2,3,\ldots.$$

8. **Computation of tail sequences and approximants.** Given the periodic continued fraction

$$\mathop{\mathrm{K}}_{n=1}^{\infty}\frac{a_n}{b_n}=\frac{2}{4}{+}\frac{-4}{1}{+}\frac{2}{4}{+}\frac{-4}{1}{+}\frac{2}{4}{+}\frac{-4}{1}{+}\cdots=\frac{2}{4}{-}\frac{4}{1}{+}\frac{2}{4}{-}\frac{4}{1}{+}\frac{2}{4}{-}\frac{4}{1}{+}\cdots.$$

(a) Find the periodic tail sequences of $\mathbf{K}(a_n/b_n)$.

(b) Find the first few terms of its critical tail sequence $\{h_n\}$.

(c) Show that

$$\frac{1}{3} \le h_{2n} \le 1 \quad \text{and} \quad 6 \le h_{2n+1} \le 10 \quad \text{for all } n \ge 2\,.$$

(d) Use the results from (a) and Corollary 2.7 on page 68 to prove that $\mathbf{K}(a_n/b_n)$ converges to 1, and to find bounds for $|f_n - 1|$.

(e) Use Corollary 2.8 on page 69 to determine the asymptotic behavior of $\{h_n\}$.

(f) Compute the first few approximants of the types $S_n(0), S_n(t_n)$ and $S_n(\tilde{t}_n)$ for the continued fraction

$$\frac{2+0.5}{4} \,-\, \frac{4+(0.5)^2}{1} \,+\, \frac{2+(0.5)^3}{4} \,-\, \frac{4+(0.5)^4}{1} \,+\, \frac{2+(0.5)^5}{4} \,-\cdots$$

where $\{t_n\}$ and $\{\tilde{t}_n\}$ are the periodic tail sequences of $\mathbf{K}(a_n/b_n)$ from (a). Compare these sequences of approximants.

(g) Construct the Bauer-Muir transform of the continued fraction in (f) with respect to $\{t_n\}$ and with respect to $\{\tilde{t}_n\}$.

9. **Tail sequences.** In each of the following cases, find a tail sequence for the given continued fraction, and use this to find its value.

(a) $\displaystyle \mathop{\mathrm{K}}_{n=1}^{\infty} \frac{(x+n)(x+n+2)}{1}\,.$

(b) $\displaystyle \mathop{\mathrm{K}}_{n=1}^{\infty} \frac{(x+n)^2}{-1}\,.$

(c) $\displaystyle \mathop{\mathrm{K}}_{n=1}^{\infty} \frac{n^2-x^2}{2x-1}\,.$

(d) $\displaystyle \mathop{\mathrm{K}}_{n=1}^{\infty} \frac{1}{b_n}$ where $\displaystyle b_n = \frac{(n+a)^2+n+a-1}{n+a+1}\,.$

10. **Tail sequences.** Prove that if the continued fraction

$$1 + \frac{a_1}{1} + \frac{(a_1-z)(a_1+1)}{z} + \frac{a_1a_2}{1} + \frac{(a_2-z)(a_2+1)}{z} + \frac{a_2a_3}{1} + \cdots$$

converges, then its value is either 0 or $1+z$. (Perron [Perr57], p 9.)

11. **Tail sequences.** Prove that if the continued fraction

$$a - \frac{a(a+z)}{a+b+z+1} - \frac{(b+1)(b+z+1)}{a+b+z+2} - \frac{(a+1)(a+z+1)}{a+b+z+3} - \frac{(b+2)(b+z+2)}{a+b+z+4} - \frac{(a+2)(a+z+2)}{a+b+z+5} - \cdots$$

converges, where $a, b, z \in \mathbb{C}$ are chosen such that the terms are $\neq 0$, then it converges to either 0 or $-z$. (Perron [Perr57], p 25.)

12. **Tail sequence.** Let $t_{2n} := 1$ and $t_{2n+1} := 2^{2n+1} + 1$ for $n = 0, 1, 2, \ldots$, and let

$$\underset{n=1}{\overset{\infty}{\mathrm{K}}} \frac{a_n}{1} := \frac{4}{1} + \frac{6}{1} + \frac{10}{1} + \frac{18}{1} + \cdots + \frac{2^n + 2}{1} + \cdots .$$

   (a) Prove that $\{t_n\}$ is a tail sequence for $\mathbf{K}(a_n/1)$.

   (b) Prove that $\mathbf{K}(a_n/1)$ diverges.

   (c) Use (1.5.7) on page 66 to prove that the even part of $\mathbf{K}(a_n/1)$ converges to 1 and the odd part to 2.

   (Part (c) was proved by Angell and Hirschhorn ([AnHi05]) in a completely different way.)

13. **Continued fraction identity.** Determine the continued fraction $\mathbf{K}(a_n/1)$ which has the tail sequence $\{1/(n+1)\}_{n=0}^{\infty}$, and prove that $\mathbf{K}(a_n/1)$ converges to 1.

14. **Corresponding element sets.** For which $b_n > 0$ is $V := \{w \in \mathbb{C};\ |w-1| < 1\}$ a value set for $\mathbf{K}(1/b_n)$? For which $a_n > 0$ is V a value set for $\mathbf{K}(a_n/1)$?

15. **Corresponding element sets.** For given $p \in \mathbb{R}$, let V be the half plane $V = \{w \in \mathbb{C};\ \mathrm{Re}(w) > p\}$. For which values of p do there exist continued fractions $\mathbf{K}(a_n/1)$ which has V as a value set? How about continued fractions $\mathbf{K}(1/b_n)$?

16. **Restrained sequences.** For given $d \in \mathbb{C}$, let

$$\tau_n(w) = \frac{1 + (2 + \frac{1}{n})w}{2 + (d - \frac{1}{n})w} \quad \text{for } n = 1, 2, 3, \ldots .$$

   (a) For which pairs (n, d) are $\tau_n \in \mathcal{M}$? Assume in the following that all $\tau_n \in \mathcal{M}$.

   (b) Find the d-values for which $\{\tau_n\}$ is a restrained sequence. Then find an exceptional sequence $\{w_n^\dagger\}$ and a generic sequence $\{z_n\}$ for $\{\tau_n\}$.

   (c) Give an illustration of Theorem 2.1 on page 61 by explicit computation of τ_n^{-1}.

   (d) What can be said about $\{\tau_n\}$ for d-values different from the ones in (b)?

17. ♠ **Totally non-restrained sequence.** In Theorem 2.2 on page 62 we saw that if $\tau_n \to \tau$ in $\mathcal{M}$, then $\sigma_n := \tau_{n-1}^{-1} \circ \tau_n \to I$. For

$$\tau_n := \frac{a_n w + b_n}{c_n w + d_n}$$

   with $a_n \to a$, $b_n \to b$, $c_n \to c$, $d_n \to d$ and $ad - bc \neq 0$ for $a, b, c, d \in \mathbb{C}$, compute σ_n explicitly and verify that $\sigma_n \to I$.

18. **Periodic continued fractions.** For arbitrary $p > 0$ we are given the 3-periodic continued fraction

$$\underset{n=1}{\overset{\infty}{\mathrm{K}}} \frac{a_n}{1} = \frac{p}{1} + \frac{1}{1} - \frac{1}{1} + \frac{p}{1} + \frac{1}{1} - \frac{1}{1} + \frac{p}{1} + \cdots$$

(a) Find its 3-periodic tail sequences.

(b) Find an expression for S_n. (Compute the coefficients for $n = 3m$, $n = 3m + 1$ and $n = 3m + 2$ separately.)

(c) For which values of $p > 0$ is the continued fraction

(*i*) divergent ?

(*ii*) convergent ?

(*iii*) generally convergent ? Write down an exceptional sequence.

(d) Give in particular the set of p-values where the continued fraction converges generally but not classically.

19. **Periodic continued fraction.** We return to the 3-periodic continued fraction

$$\mathop{\mathrm{K}}_{n=1}^{\infty} \frac{a_n}{1} = \frac{2}{1} + \frac{1}{1} - \frac{1}{1} + \frac{2}{1} + \frac{1}{1} - \frac{1}{1} + \frac{2}{1} + \cdots$$

from Example 2 on page 59.

(a) Write $h_9 := -S_9^{-1}(\infty)$ as a terminating continued fraction by means of formula (1.4.6) on page 64.

(b) Show that the canonical denominators $\widehat{B}_n$ of the terminating continued fraction in (b) satisfy $\widehat{B}_{3n-1} = B_{3n-1}$ for $n = 0, 1, 2$.

(c) Compute $S_n(w_n)$ for the following values of w_n. In which cases does $\{S_n(w_n)\}$ converge?

(*i*)	$w_n = 1,$	(*ii*)	$w_n = n,$
(*iii*)	$w_n = 2^{(n+2)/3},$	(*iv*)	$w_n = 1/n,$
(*v*)	$w_n = 2^{-n/3}.$		

20. **The reversed periodic continued fraction.** Let

$$\frac{a_1}{b_1} + \frac{a_2}{b_2} + \frac{a_3}{b_3} + \frac{a_1}{b_1} + \frac{a_2}{b_2} + \frac{a_3}{b_3} + \frac{a_1}{b_1} + \cdots$$

be a 3-periodic continued fraction. The reversed (or dual) continued fraction is then the 3-periodic continued fraction

$$b_3 + \frac{a_3}{b_2} + \frac{a_2}{b_1} + \frac{a_1}{b_3} + \frac{a_3}{b_2} + \frac{a_2}{b_1} + \frac{a_1}{b_3} + \frac{a_3}{b_2} + \cdots .$$

Prove that the canonical denominators B_{3n-1} are the same for each $n \in \mathbb{N}$ for the two continued fractions for all n. (Hint: use formula (1.4.6) on page 64.)

21. ♠ **The reversed periodic continued fraction.** (Galois [Galo28].) Let $\mathbf{K}(a_n/b_n)$ be a p-periodic continued fraction for a fixed $p \in \mathbb{N}$. Prove that for each $n \in \mathbb{N}$, the canonical denominators B_{np-1} are the same for this continued fraction and its reversed continued fraction

$$b_p + \frac{a_p}{b_{p-1}} + \frac{a_{p-1}}{b_{p-2}} + \cdots + \frac{a_2}{b_1} + \frac{a_1}{b_p} + \cdots .$$

22. ♠ **General convergence of periodic continued fraction.** Give an example other than the one in Example 2 on page 59 of a periodic continued fraction $\mathbf{K}(a_n/1)$ which converges generally but not classically. Why do you need to have a period of length ≥ 3?

23. ♠ **From series to continued fraction.** (Euler [Euler48].) Prove that the continued fraction
$$\rho_0 + \frac{\rho_1}{1} \, \frac{}{-} \, \frac{\rho_2}{1+\rho_2} \, \frac{}{-} \, \frac{\rho_3}{1+\rho_3} \, \frac{}{-} \cdots \qquad \text{where all } \rho_k \neq 0$$
has canonical numerators and denominators given by
$$A_n = \rho_0 + \sum_{k=1}^{n} \left(\prod_{j=1}^{k} \rho_j \right), \qquad B_n = 1 \quad \text{for } n = 0, 1, 2, \ldots .$$

24. **Equivalence transformation.** Prove that $\mathbf{K}(a_n/(-b_n)) \sim -\mathbf{K}(a_n/b_n)$.

25. **Equivalence transformation.** (Stern [Stern32].) Prove that
$$\frac{b_1}{b_1} \, \frac{}{+} \, \frac{b_2}{b_2} \, \frac{}{+} \, \frac{b_3}{b_3} \, \frac{}{+} \cdots \sim \frac{b_0}{b_0} \, \frac{}{+} \, \frac{b_0}{b_1} \, \frac{}{+} \, \frac{b_1}{b_2} \, \frac{}{+} \cdots \qquad \text{when all } b_n \neq 0.$$

26. ♠ **Shift transformation.** Show that the canonical contraction of $b_0 + \mathbf{K}(a_n/b_n)$ with canonical numerators and denominators given by $C_n := A_{n+1}$ and $D_n := B_{n+1}$ for $n \geq 0$ exists if and only if $b_1 \neq 0$, and that then it is given by
$$\frac{b_0 b_1 + a_1}{b_1} - \frac{a_1 a_2 / b_1}{(b_1 b_2 + a_2)/b_1} \, \frac{}{+} \, \frac{a_3}{b_3} \, \frac{}{+} \, \frac{a_4}{b_4} \, \frac{}{+} \cdots . \qquad (*)$$
Show further that if $\{t_n\}_{n=0}^{\infty}$ is a tail sequence for $b_0 + \mathbf{K}(a_n/b_n)$ with $t_0, t_1 \neq \infty$, then $-t_0 t_1, t_2, t_3 \ldots$ is a tail sequence for $(*)$.

27. ♠ **Shift transformation.** Let $N > 1$ be a fixed integer. Show that the canonical contraction of $b_0 + \mathbf{K}(a_n/b_n)$ with
$$C_n = A_n, \quad D_n = B_n \quad \text{for } n = 0, 1, 2, \ldots, N-1$$
and
$$C_n = A_{n+1}, \quad D_n = B_{n+1} \quad \text{for } n = N, N+1, N+2, \ldots$$
exists if and only if $b_{N+1} \neq 0$, and show that then it is given by
$$b_0 + \frac{a_1}{b_1} \, \frac{}{+} \cdots \frac{}{+} \, \frac{a_{N-1}}{b_{N-1}} \, \frac{}{+} \, \frac{b_{N+1} a_N}{b_{N+1} b_N + a_{N+1}}$$
$$\frac{}{+} \, \frac{-a_{N+1} a_{N+2} / b_{N+1}}{(b_{N+2} b_{N+1} + a_{N+2})/b_{N+1}} \, \frac{}{+} \, \frac{a_{N+3}}{b_{N+3}} \, \frac{}{+} \cdots .$$
(Perron [Perr57], p 16.)
Show further that if $\{t_n\}_{n=0}^{\infty}$ is a tail sequence for $b_0 + \mathbf{K}(a_n/b_n)$ with $t_N \neq \infty$ and $t_{N+1} \neq \infty$, then $t_0, t_1, \ldots, t_{N-1}, -t_N t_{N+1}, t_{N+2}, t_{N+3}, \ldots$ is a tail sequence for this contraction.

28. ♠ **The Wall transformations.** Let $\{f_n\}_{n=0}^{\infty}$ be the sequence of classical approximants for a given continued fraction $\mathbf{K}(a_n/1)$. Prove the following statements:

(a) $f_1, f_2, f_4, f_5, f_7, f_8, \ldots, f_{3n+1}, f_{3n+2}, \ldots$ is the sequence of classical approximants for the continued fraction

$$a_1 - \frac{a_1 a_2}{1+a_2} + \frac{a_3}{1+a_4} - \frac{a_4 a_5}{1+a_5} + \frac{a_6}{1+a_7} - \frac{a_7 a_8}{1+a_8} + \frac{a_9}{1+a_{10}} - \cdots$$

(b) $f_0, f_2, f_3, f_5, f_6, f_8, \ldots, f_{3n}, f_{3n+2}, \ldots$ is the sequence of classical approximants for the continued fraction

$$\frac{a_1}{1+a_2} - \frac{a_2 a_3}{1+a_3} + \frac{a_4}{1+a_5} - \frac{a_5 a_6}{1+a_6} + \frac{a_7}{1+a_8} - \frac{a_8 a_9}{1+a_9} + \cdots$$

(c) $f_0, f_1, f_3, f_4, f_6, f_7, \ldots, f_{3n}, f_{3n+1}, \ldots$ is the sequence of classical approximants for the continued fraction

$$\frac{a_1}{1} + \frac{a_2}{1+a_3} - \frac{a_3 a_4}{1+a_4} + \frac{a_5}{1+a_6} - \frac{a_6 a_7}{1+a_7} + \frac{a_8}{1+a_9} - \cdots$$

(d) If two of the continued fractions in (a)–(c) converge, then also the third one converges.

(e) $K(a_n/1)$ converges if and only if the three continued fractions in (a)–(c) converge.

(Wall ([Wall57]), where also a number of similar results are presented.)

29. ♠ **Extensions.** Show that

$$\frac{a_1}{b_1 - a_2} + \frac{a_2}{1} - \frac{1}{1 + b_2 - a_3} + \frac{a_3}{1} - \frac{1}{1 + b_3 - a_4} + \frac{a_4}{1} - \cdots$$

and

$$\frac{a_1}{b_1 - 1} - \frac{1}{1} + \frac{a_2}{1 + b_2 - a_2} - \frac{1}{1} + \frac{a_3}{1 + b_3 - a_3} - \frac{1}{1} + \cdots$$

are extensions of $\mathbf{K}(a_n/b_n)$. (McLaughlin and Wyshinski [McWy07]).

30. ♠ **Continued fractions with canonical denominators equal to 1.** Let $b_0 + \mathbf{K}(a_n/b_n)$ have critical tail sequence $\{h_n\}$ (i.e., $h_n = -S_n^{-1}(\infty)$) with $h_n \neq \infty$ for $n \geq 1$. Prove that the continued fraction

$$b_0 + \frac{a_1/h_1}{1} + \frac{a_2/h_1 h_2}{b_2/h_2} + \frac{a_3/h_2 h_3}{b_3/h_3} + \cdots$$

is equivalent to $b_0 + \mathbf{K}(a_n/b_n)$ and has canonical denominators $B_n = 1$ for $n \geq 0$. (Euler.)

31. ♠ **Extension of continued fraction.** (Perron [Perr57], p15.) Let $b_0 + \mathbf{K}(a_n/b_n)$ have classical approximants $S_n(0) = f_n$, let $N \in \mathbb{N}$ with $N \geq 2$ and let $g \in \widehat{\mathbb{C}}$ be chosen such that

$$\rho := -S_N^{-1}(g) = \frac{A_N - B_N g}{A_{N-1} - B_{N-1} g} \neq 0, \infty.$$

Prove that

$$b_0 + \frac{a_1}{b_1} + \cdots + \frac{a_{N-1}}{b_{N-1}} + \frac{a_N}{b_N - \rho} + \frac{\rho}{1} - \frac{a_{N+1}/\rho}{b_{N+1} + a_{N+1}/\rho} + \frac{a_{N+2}}{b_{N+2}} + \cdots$$

is an extension of $b_0 + \mathbf{K}(a_n/b_n)$ with classical approximants

$$f_n^* = \begin{cases} f_n & \text{for } n = 0, 1, \ldots, N-1\,, \\ g & \text{for } n = N\,, \\ f_{n-1} & \text{for } n = N+1, N+2, \ldots\,. \end{cases}$$

32. ♠ **The Khovanskii transform.** The Khovanskii transform ([Khov63], p 23) of $\mathbf{K}(a_n/1)$ is given by

$$\frac{a_1}{1+2a_2} - \frac{a_2}{1} - \frac{a_3}{1+2a_3+2a_4} - \frac{a_4}{1} - \frac{a_5}{1+2a_5+2a_6} - \cdots - \frac{a_{2n}}{1} - \frac{a_{2n+1}}{1+2a_{2n+1}+2a_{2n+2}} - \cdots\,.$$

Prove that if both $\mathbf{K}(a_n/1)$ and its Khovanskii transform converge (in the classical sense), then they converge to the same value.

33. **The Bauer-Muir transform.**

(a) Find the Bauer-Muir transform of

$$z^2 - 1 + \frac{2^2}{1} + \frac{2^2}{z^2-1} + \frac{4^2}{1} + \frac{4^2}{z^2-1} + \frac{6^2}{1} + \frac{6^2}{z^2-1} + \cdots$$

with respect to $\{w_n\}$ given by

$$w_n := \begin{cases} \dfrac{n+1}{z+1} - \dfrac{1}{2} & \text{if } n \text{ is odd,} \\ n(z+1) + \frac{1}{2}(3+2z-z^2) & \text{if } n \text{ is even.} \end{cases}$$

(b) Assume that the continued fraction in (a) and its Bauer-Muir transform converge to the same value $f(z)$ for $z > 1$. Find a functional equation for $f(z)$.

34. **The Bauer-Muir transform.** Assume that the two continued fractions

$$\frac{a_1+h}{1} + \frac{a_1}{b} + \frac{a_2+h}{1} + \frac{a_2}{b} + \cdots$$

$$h + \frac{a_1}{1} + \frac{a_1+h}{b} + \frac{a_2}{1} + \frac{a_2+h}{b} + \cdots$$

converge. Prove that then they converge to the same value. (Ramanujan [Bern89], p 122), ([Jaco90]).

Chapter 3

Convergence criteria

The convergence theory for continued fractions relies on some basic concepts and ideas. The first part of this chapter is devoted to some of these tools.

Next we go on to present some classical convergence theorems. They are classical in two meanings of the word. For one thing they are old, well-proven results. But they are also classical in the sense that they aim for classical convergence. The proofs we offer are not always the classical ones, though. We have chosen to see the theorems in a more modern setting whenever convenient. In particular we use value sets to prove uniform convergence of $\{S_n(w)\}$ both with respect to w and with respect to a family of continued fractions. This allows us to produce some newer results as well.

Our convergence criteria are mainly stated as conditions on the elements $\{a_n\}$ and $\{b_n\}$ of $\mathbf{K}(a_n/b_n)$. Even if these conditions are met only from some n on, $\mathbf{K}(a_n/b_n)$ still converges, since a tail of $\mathbf{K}(a_n/b_n)$ converges.

Let $\mathcal{F}$ be a family of continued fractions which satisfy a given convergence criterion. In applications it is often vital to have reliable truncation error bounds for continued fractions from $\mathcal{F}$; i.e., bounds for the error $|f - f_n|$ in the approximation $f \approx f_n$. We have collected such bounds, valid for a family characterized by a convergence criterion. They are useful because of their generality and their simplicity. For bounds valid in more special (i.e., smaller) families, we refer to Chapter 5.

3.1 Tools

3.1.1 The Stern-Stolz Divergence Theorem

Stern ([Stern48]) and Stolz ([Stolz86]) proved independently that a continued fraction $\mathbf{K}(1/b_n)$ diverges in the classical sense if $\sum |b_n| < \infty$. Later on, von Koch ([Koch95a], [Koch95b]) proved that $\{A_{2n+m}\}_n$ and $\{B_{2n+m}\}_n$ converge to finite values for which (1.1.1) holds in this case. From their proofs we can even extract a little more information. We say that a sequence $\{c_n\}$ *converges absolutely* if $\sum |c_n - c_{n-1}| < \infty$. Absolute convergence implies convergence to a finite value since $\sum_{k=1}^{n}(c_k - c_{k-1}) = c_n - c_0$. We also say that a continued fraction converges absolutely if its sequence of classical approximants converges absolutely.

Theorem 3.1. (The Stern-Stolz Divergence Theorem.) *If* $\sum |b_n| < \infty$, *then the continued fraction* $\mathbf{K}(1/b_n)$ *diverges generally, the sequences* $\{A_{2n+m}\}_n$ *and* $\{B_{2n+m}\}_n$ *converge absolutely to finite values* $\mathcal{A}_m$ *and* $\mathcal{B}_m$ *respectively (for* $m = 0, 1$*), and*

$$\mathcal{A}_1\mathcal{B}_0 - \mathcal{A}_0\mathcal{B}_1 = 1. \tag{1.1.1}$$

Proof : We shall use the classical proof of this theorem. It is related to Example 19 on page 87. Let $\sum |b_n| < \infty$. Now, $\{A_n\}$ and $\{B_n\}$ are solutions of the recurrence relation

$$X_n = b_n X_{n-1} + X_{n-2} \quad \text{for } n = 1, 2, 3, \dots . \tag{1.1.2}$$

For every solution $\{X_n\}$ of this relation, $|X_n| \le |b_n| \cdot |X_{n-1}| + |X_{n-2}| \le |b_n| \cdot |X_{n-1}| + (|b_{n-1}| + 1)|X_{n-2}|$, so if $|X_m| \le \lambda(|b_1| + 1) \cdots (|b_m| + 1)$ for some $\lambda > 0$ for $m := n - 1$ and $m := n - 2$, then

$$|X_n| \le (|b_n| + 1)\lambda \prod_{k=1}^{n-1}(|b_k| + 1) = \lambda \prod_{k=1}^{n}(|b_k| + 1).$$

It follows therefore by induction that

$$|X_n| \le \max\{|X_{-1}|, |X_0|\} \cdot (|b_1| + 1)(|b_2| + 1) \cdots (|b_n| + 1).$$

Since $\prod_{n=1}^{\infty}(1 + |b_n|) < \infty$ if and only if $\sum_{n=1}^{\infty} |b_n| < \infty$, this means that $\{A_n\}$ and $\{B_n\}$ are bounded under our conditions. Therefore the two series $\sum b_n A_{n-1}$ and $\sum b_n B_{n-1}$ converge absolutely. Since solutions $\{X_n\}$ of (1.1.2) satisfy $X_n - X_{n-2} = b_n X_{n-1}$, this means that $\sum |A_n - A_{n-2}| < \infty$ and $\sum |B_n - B_{n-2}| < \infty$. In other words, $\{A_{2n}\}$, $\{A_{2n+1}\}$, $\{B_{2n}\}$ and $\{B_{2n+1}\}$ converge absolutely to finite values.

The identity (1.1.1) follows since by the determinant formula on page 7,

$$A_{2n-1}B_{2n} - A_{2n}B_{2n-1} = (-1)^{2n} = 1 \quad \text{for all } n\,.$$

Finally, since

$$\lim_{n\to\infty} S_{2n}(w) = \frac{\mathcal{A}_1 w + \mathcal{A}_0}{\mathcal{B}_1 w + \mathcal{B}_0}\,, \qquad \lim_{n\to\infty} S_{2n+1}(w) = \frac{\mathcal{A}_0 w + \mathcal{A}_1}{\mathcal{B}_0 w + \mathcal{B}_1}\,, \tag{1.1.3}$$

the sequence $\{S_n\}$ is totally non-restrained, and the general divergence follows. □

Remarks:

1. Since $\mathbf{K}(1/b_n)$ diverges generally, it definitely diverges in the classical sense. Indeed, $\lim S_{2n}(w)$ and $\lim S_{2n+1}(w)$ exist for every $w \in \widehat{\mathbb{C}}$, but their limits depend totally on w. In particular,

$$\lim_{n\to\infty} S_{2n}(0) = \frac{\mathcal{A}_0}{\mathcal{B}_0} \quad\text{and}\quad \lim_{n\to\infty} S_{2n+1}(0) = \lim_{n\to\infty} S_{2n}(\infty) = \frac{\mathcal{A}_1}{\mathcal{B}_1}$$

where $\mathcal{A}_0/\mathcal{B}_0 \neq \mathcal{A}_1/\mathcal{B}_1$ by (1.1.1). ($\mathcal{A}_m/\mathcal{B}_m$ is well defined in $\widehat{\mathbb{C}}$ since $\mathcal{B}_m = 0$ implies that $\mathcal{A}_m \neq 0$ by (1.1.1).)

2. If for $m = 0$ or $m = 1$ all $B_{2n+m} \neq 0$ and $\mathcal{B}_m \neq 0$, then also $\{S_{2n+m}(0)\}_n$ converges absolutely since by the determinant formula

$$|S_n(0) - S_{n-2}(0)| = \left|\frac{A_n B_{n-2} - A_{n-2} B_n}{B_n B_{n-2}}\right| = \left|\frac{b_n}{B_n B_{n-2}}\right|.$$

$\mathcal{B}_m$ is in particular $\neq 0$ if $\lim_{n\to\infty} S_{2n+m}(0) \neq \infty$ (see (1.1.3)).

3. An equivalence transformation does not change the classical approximants of a continued fraction $\mathbf{K}(a_n/b_n)$. The equivalence transformation in Corollary 2.15 on page 77 brings $\mathbf{K}(a_n/b_n)$ to the form $\mathbf{K}(1/d_n)$. Hence, $\mathbf{K}(a_n/b_n)$ diverges if

$$\mathcal{S} := \sum_{n=1}^{\infty} |d_n| = \sum_{n=1}^{\infty} \left| b_n \prod_{k=1}^{n} a_k^{(-1)^{n-k+1}} \right| < \infty\,, \tag{1.1.4}$$

and its sequence of even and sequence of odd classical approximants both converge to distinct values. The series $\mathcal{S}$ in (1.1.4) is called the *Stern-Stolz Series* of $\mathbf{K}(a_n/b_n)$. It is invariant under equivalence transformations of $\mathbf{K}(a_n/b_n)$, and it can also be written

$$\mathcal{S} = \sum_{n=1}^{\infty} \left| b_{2n} \frac{a_1 a_3 \cdots a_{2n-1}}{a_2 a_4 \cdots a_{2n}} \right| + \sum_{n=1}^{\infty} \left| b_{2n+1} \frac{a_2 a_4 \cdots a_{2n}}{a_1 a_3 \cdots a_{2n+1}} \right|. \tag{1.1.5}$$

4. In principle it may occur that $\mathbf{K}(a_n/b_n)$ converges generally even if its Stern-Stolz Series (1.1.4) converges and thus $\mathbf{K}(a_n/b_n)$ diverges in the classical sense (Theorem 2.16 on page 81). However, by Theorem 2.17 on page 82, this problem can not occur if

$$\inf |r_n| > 0 \;\text{ and }\; \sup |r_n| < \infty \;\text{ for }\; r_n := \Pi_{k=1}^{n} a_k^{(-1)^{n-k+1}}. \tag{1.1.6}$$

5. A continued fraction of the form $\mathbf{K}(a_n/1)$ diverges generally if its Stern-Stolz Series converges. This follows since $S_n(-1) = S_{n-2}(0)$, which means that if $\{S_{2n+m}(0)\}_n$ converges to f, then $\{S_{2n+m}\}_n$ converges generally to f. But $\{S_{2n}\}_n$ and $\{S_{2n+1}\}_n$ converge generally to distinct values in this case.

6. If the limits $\mathcal{A} := \lim A_n$ and $\mathcal{B} := \lim B_n$ both exist and are finite, we say that $\mathbf{K}(a_n/b_n)$ *converges separately.* Hence, if all $b_n \neq 0$ and $\sum |b_n| < \infty$, then the even and odd parts of $\mathbf{K}(1/b_n)$ converge separately. (The condition $b_n \neq 0$ is needed for the existence of the even and odd parts.) Indeed, Theorem 3.1 is actually a *convergence theorem* in this sense.

Condition (1.1.4) can be difficult to check at times. Then the following theorem ([Prin99b]) may come in handy:

Theorem 3.2. *The Stern-Stolz Series of* $\mathbf{K}(a_n/b_n)$ *has sum* ∞ *if at least one of the following three conditions hold:*

$$(i) \qquad \sum_{n=2}^{\infty} \sqrt{|b_n b_{n-1}/a_n|} = \infty\,, \tag{1.1.7}$$

$$(ii) \qquad \liminf_{n\to\infty} \left|\frac{a_n}{b_n b_{n-1}}\right| < \infty, \tag{1.1.8}$$

$$(iii) \qquad \sum_{n=2}^{\infty} \frac{|b_n b_{n-1}|}{n|a_n|} = \infty. \tag{1.1.9}$$

Proof : (i): The series in (1.1.7) and the Stern-Stolz Series are invariant under equivalence transformations, so it suffices to consider continued fractions $\mathbf{K}(1/b_n)$ and to prove that

$$\sum_{n=2}^{\infty} \sqrt{|b_n b_{n-1}|} = \infty \qquad \Longrightarrow \qquad \sum_{n=1}^{\infty} |b_n| = \infty\,.$$

This implication follows easily since $\sqrt{pq} \leq \frac{1}{2}(p+q)$ for positive numbers p and q, and thus

$$\sum_{n=2}^{\infty} \sqrt{|b_n b_{n-1}|} \leq \sum_{n=2}^{\infty} \frac{1}{2}(|b_n| + |b_{n-1}|) \leq \sum_{n=1}^{\infty} |b_n|.$$

(ii): This follows immediately since (1.1.8) $\Rightarrow$ (1.1.7).

(iii): It suffices to prove that (1.1.9) $\Rightarrow$ (1.1.7), or rather that

$$\sum_{n=2}^{\infty} \frac{q_n}{n} = \infty \quad \Longrightarrow \quad \sum_{n=2}^{\infty} \sqrt{q_n} = \infty$$

for an arbitrary sequence $\{q_n\}$ with $q_n \geq 0$. If $\limsup q_n > 0$, then clearly $\sum \sqrt{q_n} = \infty$. Let $q_n \to 0$. Then $q_n \leq n^2$ from some n on, and thus $q_n/n \leq q_n/\sqrt{q_n} = \sqrt{q_n}$, and the implication follows. □

3.1.2 The Lane-Wall Characterization

The Stern-Stolz Divergence Theorem gives sufficient conditions for divergence of classical approximants. But how about convergence? When is divergence of the Stern-Stolz Series sufficient for convergence of $\mathbf{K}(a_n/b_n)$? The following theorem by Lane and Wall ([LaWa49]), gives a very useful answer:

Theorem 3.3. (The Lane-Wall Characterization.) *Let* $\mathbf{K}(a_n/b_n)$ *with classical approximants* $f_n = S_n(0) \neq \infty$ *satisfy*

$$\sum_{n=1}^{\infty} |f_{n+1} - f_{n-1}| < \infty. \tag{1.2.1}$$

Then $\mathbf{K}(a_n/b_n)$ *converges if and only if*

$$\sum_{n=1}^{\infty} \left| b_n \prod_{k=1}^{n} a_k^{(-1)^{n-k+1}} \right| = \infty. \tag{1.2.2}$$

Proof : Since the approximants f_n and the Stern-Stolz Series (1.2.2) are invariant under equivalence transformations, we may assume that $\mathbf{K}(a_n/b_n)$ has the form $\mathbf{K}(1/b_n)$. By the Stern-Stolz Divergence Theorem we know that $\mathbf{K}(1/b_n)$ diverges if $\sum |b_n| < \infty$. Assume that $\sum |b_n| = \infty$. Since $\{f_{2n}\}$ and $\{f_{2n+1}\}$ converge absolutely, the limit L_0 of $\{f_{2n}\}$ and the limit L_1 of $\{f_{2n+1}\}$ exist and are finite. We want to prove that they coincide.

Assume that $L_0 \neq L_1$. Then $|f_{n+1} - f_n| \to |L_0 - L_1| > 0$, and thus

$$\sum_{n=1}^{\infty} |\delta_n| < \infty \quad \text{for} \quad \delta_n := -\frac{f_{n+1} - f_{n-1}}{f_{n+1} - f_n}$$

since $f_{n+1} \neq f_n$ by Theorem 1.3A on page 9. In particular $\delta_n \neq \infty, -1$. The determinant formula and the recurrence relation (1.1.2) on page 100 for $\{A_n\}$ and

$\{B_n\}$ show that

$$\delta_n = -\frac{\dfrac{A_{n+1}}{B_{n+1}} - \dfrac{A_{n-1}}{B_{n-1}}}{\dfrac{A_{n+1}}{B_{n+1}} - \dfrac{A_n}{B_n}} = -\frac{A_{n+1}B_{n-1} - A_{n-1}B_{n+1}}{A_{n+1}B_n - A_nB_{n+1}} \cdot \frac{B_n}{B_{n-1}}$$
$$= -b_{n+1}\frac{A_nB_{n-1} - A_{n-1}B_n}{A_{n+1}B_n - A_nB_{n+1}} \cdot \frac{B_n}{B_{n-1}} = b_{n+1}\frac{B_n}{B_{n-1}} \tag{1.2.3}$$

where all $B_n \neq 0$ since all $f_n \neq \infty$. That is,

$$\delta_n B_{n-1} = b_{n+1}B_n \quad \text{for all } n \in \mathbb{N}. \tag{1.2.4}$$

We shall prove that $\sum |\delta_n| < \infty$ implies $\sum |b_n| < \infty$, a contradiction, and thus $L_0 = L_1$. We first observe that

$$\frac{B_{n-1}}{B_n} = \frac{B_{n-1}}{b_nB_{n-1} + B_{n-2}} = \frac{B_{n-1}}{\delta_{n-1}B_{n-2} + B_{n-2}} = \frac{1}{\delta_{n-1}+1} \cdot \frac{1}{B_{n-2}/B_{n-1}}. \tag{1.2.5}$$

Therefore it follows from (1.2.4) that

$$b_{n+1} = \delta_n\frac{B_{n-1}}{B_n} = \frac{\delta_n}{\delta_{n-1}+1} \cdot \frac{1}{B_{n-2}/B_{n-1}} = \frac{\delta_n(\delta_{n-2}+1)}{\delta_{n-1}+1} \cdot \frac{B_{n-3}}{B_{n-2}}$$
$$= \frac{\delta_n(\delta_{n-2}+1)}{(\delta_{n-1}+1)(\delta_{n-3}+1)} \cdot \frac{1}{B_{n-4}/B_{n-3}} = \frac{\delta_n(\delta_{n-2}+1)(\delta_{n-4}+1)}{(\delta_{n-1}+1)(\delta_{n-3}+1)} \cdot \frac{B_{n-5}}{B_{n-4}}$$

and so on. That is,

$$b_{2n+1} = \frac{\delta_{2n}(\delta_{2n-2}+1)(\delta_{2n-4}+1)\cdots(\delta_2+1)}{(\delta_{2n-1}+1)(\delta_{2n-3}+1)\cdots(\delta_3+1)} \cdot \frac{B_1}{B_2} \tag{1.2.6}$$

and

$$b_{2n+2} = \frac{\delta_{2n+1}(\delta_{2n-1}+1)(\delta_{2n-3}+1)\cdots(\delta_1+1)}{(\delta_{2n}+1)(\delta_{2n-2}+1)\cdots(\delta_2+1)} \cdot \frac{B_0}{B_1}. \tag{1.2.7}$$

Since all $\delta_n \neq -1, \infty$ and all $B_n \neq 0, \infty$, this shows that if $\sum |\delta_n| < \infty$, then $\sum |b_n| < \infty$. □

Remark. If $f_n \neq \infty$ for $n \geq m \in \mathbb{N}$ only, and $\sum_{n=m+1}^{\infty} |f_{n+1} - f_n| < \infty$, then the conclusion of Theorem 3.3 still holds. The proof only needs the minor modification that the process of using (1.2.5) to produce (1.2.6)-(1.2.7) stops at B_m/B_{m+1} and B_{m+1}/B_{m+2}.

The absolute convergence in (1.2.1) is vital. As Wall proved in [Wall56], there exists a divergent continued fraction for which the even and odd parts converge to distinct values and (1.2.2) holds:

Example 1. Let $\mathbf{K}(a_n/1)$ be given by

$$a_n := (-1)^{n-1}n^2 \quad \text{for } n = 1, 2, 3, \dots .$$

Then $\mathbf{K}(a_n/1) \sim \mathbf{K}(1/b_n)$ where $1/b_1 := a_1 = 1$ and $1/b_{n-1}b_n := a_n$ for $n \geq 2$. Since the geometric mean $\sqrt{pq}$ of two non-negative real numbers is less that or equal to their arithmetic mean $\frac{1}{2}(p+q)$, it follows that

$$\sqrt{|b_{n-1}b_n|} = \tfrac{1}{n} \leq \tfrac{1}{2}(|b_{n-1}| + |b_n|),$$

and thus

$$\sum_{n=1}^{\infty} |b_n| = \frac{|b_1|}{2} + \sum_{n=2}^{\infty} \frac{1}{2}(|b_{n-1}| + |b_n|) \geq \frac{1}{2} + \sum_{n=2}^{\infty} \frac{1}{n} = \infty;$$

i.e., (1.2.2) holds for $\mathbf{K}(a_n/1)$. The even part of $\mathbf{K}(a_n/1)$ is the continued fraction

$$\frac{a_1}{1+a_2} - \frac{a_2a_3}{1+a_3+a_4} - \frac{a_4a_5}{1+a_5+a_6} - \cdots - \frac{a_{2n-2}a_{2n-1}}{1+a_{2n-1}+a_{2n}} - \cdots \sim \mathop{\mathrm{K}}_{n=1}^{\infty} \frac{c_n}{1}$$

where

$$c_1 := \frac{a_1}{1+a_2} = -\frac{1}{3}, \quad c_2 := \frac{-a_2a_3}{(1+a_2)(1+a_3+a_4)} = 2,$$

$$c_n := \frac{-a_{2n-2}a_{2n-1}}{(1+a_{2n-3}+a_{2n-2})(1+a_{2n-1}+a_{2n})} = \frac{(n-1)^2(2n-1)}{2n-3}$$

for $n \geq 3$. Similarly, the odd part of $\mathbf{K}(a_n/1)$ is

$$a_1 - \frac{a_1a_2}{1+a_2+a_3} - \frac{a_3a_4}{1+a_4+a_5} - \frac{a_5a_6}{1+a_6+a_7} - \cdots \sim a_1 + \mathop{\mathrm{K}}_{n=1}^{\infty} \frac{d_n}{1}$$

where

$$d_1 := \frac{-a_1a_2}{1+a_2+a_3} = \frac{2}{3},$$

$$d_n := \frac{-a_{2n-1}a_{2n}}{(1+a_{2n-2}+a_{2n-1})(1+a_{2n}+a_{2n+1})} = \frac{n^2(2n-1)}{2n+1}.$$

Both $\mathbf{K}(c_n/1)$ and $\mathbf{K}(d_n/1)$ converge, as you are asked to prove in Problem 3 on page 166. Indeed, since all $d_n > 0$, it follows that $\mathbf{K}(d_n/1)$ converges to a positive number. Hence the odd part of $\mathbf{K}(a_n/1)$ converges to a number $> a_1 = 1$. However, $\mathbf{K}(c_n/1)$ converges to a number $f := c_1/(1+f^{(1)})$ where $f^{(1)} \geq 0$, and thus $f < 0$. Hence the even and odd parts of $\mathbf{K}(a_n/1)$ converge to distinct values. ◇

On the other hand, Wall ([Wall56]) has proved the following useful result:

Theorem 3.4. *Let the even and odd approximants $S_n(0)$ for $\mathbf{K}(a_n/1)$ converge. If $\liminf |a_n| < \infty$, then $\mathbf{K}(a_n/1)$ converges.*

Proof : Let $L_0 := \lim S_{2n}(0)$ and $L_1 := \lim S_{2n+1}(0)$, and assume that $L_0 \neq L_1$. Without loss of generality we assume that $L_1 \neq \infty$ and $L_2 \neq \infty$. (Otherwise we consider a continued fraction $a/(1 + \mathbf{K}(a_n/1))$ (or $\tilde{a}/(1 + a/(1 + \mathbf{K}(a_n/1)))$ if necessary) for appropriately chosen $a, \tilde{a} \neq 0$.) Since $S_n(-1) = S_{n-2}(0)$, the two sequences $\{S_{2n+m}\}_n$ for $m = 0, 1$ converge generally to L_m, respectively, with exceptional sequences $\{\zeta_{2n+m}\}_n$. Let δ_n be as defined in the previous proof. Then $\delta_n \to 0$, and by (1.2.3)

$$\delta_n := -\frac{f_{n+1} - f_{n-1}}{f_{n+1} - f_n} = \frac{1}{a_{n+1}} \cdot \frac{B_n}{B_{n-1}} = -\frac{\zeta_n}{a_{n+1}} \to 0.$$

Since $\delta_n \to 0$ and $\liminf |a_n| < \infty$, there therefore exists a subsequence $\{\zeta_{n_k}\}$ converging to 0. Without loss of generality we may assume that all n_k are either even or odd numbers, say n_k are even. Then $\lim S_{n_k}(\infty) = L_0$. However, $S_{n_k}(\infty) = S_{n_k-1}(0)$ which converges to L_1 as $k \to \infty$. This is a contradiction. Hence $L_0 = L_1$. □

3.1.3 Truncation error bounds

Let $\mathbf{K}(a_n/b_n)$ converge generally to some finite value $f \in \mathbb{C}$. The approximants $S_n(w_n)$ can be used to approximate f. But we need to control the *truncation error* $(f - S_n(w_n))$. This is done by deriving reliable *truncation error bounds* λ_n; i.e., positive numbers λ_n such that

$$|f - S_n(w_n)| \leq \lambda_n. \tag{1.3.1}$$

We have already seen three methods to establish such bounds:

1. In formula (1.6.6) on page 71 we observed that

   $$|f - S_n(w_n)| \leq \operatorname{diam}(K_n) \quad \text{for } w_n \in V_n \tag{1.3.2}$$

   whenever $\mathbf{K}(a_n/b_n)$ converges generally to f and $w_n \in V_n$ with $\operatorname{diam}_{\mathrm{m}}(V_n) \geq \varepsilon > 0$ for all n. Here $\{V_n\}$ are value sets for $\mathbf{K}(a_n/b_n)$ and $K_n := S_n(V_n)$. Therefore, if we find bounds for $\operatorname{diam}(K_n)$, then we have quite general bounds for the absolute error $|f - S_n(w_n)|$. More generally, (1.3.2) implies that

   $$|S_{n+m}(w_{n+m}) - S_n(w_n)| \leq \operatorname{diam}(K_n) \quad \text{for } m, n \in \mathbb{N} \tag{1.3.3}$$

   whenever $w_k \in V_k$ for $k = n, m+n$, since then $S_{n+m}(w_{n+m}) \in S_{n+m}(V_{n+m}) \subseteq S_n(V_n) = K_n$. This still holds if $\mathbf{K}(a_n/b_n)$ diverges.

2. By Corollary 2.7 on page 68

   $$f - f_n = t_0\Big(\frac{1}{\Sigma_n} - \frac{1}{\Sigma_\infty}\Big) \tag{1.3.4}$$

when $\mathbf{K}(a_n/b_n)$ converges to f, where

$$\Sigma_n := \sum_{k=0}^{n} P_k \to \Sigma_\infty \in \widehat{\mathbb{C}} \quad \text{for} \quad P_k := \prod_{j=1}^{k} \frac{b_j + t_j}{-t_j} \tag{1.3.5}$$

for an arbitrary tail sequence $\{t_n\}$ for $\mathbf{K}(a_n/b_n)$ with all $t_n \neq \infty$. Of course, $f_{n+m} - f_n = t_0(\Sigma_n^{-1} - \Sigma_{n+m}^{-1})$ holds even if $\mathbf{K}(a_n/b_n)$ diverges.

3. From the invariance of the cross ratio

$$\frac{S_n(u) - S_n(z)}{S_n(u) - S_n(w)} \cdot \frac{S_n(v) - S_n(w)}{S_n(v) - S_n(z)} = \frac{u - z}{u - w} \cdot \frac{v - w}{v - z} \tag{1.3.6}$$

with $u := 0$, $z := f_k^{(n)}$ (the kth classical approximant of the nth tail), $w := \infty$ and $v := \zeta_n := S_n^{-1}(\infty)$, we obtain

$$\frac{f_n - f_{n+k}}{f_n - f_{n-1}} \cdot 1 = \frac{-f_k^{(n)}}{\zeta_n - f_k^{(n)}}$$

when $f_k^{(n)} \neq 0, \infty$ and $\zeta_n \neq 0, \infty$, and thus $f_n, f_{n-1} \neq \infty$. I.e.,

$$f_{n+k} - f_n = \frac{-f_k^{(n)}}{f_k^{(n)} - \zeta_n}(f_n - f_{n-1}) . \tag{1.3.7}$$

Similarly, if $\mathbf{K}(a_n/b_n)$ converges to a finite value f and $\zeta_n, f^{(n)} \neq 0, \infty$, then

$$f - f_n = \frac{-f^{(n)}}{f^{(n)} - \zeta_n}(f_n - f_{n-1}). \tag{1.3.8}$$

If $\mathbf{K}(a_n/b_n)$ converges to $f = \infty$, it is natural to use

$$\frac{1}{f} - \frac{1}{f_n} = -\frac{1}{f_n} \quad \text{or} \quad \frac{1}{f} - \frac{1}{S_n(w_n)} = -\frac{1}{S_n(w_n)}$$

as a measure for the truncation error. Bounds for this error can be found from the truncation error of a continued fraction $a/(b + \mathbf{K}(a_n/b_n))$ which then converges to 0, regardless of the choice of $a \neq 0$ and b.

Now, a lot of the truncation error bounds derived up through the ages concern classical approximants. On the other hand we have seen that a clever choice of approximants $S_n(w_n)$ can give faster convergence, so we really want bounds for $|f - S_n(w_n)|$. Fortunately we have the following method to adjust existing truncation error bounds to this newer situation:

Theorem 3.5. *Let $\mathbf{K}(a_n/b_n)$ converge generally to $f \neq \infty$. If both $f^{(n)} \neq \infty$ and $\zeta_n \neq \infty$, then*

$$f - S_n(w_n) = \frac{f^{(n)} - w_n}{\zeta_n - w_n}(f - f_{n-1}). \tag{1.3.9}$$

Proof : Also this follows from (1.3.6), but this time we use $u := f^{(n)}$, $z := w_n$, $w := \infty$ and $v := \zeta_n$ if these are distinct points. Now, $f^{(n)} \neq \zeta_n$ since $f = S_n(f^{(n)}) \neq \infty$ and $S_n(\zeta_n) = \infty$. If $w_n = \zeta_n$ or $w_n = f^{(n)}$, the identity (1.3.9) reduces to $\infty = \infty$ or $0 = 0$. Finally, if $w_n = \infty$, the identity reduces to $f - f_{n-1} = f - f_{n-1}$. □

Equation (1.3.9) also shows that to get fast convergence, a sensible choice for w_n is a choice that makes $|\kappa(w_n)|$ small, preferably $\kappa(w_n) \to 0$ for

$$\kappa(w_n) := \frac{f^{(n)} - w_n}{\zeta_n - w_n}. \tag{1.3.10}$$

That is, we want w_n to be close to the tail value $f^{(n)}$ and far away from the critical tail ζ_n.

We distinguish between *a priori bounds* which can be computed in advance, before we compute the approximants, and *a posteriori bounds* which are expressed in terms of the approximants. The bounds (1.3.2) and (1.3.4) are normally a priori bounds, whereas (1.3.8) is a typical a posteriori bound. A posteriori bounds are normally tighter than a priori bounds. A priori bounds on the other hand make it possible to estimate in advance the order of the approximant and thus the number of digits needed in the computation to reach a wanted degree of approximation.

Example 2. Let $0 < g < \frac{1}{2}$ be given. Then the real interval $V := [-g, g]$ is a simple value set for every continued fraction $\mathbf{K}(a_n/1)$ from $E := [-g(1-g), g(1-g)]$. Let $\mathbf{K}(a_n/1)$ be a convergent continued fraction from E. By (1.4.6) on page 64 we know that

$$-\zeta_n = h_n = 1 + \frac{a_n}{1} \, \frac{}{+} \, \frac{a_{n-1}}{1} \, \frac{}{+} \cdots \frac{}{+} \, \frac{a_2}{1}.$$

Since $0 \in V$ and $a_k/(1+V) \subseteq V$ for all k, it follows that $h_n \in 1 + V$; i.e., $\zeta_n \in -1 - V = [-1-g, -1+g]$. It is also clear that $f^{(n)} \in V$. Therefore $|f^{(n)}| \leq g$ and $|\zeta_n - f^{(n)}| \geq 1 - 2g$, and thus by (1.3.8)

$$|f - f_n| \leq \frac{g}{1-2g} |f_n - f_{n-1}| \quad \text{for } n = 1, 2, 3, \ldots.$$

◇

3.1.4 Mapping with linear fractional transformations

A circle C on the Riemann sphere $\widehat{\mathbb{C}}$ is also a circle in the complex plane $\mathbb{C}$ if $\infty \notin C$, but if $\infty \in C$, then $C \setminus \{\infty\}$ is a straight line in $\mathbb{C}$. Therefore circles and lines on $\mathbb{C}$ are also called *generalized circles* as a common name.

Linear fractional transformations $\tau \in \mathcal{M}$ have beautiful mapping properties. Indeed, they

1. map circles on $\widehat{\mathbb{C}}$ onto circlcs on $\widehat{\mathbb{C}}$,

2. map points symmetric with respect to a generalized circle C in $\widehat{\mathbb{C}}$ onto points symmetric with respect to $\tau(C)$. (Two points u and v are symmetric with respect to the circle with center $\Gamma \in \mathbb{C}$ and radius $\rho > 0$ if $(u-\Gamma)(\overline{v}-\overline{\Gamma}) = \rho^2$. (Two points u and v are symmetric with respect to the line L if L is the perpendicular bisector of the line segment connecting u and v.)

We shall focus on the family $\mathfrak{V}$ of closed sets V on $\widehat{\mathbb{C}}$ where ∂V (the boundary of V) is a circle on $\widehat{\mathbb{C}}$. We distinguish between the cases where

$\infty \notin V$: We say that V is a *closed (circular) disk*, and we write $V = \mathfrak{B}(\Gamma, \rho) := \{w \in \mathbb{C};\ |w - \Gamma| \leq \rho\}$ where $\Gamma \in \mathbb{C}$ is the center and $\rho > 0$ is the (euclidean) radius of V.

$\infty \in \partial V$: We say that V is a *closed half plane*, and we write $V = \xi + e^{i\alpha}\overline{\mathbb{H}}$ where $\xi \in \mathbb{C}$, $\alpha \in \mathbb{R}$ (the set of real numbers), $\mathbb{H} := \{w \in \mathbb{C};\ \operatorname{Re} w > 0\}$, and $\overline{\mathbb{H}}$ is its closure in $\widehat{\mathbb{C}}$.

$\infty \in V^\circ$ (the interior of V): We say that V is *the complement of* (or *the exterior of*) *an open disk*, and we write $V = \mathfrak{B}(\Gamma, -\rho) := \{w \in \widehat{\mathbb{C}};\ |w - \Gamma| \geq \rho\}$ with $\rho > 0$, when V is the exterior of $\mathfrak{B}(\Gamma, \rho)^\circ$.

Let $\tau \in \mathcal{M}$ be given by

$$\tau(w) := \frac{aw+b}{cw+d} \quad \text{with} \quad \Delta := ad - bc \neq 0. \tag{1.4.1}$$

We want to describe $\tau(V)$ for $V \in \mathfrak{V}$. The easy case is the case where $c = 0$. Then τ can be written $\tau(w) = \frac{a}{d}w + \frac{b}{d}$, and therefore

$$c = 0 \quad \Longrightarrow \quad \begin{cases} \tau(\mathfrak{B}(\Gamma, \mu)) = \mathfrak{B}(\frac{a}{d}\Gamma + \frac{b}{d}, |\frac{a}{d}|\mu)\,, \\ \tau(\xi + e^{i\alpha}\overline{\mathbb{H}}) = \tau(\xi) + e^{i(\alpha+\beta)}\overline{\mathbb{H}} \text{ where } \beta := \arg\frac{a}{d}\,. \end{cases} \tag{1.4.2}$$

Otherwise we have:

Lemma 3.6. *Let τ be given by (1.4.1) with $c \neq 0$, and let $V \in \mathfrak{V}$. If $-\frac{d}{c} \notin \partial V$, then $\tau(V) = \mathfrak{B}(\Gamma_1, \mu_1)$ where*

$$\Gamma_1 := \frac{a}{c} - \frac{\Delta(\overline{c}\overline{\Gamma} + \overline{d})/c}{|c\Gamma + d|^2 - |c\mu|^2} \quad \textit{and} \quad \mu_1 := \frac{\mu|\Delta|}{|c\Gamma + d|^2 - |c\mu|^2} \tag{1.4.3}$$
$$\textit{if } V = \mathfrak{B}(\Gamma, \mu) \textit{ with } \Gamma \in \mathbb{C} \textit{ and } \mu \in \mathbb{R} \setminus \{0\}$$

and

$$\Gamma_1 := \frac{a}{c} - \frac{\frac{1}{2}\Delta\, e^{-i\alpha}/c^2}{\mathrm{Re}[(\xi + \frac{d}{c})e^{-i\alpha}]} \quad \textit{and} \quad |\mu_1| := \left|\frac{\frac{1}{2}\Delta/c^2}{\mathrm{Re}[(\xi + \frac{d}{c})e^{-i\alpha}]}\right| \tag{1.4.4}$$
$$\textit{if } V = \xi + e^{i\alpha}\mathbb{H} \textit{ with } \xi \in \mathbb{C} \textit{ and } \alpha \in \mathbb{R}.$$

Proof: This time τ can be written

$$\tau(w) = \frac{a}{c} - \frac{\Delta/c^2}{w + \beta} \quad \text{where } \beta := \frac{d}{c}. \tag{1.4.5}$$

Since $-\beta \notin \partial V$; i.e., $0 \notin \partial(\beta + V)$, the case where $\tau(V)$ is a half plane is ruled out.

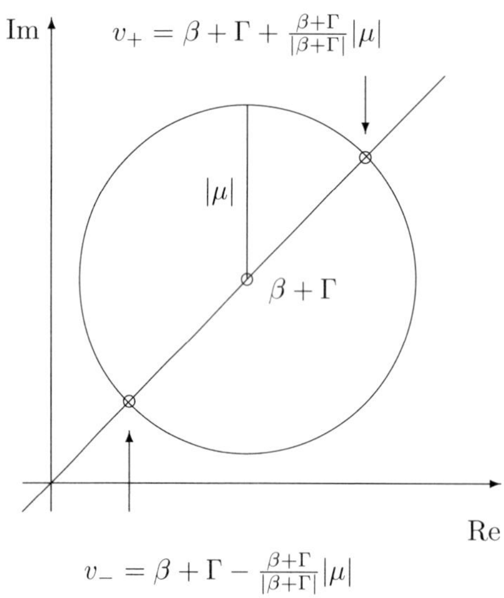

Let first $V := \mathfrak{B}(\Gamma, \mu)$. If $\beta + \Gamma = 0$, then the boundary of $(\beta + V)$ is a circle with center at the origin and radius $|\mu|$. Therefore $1/\partial(\beta + V)$ is a circle with center at the origin and radius $1/|\mu|$. Therefore $\frac{\Delta}{c^2}/\partial(\beta + V)$ is a circle with center at the origin and radius $|\frac{\Delta}{c^2}/\mu|$. Therefore V is mapped onto $\mathfrak{B}(\frac{a}{c}, -|\frac{\Delta}{c^2}|/\mu)$, as claimed in (1.4.3).

Let $\beta + \Gamma \neq 0$. Since

$v_\pm := (\beta + \Gamma)(1 \pm |\mu|/|\beta + \Gamma|)$

are the two points on $\partial(\beta + V)$ furthest away from 0 and closest to 0 respectively (see the picture), it follows that the circle $1/(\beta + \partial V)$ has center $\widehat{\Gamma}$ and radius r given by

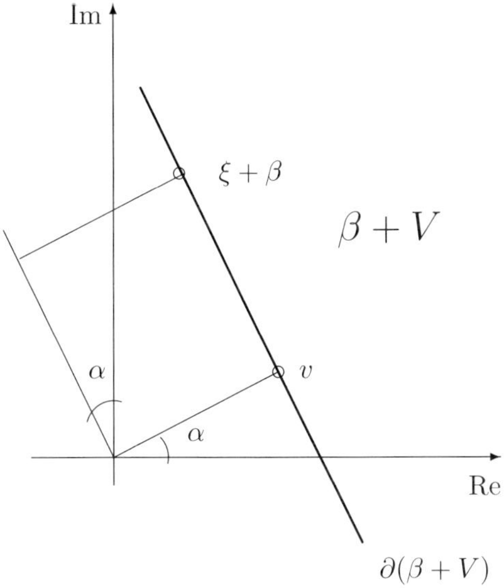

$$\widehat{\Gamma} := \frac{1}{2}\Big(\frac{1}{v_-}+\frac{1}{v_+}\Big) = \frac{\overline{\beta}+\overline{\Gamma}}{|\beta+\Gamma|^2-\mu^2}$$
$$r := \frac{1}{2}\Big|\frac{1}{v_-}-\frac{1}{v_+}\Big| = \Big|\frac{\mu}{|\beta+\Gamma|^2-\mu^2}\Big|.$$

Since $\tau(V)$ is bounded if and only if $0 \notin \beta+V$, this proves (1.4.3).

Next, let $V := \xi+e^{i\alpha}\overline{\mathbb{H}}$. Then $V+\beta = \xi+\beta+e^{i\alpha}\overline{\mathbb{H}}$. The point on $\partial(V+\beta)$ closest to the origin is $v := e^{i\alpha}\mathrm{Re}[(\xi+\beta)e^{-i\alpha}]$. Therefore $1/\partial(V+\beta)$ is the circle through the origin with center $\frac{1}{2v}$ and radius $\frac{1}{2|v|}$. Hence the expression for $\tau(V)$ follows from (1.4.5). □

We conclude this section with some useful observations ([Lore94a]), strongly inspired by work of Jones and Thron. Here $\mathrm{rad}\, V$ denotes the radius of a disk V.

Lemma 3.7. *Let $\mathcal{T}_n := \tau_1 \circ \tau_2 \circ \cdots \circ \tau_n$ for all $n \in \mathbb{N}$ where all $\tau_n \in \mathcal{M}$ map the unit disk $\mathbb{D}$ into itself with*

$$\limsup_{n\to\infty} \mathrm{rad}\, \tau_n(\mathbb{D}) < 1. \tag{1.4.6}$$

Then $\{\mathcal{T}_n\}$ is restrained with exceptional sequence $\{\mathcal{T}_n^{-1}(\infty)\}$ from $\widehat{\mathbb{C}}\setminus\overline{\mathbb{D}}$. If moreover $\lim \mathrm{rad}\, \mathcal{T}_n(\overline{\mathbb{D}}) > 0$, then $\sum \mathfrak{m}(|\mathcal{T}_n^{-1}(\infty)|, 1) < \infty$, and $\{\mathcal{T}_n\}$ has an exceptional sequence from $\partial\mathbb{D}$.

Proof : Let Γ_n and R_n be the center and radius of the disk $\mathcal{T}_n(\overline{\mathbb{D}})$. The nestedness $\mathcal{T}_{n+1}(\overline{\mathbb{D}}) = \mathcal{T}_n(\tau_{n+1}(\overline{\mathbb{D}})) \subseteq \mathcal{T}_n(\overline{\mathbb{D}})$ shows that $\{R_n\}$ is a non-increasing sequence of positive numbers. Hence $R_n \to R \geq 0$. If $R=0$, the limit point case, then the sequence $\{\mathcal{T}_n\}$ is clearly restrained. Indeed, it converges generally to the limit point $\gamma \in \lim \mathcal{T}_n(\overline{\mathbb{D}}) \subseteq \overline{\mathbb{D}}$.

Assume that $R>0$, the limit circle case. A linear fractional transformation mapping $\mathbb{D}$ onto $\mathbb{D}$ can always be written on the form $e^{i\alpha}(w-q)/(1-\overline{q}w)$ for some real constant α and complex constant $q \in \mathbb{D}$. Therefore $\mathcal{T}_n$ can be written

$$\mathcal{T}_n(w) = \Gamma_n + P_n\frac{w-Q_n}{1-\overline{Q}_n w} = \frac{(\Gamma_n - P_nQ_n)+(P_n-\overline{Q}_n\Gamma_n)w}{1-\overline{Q}_n w} \tag{1.4.7}$$

where $|P_n| = R_n$ and $|Q_n| < 1$. Similarly, τ_n can be written

$$\tau_n(w) = \gamma_n + p_n \frac{w - q_n}{1 - \overline{q}_n w} \quad \text{for } n = 1, 2, 3, \dots,$$

where γ_n and $r_n := |p_n|$ are the center and radius of $\tau_n(\mathbb{D})$ and $|q_n| < 1$. By (1.4.6) $\limsup r_n < 1$; i.e., $r_n \le r$ from some n on for some $r < 1$. Since $\tau_n(\mathbb{D}) \subseteq \mathbb{D}$, we have $|\gamma_n| \le 1 - r_n$. From (1.4.3) it follows that the radius R_{n+1} of $\mathcal{T}_{n+1}(\mathbb{D}) = \mathcal{T}_n(\tau_{n+1}(\mathbb{D}))$ is equal to

$$\begin{aligned} R_{n+1} &= \frac{|(\Gamma_n - P_n Q_n)(-\overline{Q}_n) - (P_n - \overline{Q}_n \Gamma_n) \cdot 1|}{|1 - \overline{Q}_n \gamma_{n+1}|^2 - |Q_n|^2 r_{n+1}^2} r_{n+1} \\ &= \frac{R_n(1 - |Q_n|^2) r_{n+1}}{|1 - \overline{Q}_n \gamma_{n+1}|^2 - |Q_n|^2 r_{n+1}^2}. \end{aligned}$$

Therefore

$$\begin{aligned} \frac{R_{n+1}}{R_n} &\le \frac{(1 - |Q_n|^2) r_{n+1}}{(1 - |Q_n|(1 - r_{n+1}))^2 - |Q_n|^2 r_{n+1}^2} \\ &= \frac{(1 + |Q_n|) r_{n+1}}{1 - |Q_n| + 2|Q_n| r_{n+1}} = 1 - \frac{(1 - r_{n+1})(1 - |Q_n|)}{1 - |Q_n| + 2|Q_n| r_{n+1}} \\ &\le 1 - \frac{(1 - r)(1 - |Q_n|)}{1 - |Q_n| + 2r|Q_n|} =: 1 - \delta_n \end{aligned}$$

where $\delta_n > 0$. Since $\prod_{n=1}^{\infty}(R_{n+1}/R_n) = R/R_1 > 0$, it follows that $\sum \delta_n < \infty$; i.e., $\sum(1 - |Q_n|) < \infty$. Since $\mathcal{T}_n^{-1}(\infty) = 1/\overline{Q}_n$ where $|Q_n| \to 1$, this proves that $\mathcal{T}_n^{-1}(\infty) \ne \infty$ for $n \ge n_0$ for some $n_0 \in \mathbb{N}$ and that $\sum_{n=n_0}^{\infty}(|\mathcal{T}_n^{-1}(\infty)| - 1) < \infty$, or, equivalently, $\sum_{n=1}^{\infty} \mathfrak{m}(|\mathcal{T}_n^{-1}(\infty)|, 1) < \infty$. From the expression (1.4.7) for $\mathcal{T}_n$ it follows that its derivative is

$$\mathcal{T}_n'(w) = P_n \frac{1 - |Q_n|^2}{(1 - \overline{Q}_n w)^2} \tag{1.4.8}$$

where $|P_n| = R_n \to R < 1$ and $|Q_n| \to 1$. Hence $\mathcal{T}_n'(w) \to 0$ for every $w \in \mathbb{D}$, and $\{\mathcal{T}_n\}$ is restrained.

It remains to prove that $\{\mathcal{T}_n^{-1}(\infty)\}$ is an exceptional sequence. But this follows trivially from (1.4.8) since $w := 1/\overline{Q}_n = \mathcal{T}_n^{-1}(\infty)$ gives $\mathcal{T}_n'(w) = \infty$. Since $|\mathcal{T}_n^{-1}(\infty)| \to 1$, there also exist exceptional sequences $\{w_n^\dagger\}$ with $|w_n^\dagger| = 1$ for all n; i.e., $w_n^\dagger \in \partial\mathbb{D}$. □

Lemma 3.8. *Let $\mathcal{T}_n$ and τ_n be as in Lemma 3.7, and assume there exists a sequence $\{w_n\} \subseteq \widehat{\mathbb{C}}$ such that*

$$\liminf \big||w_n| - 1\big| > 0 \quad \textit{and} \quad \liminf \big||\tau_n(w_n)| - 1\big| > 0. \tag{1.4.9}$$

Then $\{\mathcal{T}_n\}$ converges generally to some constant $\gamma \in \overline{\mathbb{D}}$ with an exceptional sequence $\{w_n^\dagger\}$ with $w_n^\dagger \in \widehat{\mathbb{C}} \setminus \overline{\mathbb{D}}$. If the limit point case fails to occur for $\{\mathcal{T}_n(\overline{\mathbb{D}})\}$, then $\{\mathcal{T}_n(w)\}$ converges absolutely to γ for every $w \in \mathbb{D}$.

Proof : We use the notation from the proof of Lemma 3.7. If the limit point case occurs for $\mathcal{T}_n(\overline{\mathbb{D}})$, then the general convergence is clear. Assume that $R > 0$. Assume there exists such a sequence $\{w_n\}$ bounded away from the unit circle $\partial\mathbb{D}$, with the property that also $\{\tau_n(w_n)\}$ is bounded away from $\partial\mathbb{D}$, as required in (1.4.9). From (1.4.7)

$$|\mathcal{T}_n(u) - \mathcal{T}_n(v)| = \begin{cases} R_n \dfrac{(1-|Q_n|^2)|u-v|}{|1-\overline{Q}_n u| \cdot |1-\overline{Q}_n v|} & \text{if } u, v \neq \infty, \\ R_n \dfrac{1-|Q_n|^2}{|Q_n| \cdot |1-\overline{Q}_n v|} & \text{if } u = \infty, \\ 0 & \text{if } u = v = \infty. \end{cases} \tag{1.4.10}$$

Since $|Q_n| \to 1$ in this case, there therefore exists a constant $A > 0$ such that

$$|\mathcal{T}_{n+1}(w_{n+1}) - \mathcal{T}_n(w_n)| = |\mathcal{T}_n(\tau_{n+1}(w_{n+1})) - \mathcal{T}_n(w_n)| < A(1-|Q_n|)$$

from some n on, say $n \geq n_0$. Since $\sum(1-|Q_n|) < \infty$ under our conditions, this means that $\sum_{n=n_0}^{\infty} |\mathcal{T}_{n+1}(w_{n+1}) - \mathcal{T}_n(w_n)| < \infty$; i.e., $\{\mathcal{T}_n(w_n)\}_{n_0}^{\infty}$ converges absolutely to some $\gamma \in \overline{\mathbb{D}}$.

Let $w \in \mathbb{D}$ be arbitrarily chosen. Then $(1-\overline{Q}_n w)$ is bounded away from 0. Hence also $|\mathcal{T}_n(w_n) - \mathcal{T}_n(w)| \leq A(1-|Q_n|)$ for some $A > 0$ from some n on. Therefore $\{\mathcal{T}_n(w)\}$ converges absolutely to γ for every $w \in \mathbb{D}$, and thus $\{\mathcal{T}_n\}$ converges generally to γ. □

Remark: The conditions in Lemma 3.8 are in particular satisfied if all τ_n map $\mathbb{D}$ into itself and $|\tau_n(w_0)| \leq r < 1$ for all n for some given point $w_0 \notin \overline{\mathbb{D}}$, since then $\operatorname{rad}(\tau_n(\mathbb{D})) \leq \frac{1+r}{2}$ for all n. This generalizes a result by Hillam and Thron ([HiTh65]) who required that $\tau_n(w) = z$ for all n for some $w \in \widehat{\mathbb{C}} \setminus \overline{\mathbb{D}}$ and $z \in \mathbb{D}$.

One application of Lemma 3.8 is to derive convergence criteria for continued fractions $\mathbf{K}(a_n/b_n)$ with value sets $\{V_n\}_{n=0}^{\infty}$ where $V_n \in \mathfrak{V}$ (i.e., ∂V_n is a circle on $\widehat{\mathbb{C}}$). The idea is to define linear fractional transformations φ_n such that $\varphi_n(\overline{\mathbb{D}}) = V_n$. Then

$$\tau_n := \varphi_{n-1}^{-1} \circ s_n \circ \varphi_n$$

maps $\mathbb{D}$ into $\mathbb{D}$. Moreover, $s_n = \varphi_{n-1} \circ \tau_n \circ \varphi^{-1}$, so

$$S_n := s_1 \circ s_2 \circ \cdots \circ s_n = \varphi_0^{-1} \circ \tau_1 \circ \tau_2 \circ \cdots \circ \tau_n \circ \varphi_n = \varphi_0^{-1} \circ \mathcal{T}_n \circ \varphi_n.$$

Therefore, if $\{\varphi_n\}$ is totally non-restrained, then $\{S_n\}$ converges generally if and only if $\{\mathcal{T}_n\}$ converges generally. Hence the following result is for instance a consequence of Lemma 3.8:

Corollary 3.9. *Let $\{\varphi_n\}_{n=0}^{\infty}$ be a totally non-restrained sequence of linear fractional transformations, and let $\{V_n\}_{n=0}^{\infty}$ given by $V_n := \varphi_n(\overline{\mathbb{D}})$ be a sequence of value sets for the continued fraction $\mathbf{K}(a_n/b_n)$. If*

$$\limsup_{n\to\infty} \operatorname{rad} \varphi_{n-1}^{-1} \circ s_n \circ \varphi_n(\mathbb{D}) < 1 \tag{1.4.11}$$

and there exists a sequence $\{w_n\}$ from $\widehat{\mathbb{C}}$ such that

$$\liminf_{n\to\infty} \operatorname{dist}_{\mathfrak{m}}(w_n, \partial V_n) > 0 \ \ \text{and} \ \ \liminf_{n\to\infty} \operatorname{dist}_{\mathfrak{m}}(s_n(w_n), \partial V_{n-1}) > 0, \tag{1.4.12}$$

then $\mathbf{K}(a_n/b_n)$ converges generally to some value $f \in V_0$ with exceptional sequence $\{w_n^\dagger\}$ with $w_n^\dagger \in \widehat{\mathbb{C}} \setminus V_n$ for all n.

Here $\operatorname{dist}_{\mathfrak{m}}(w, V) := \inf\{\mathfrak{m}(w, v);\, v \in V\}$ denotes the chordal distance between a point $w \in \widehat{\mathbb{C}}$ and a set $V \subseteq \widehat{\mathbb{C}}$.

3.1.5 The Stieltjes-Vitali Theorem

Convergence of a continued fraction is defined as convergence of the sequence $\{S_n\}$ from $\mathcal{M}$, either evaluated at the origin or in the sense of general convergence. There is a useful theorem on convergence of sequences of holomorphic functions, generally known as *Vitali's Theorem*, ([Vita35]). However, Stieltjes presented the idea earlier in his important 1894-paper ([Stie94]). Therefore we follow Jones and Thron ([JoTh80], p 89) in naming the theorem:

Theorem 3.10. (The Stieltjes-Vitali Theorem.) *Let $\{F_n\}$ be a sequence of holomorphic functions in a region D, such that*

(i) there exist two distinct values $p, q \in \mathbb{C}$ for which $F_n(z) \neq p$ and $F_n(z) \neq q$ for all $n \in \mathbb{N}$ and all $z \in D$, and

(ii) $\{F_n(z)\}$ converges to a finite value for each z in a set $D^* \subseteq D$ which has at least one accumulation point in D.

Then $\{F_n(z)\}$ converges locally uniformly in D to a holomorphic function.

Here region is meant in the strict sense; i.e., an open, connected set $\subseteq \mathbb{C}$. Moreover, D^* needs to contain infinitely many points since it shall have an accumulation point. For the proof of this theorem we refer to text books on functions of a complex variable, for instance ([Hille62], p 248–251).

3.1.6 A simple estimate

In [Lange99b] the following estimate was proved:

Lemma 3.11. *For given positive constants a, b with $a < b+1$,*

$$\prod_{k=1}^{n} \frac{b+k-a}{b+k} < \Big(\frac{b+1}{b+n+1}\Big)^a \qquad \text{for } n > 0. \tag{1.6.1}$$

Proof : For $u := (b+k+1)/(b+k)$ we consider the integral

$$\begin{aligned} I &:= \int_1^u t^{-a-2}(u-t)\,dt = \Big[\frac{t^{-a-1}}{-a-1}\,u - \frac{t^{-a}}{-a}\Big]_1^u \\ &= -\tfrac{1}{a+1}(u^{-a}-u) + \tfrac{1}{a}(u^{-a}-1) \\ &= \frac{1}{a(a+1)}\Big[\Big(\frac{b+k}{b+k+1}\Big)^a - \frac{b+k-a}{b+k}\Big]. \end{aligned}$$

It is clear from the definition of I that $I > 0$. Hence

$$\frac{b+k-a}{b+k} < \Big(\frac{b+k}{b+k+1}\Big)^a,$$

and (1.6.1) follows by taking products on both sides. □

3.2 Classical convergence theorems

3.2.1 Positive continued fractions

A continued fraction $\mathbf{K}(a_n/b_n)$ with all $a_n > 0$ and $b_n \geq 0$ is called a *positive* continued fraction. (Note that we allow $b_n = 0$.) It has the following useful property:

Theorem 3.12. *Let $\mathbf{K}(a_n/b_n)$ be a positive continued fraction. Then*

$$S_2(0) \leq S_4(0) \leq S_6(0) \leq \cdots \leq S_5(0) \leq S_3(0) \leq S_1(0)\,. \tag{2.1.1}$$

If moreover $b_1 > 0$, then $\{S_{2n}(0)\}$ and $\{S_{2n+1}(0)\}$ converge absolutely to finite, non-negative values. If all $b_n > 0$, then (2.1.1) holds with strict inequalities.

Proof : Let first $b_1 > 0$. Then $B_n > 0$ for $n \geq 1$ by the recurrence relation. Therefore the Euler-Minding series

$$\sum_{n=1}^{\infty}(f_n - f_{n-1}) \quad \text{where} \quad f_n - f_{n-1} = (-1)^{n+1}\frac{\prod_{k=1}^{n} a_k}{B_n B_{n-1}} \tag{2.1.2}$$

is an alternating series. Since

$$\left|\frac{f_n - f_{n-1}}{f_{n-1} - f_{n-2}}\right| = \frac{a_n B_{n-2}}{B_n} = \frac{B_n - b_n B_{n-1}}{B_n} \leq 1, \tag{2.1.3}$$

we find that $\{|f_n - f_{n-1}|\}$ is non-increasing, and thus (2.1.1) follows since $S_n(0) = f_n = \sum_{m=1}^{n}(f_m - f_{m-1})$. The convergence of $\{f_{2n}\}$ and $\{f_{2n+1}\}$ is therefore clear. Indeed, since $(f_{2n+2} - f_{2n}) \geq 0$ and $(f_{2n+1} - f_{2n-1}) \leq 0$ for all n, both $\{f_{2n}\}$ and $\{f_{2n+1}\}$ converge absolutely.

Next let $b_1 = 0$. If all $b_n = 0$, then $S_{2n}(0) = 0$ and $S_{2n+1}(0) = \infty$ for all n, and (2.1.1) holds trivially. Otherwise, let b_{n_0+1} be the first non-zero b_n. Then (2.1.1) holds for the n_0th tail $\mathbf{K}_{n=n_0+1}^{\infty}(a_n/b_n)$ of $\mathbf{K}(a_n/b_n)$. Since

$$\mathbf{K}(a_n/b_n) = S_{n_0}(\mathbf{K}_{n=n_0+1}^{\infty}(a_n/b_n)) \tag{2.1.4}$$

where

$$S_{n_0}(w) = \frac{a_1}{0} \, + \, \frac{a_2}{0} \, + \cdots + \, \frac{a_{n_0}}{w} = \begin{cases} \dfrac{a_1 a_3 \cdots a_{n_0}}{a_2 a_4 \cdots a_{n_0-1} w} & \text{if } n_0 \text{ is odd} \\[2ex] \dfrac{a_1 a_3 \cdots a_{n_0-1}}{a_2 a_4 \cdots a_{n_0}} w & \text{if } n_0 \text{ is even} \end{cases}$$

(2.1.1) holds also now. □

The following classical result due to Seidel ([Seid46]) and Stern ([Stern48]) is a direct consequence of Theorem 3.12 combined with the Stern-Stolz Divergence Theorem on page 100 and the Lane-Wall Characterization on page 103:

Theorem 3.13. *A positive continued fraction* $\mathbf{K}(1/b_n)$ *converges if and only if* $\sum b_n = \infty$. *If* $\sum b_n < \infty$, *then* $\mathbf{K}(1/b_n)$ *diverges generally.*

Proof : Let first $\sum b_n = \infty$. Without loss of generality we assume that $b_1 > 0$. (Otherwise we consider a tail of $\mathbf{K}(1/b_n)$.) Then $\{f_{2n}\}$ and $\{f_{2n+1}\}$ converge absolutely (Theorem 3.12), and thus $\mathbf{K}(1/b_n)$ converges (the Lane-Wall Characterization).

Next let $\sum b_n < \infty$. The general divergence follows then from the Stern-Stolz Divergence Theorem. □

Seidel and Stern required that all $b_n > 0$. That $b_n \geq 0$ suffices was proved by Broman ([Brom77]). An equivalence transformation gives the formulation:

Theorem 3.14. (The Seidel-Stern Theorem.) *A positive continued fraction* $\mathbf{K}(a_n/b_n)$ *converges if and only if its Stern-Stolz Series diverges to* ∞; *i.e., if and only if*

$$\mathcal{S} := \sum_{n=1}^{\infty} b_n \prod_{k=1}^{n} a_k^{(-1)^{n+k+1}} = \infty. \tag{2.1.5}$$

Positive continued fractions and tail sequences.

It is clear that if the positive continued fraction $\mathbf{K}(a_n/b_n)$ with all $b_n > 0$ converges generally, then its sequence $\{f^{(n)}\}$ of tail values is a positive sequence. What is more interesting is that a kind of converse result is true. The following was published by Khrushchev ([Khru06b]) under the name of Markov's Theorem:

Theorem 3.15. *Let* $\{t_n\}$ *be a positive tail sequence for the generally convergent positive continued fraction* $\mathbf{K}(a_n/b_n)$. *Then* $\mathbf{K}(a_n/b_n)$ *converges to* t_0 *and* $\{t_n\}$ *is its sequence of tail values.*

Proof : Since $\mathbf{K}(a_n/b_n)$ converges and all $t_n \neq 0, -b_n$, it follows from Corollary 2.7 on page 68 that $\{\Sigma_n\}$ given by

$$\Sigma_n := \sum_{k=0}^{n} P_k \text{ where } P_k := \prod_{m=1}^{k} \frac{b_m + t_m}{-t_m}$$

converges to some $\Sigma_\infty \in \widehat{\mathbb{C}}$ as $n \to \infty$. Since $|P_k| \geq 1$ for all n, the possibility that Σ_∞ is finite is ruled out. Hence, by Corollary 2.7 t_0 is the value of $\mathbf{K}(a_n/b_n)$. □

Truncation error bounds.

Let $\mathbb{R}^+$ be the set of positive numbers, and $\overline{\mathbb{R}^+}$ be its closure in $\widehat{\mathbb{C}}$. Both $\mathbb{R}^+$ and $\overline{\mathbb{R}^+}$ are simple value sets for the family of positive continued fractions. Indeed, $\overline{\mathbb{R}^+}$ is the limit set for this family. We can use this to construct truncation error bounds:

Theorem 3.16. *If $\mathbf{K}(a_n/b_n)$ is a positive continued fraction, then*

$$\operatorname{diam} S_n(\overline{\mathbb{R}^+}) = |f_n - f_{n-1}|. \tag{2.1.6}$$

Proof : $S_n(\overline{\mathbb{R}^+})$ is an interval $\subseteq \overline{\mathbb{R}^+}$ with end points $S_n(0) = f_n$ and $S_n(\infty) = f_{n-1}$. □

This means that if $\mathbf{K}(a_n/b_n)$ converges, then $(f_n - f_{n-1}) \to 0$ and

$$|f - S_n(w_n)| \leq |f_n - f_{n-1}| \quad \text{for } w_n \in \overline{\mathbb{R}^+}. \tag{2.1.7}$$

In particular the choice $w_n := \infty$ leads to

$$|f - f_{n-1}| \leq |f_n - f_{n-1}| \tag{2.1.8}$$

which also follows directly from (2.1.1). Indeed, from (2.1.1) we also know that

$$(f - f_{n-1}) \text{ is } \begin{cases} \leq 0 & \text{if } n \text{ odd,} \\ \geq 0 & \text{if } n \text{ even.} \end{cases} \tag{2.1.9}$$

We can even exploit this oscillatory character further to get

$$|f - f_n^*| \leq \frac{1}{2}(f_{2n+1} - f_{2n}) \quad \text{for} \quad f_n^* := \frac{1}{2}(f_{2n+1} + f_{2n}). \tag{2.1.10}$$

These simple a posteriori bounds secure reliable computation of positive continued fractions. We shall see later (Theorem 3.24 on page 128 and Corollary 3.41 on page 148) that the following a priori bounds hold:

Theorem 3.17. *If* $\mathbf{K}(a_n/b_n)$ *is a positive continued fraction, then*

$$\operatorname{diam} S_n(\overline{\mathbb{R}^+}) \leq 2a_1 \prod_{k=2}^{n} \frac{\sqrt{1+4a_k}-1}{\sqrt{1+4a_k}+1} \quad \text{if all } b_n = 1, \tag{2.1.11}$$

and

$$\operatorname{diam} S_n(\overline{\mathbb{R}^+}) \leq \frac{1/b_1}{\prod_{k=2}^{n}(1+b_k b_{k-1})} \quad \text{if all } a_n = 1. \tag{2.1.12}$$

Example 3. In Example 12 on page 26 we used the continued fraction

$$\operatorname{Ln} 2 = \frac{1}{1+}\,\frac{1/2}{1}\,+\,\frac{1/(2\cdot 3)}{1}\,+\,\frac{2/(2\cdot 3)}{1}\,+\,\frac{2/(2\cdot 5)}{1}\,+\,\frac{3/(2\cdot 5)}{1}\,+\,\frac{3/(2\cdot 7)}{1}\,+\cdots$$

to estimate the value of $\operatorname{Ln} 2$. The first seven approximants $f_n = S_n(0)$ were given in a table. The oscillatory character of $\{S_n(0)\}$ in this table is consistent with (2.1.1). By (2.1.8)-(2.1.9) the approximation $\operatorname{Ln}(2) \approx f_6 \approx 0.693121$ satisfies

$$0 < \operatorname{Ln} 2 - f_6 \leq (0.693152 - 0.693121) \approx 3.1 \cdot 10^{-5},$$

and by (2.1.11)

$$|\operatorname{Ln} 2 - f_6| \leq 2 \cdot \frac{\sqrt{1+2}-1}{\sqrt{1+2}+1} \cdot \frac{\sqrt{1+\frac{2}{3}}-1}{\sqrt{1+\frac{2}{3}}+1} \cdot \frac{\sqrt{1+\frac{4}{3}}-1}{\sqrt{1+\frac{4}{3}}+1} \cdot \frac{\sqrt{1+\frac{6}{5}}-1}{\sqrt{1+\frac{6}{5}}+1} \tag{2.1.13}$$

which is $\approx 2.8 \cdot 10^{-3}$. Now,

$$a_{2k} = \frac{k/2}{2k-1} = \frac{1}{4} + \frac{1/4}{2k-1}, \qquad a_{2k+1} = \frac{k/2}{2k+1} = \frac{1}{4} - \frac{1/4}{2k+1}$$

for $k \geq 1$. Hence $a_n \leq \frac{1}{4} + \frac{1/4}{5} = 0.3$ for $n \geq 6$, and so

$$\begin{aligned} |\operatorname{Ln} 2 - f_n| \leq 2 \cdot \frac{\sqrt{3}-1}{\sqrt{3}+1} \cdot \frac{\sqrt{5}-\sqrt{3}}{\sqrt{5}+\sqrt{3}} \cdot \frac{\sqrt{7}-\sqrt{3}}{\sqrt{7}+\sqrt{3}} \times \\ \times \frac{\sqrt{9}-\sqrt{5}}{\sqrt{9}+\sqrt{5}} \cdot \left(\frac{\sqrt{2.2}-1}{\sqrt{2.2}+1}\right)^{n-5} \end{aligned} \tag{2.1.14}$$

for $n \geq 4$, which is an easier bound to compute. In the table below we compare the actual value of $|\operatorname{Ln} 2 - f_n|$ to the three bounds given above.

n	f_n	$\lvert\operatorname{Ln} 2 - f_n\rvert$	(2.1.8)	(2.1.11)	(2.1.14)
10	.693147157853...	$2 \cdot 10^{-8}$	$2 \cdot 10^{-7}$	$5 \cdot 10^{-8}$	$1 \cdot 10^{-7}$
11	.693147184962...	$4 \cdot 10^{-9}$	$3 \cdot 10^{-8}$	$1 \cdot 10^{-8}$	$2 \cdot 10^{-8}$
12	.693147179886...	$7 \cdot 10^{-10}$	$5 \cdot 10^{-9}$	$2 \cdot 10^{-9}$	$4 \cdot 10^{-9}$
13	.693147180688...	$1 \cdot 10^{-10}$	$8 \cdot 10^{-10}$	$3 \cdot 10^{-10}$	$8 \cdot 10^{-10}$

The improved approximant $f_3^* := \frac{1}{2}(f_7 + f_6)$ gives for instance

$$\operatorname{Ln} 2 \approx 0.693136 \ \text{ with } \ |\operatorname{Ln} 2 - f_3^*| \le 1.5 \cdot 10^{-5}.$$

The correct value is of course $\operatorname{Ln} 2 = 0.69314718055\ldots$. ◇

If we use approximants $S_n(w_n)$, the useful oscillation property (2.1.1) may get lost. It can be saved, though, if we are a little careful:

Theorem 3.18. *Let $\mathbf{K}(a_n/b_n)$ be a positive continued fraction with $b_1 > 0$, and let $w_n \ge 0$ satisfy*

$$w_n < \frac{a_{n+1}}{b_{n+1} + w_{n+1}} \quad \text{and} \quad w_n < \cfrac{a_{n+1}}{b_{n+1} + \cfrac{a_{n+2}}{b_{n+2} + w_{n+2}}} \tag{2.1.15}$$

for all $n \in \mathbb{N}$. Then

$$S_2(w_2) < S_4(w_4) < S_6(w_6) < \cdots < S_5(w_5) < S_3(w_3) < S_1(w_1). \tag{2.1.16}$$

Proof : Since $\mathbf{K}(a_n/b_n)$ is positive with $b_1 > 0$, it follows from the recurrence relation for $\{B_n\}$ that $B_n > 0$ for $n \ge 0$. Hence $\zeta_n := S_n^{-1}(\infty) = -B_n/B_{n-1} < 0$. For $w > \zeta_n$, the derivative of S_n satisfies

$$S_n'(w) = \frac{A_{n-1}B_n - B_{n-1}A_n}{(B_{n-1}w + B_n)^2} = \frac{\prod_{k=1}^n (-a_k)}{(B_{n-1}w + B_n)^2} \begin{cases} < 0 & \text{for } n \text{ odd} \\ > 0 & \text{for } n \text{ even} \end{cases}$$

and so, for fixed n, $S_{2n}(w)$ increases and $S_{2n+1}(w)$ decreases as $w > \zeta_n$ increases. In particular this holds for $w \ge 0$. Since

$$S_{n+1}(w_{n+1}) = S_n\left(\frac{a_{n+1}}{b_{n+1} + w_{n+1}}\right),$$

the first condition in (2.1.15) implies that $S_{2n}(w_{2n}) < S_{2n+1}(w_{2n+1})$ and $S_{2n+1}(w_{2n+1}) > S_{2n+2}(w_{2n+2})$. Moreover, since

$$S_{n+2}(w_{n+2}) = S_n(v_n) \quad \text{where } v_n := \cfrac{a_{n+1}}{b_{n+1} + \cfrac{a_{n+2}}{b_{n+2} + w_{n+2}}},$$

the second condition in (2.1.15) implies that the sequence $\{S_{2n}(w_{2n})\}$ increases and $\{S_{2n+1}(w_{2n+1})\}$ decreases. This proves (2.1.16). □

Remark. It is also clear from the sign of $S_n'(w)$ that

- if both "$<$" are replaced by "$\leq$" in (2.1.15), then (2.1.16) holds with all "$<$" replaced by "$\leq$".
- if both "$<$" are replaced by "$>$" (or "$\geq$") in (2.1.15), then (2.1.16) holds with all "$<$" replaced by "$>$" (or "$\geq$").

Corollary 3.19. *Let $\mathbf{K}(a_n/1)$ be a positive continued fraction with $a_n \to a < \infty$, and let x be the non-negative solution of the equation $x = a/(1+x)$. If $(a_n - a) > 0$ and $(a_n - a) > x(a_{n+1} - a_n)$ for all n, then*

$$S_2(x) < S_4(x) < S_6(x) < \cdots < S_5(x) < S_3(x) < S_1(x)\,. \qquad (2.1.17)$$

Proof : Condition (2.1.15) holds since

$$\frac{a_{n+1}}{1+x} - x = \frac{a_{n+1} - a}{1+x} > 0$$

and

$$\frac{a_{n+1}}{1 + \frac{a_{n+2}}{1+x}} - x = \frac{a_{n+1} - a + x(a_{n+1} - a_{n+2})}{1 + x + a_{n+2}} > 0.$$

□

Remark: As in the previous remark, the inequalities in (2.1.17) are for instance reversed if $(a_n - a) < 0$ and $(a_n - a) < x(a_{n+1} - a_n)$ for all n.

Example 4. The positive continued fraction

$$\mathop{\mathrm{K}}_{n=1}^{\infty} \frac{a_n}{1} := \frac{2 + \frac{1}{1}}{1} + \frac{2 + \frac{1}{2}}{1} + \frac{2 + \frac{1}{3}}{1} + \cdots + \frac{2 + \frac{1}{n}}{1} + \cdots$$

has elements $0 < a_n \to 2$, and $x = 1$ is the positive root of the equation $x = 2/(1+x)$. Here $(a_n - 2) = \frac{1}{n} > 0$ and $(a_n - 2) = \frac{1}{n} > x(a_{n+1} - a_n) = 1 \cdot (\frac{1}{n+1} - \frac{1}{n})$. Hence (2.1.17) holds. Since $\mathbf{K}(a_n/1)$ converges by the Seidel-Stern Theorem (see Remark 3 on page 101 and Theorem 3.2(ii) on page 102), this means that

$$|f - S_n(x)| < |S_{n+1}(x) - S_n(x)| \qquad (2.1.18)$$

and

$$|f - f_n^*| \leq \tfrac{1}{2}(S_{2n+1}(x) - S_{2n}(x)) \quad \text{for } f_n^* := \tfrac{1}{2}(S_{2n+1}(x) + S_{2n}(x)).$$

In the table below we have listed the approximants $f_n := S_n(0)$ and the truncation errors $(f - f_n)$ and the expression $(f_{n+1} - f_n)$ from the truncation error bound in (2.1.8) for $5 \le n \le 10$. Moreover, $(f - S_n(x))$ and $(S_{n+1}(x) - S_n(x))$ from (2.1.18) are listed. The result is consistent with the Seidel-Stern Theorem, Corollary 3.19 and the idea of convergence acceleration from Example 3 on page 11. The value of $\mathbf{K}(a_n/1)$ is 1.37587055 correctly rounded to 8 decimals.

n	f_n	$f - f_n$	$f_{n+1} - f_n$	$f - S_n(x)$	$S_{n+1}(x) - S_n(x)$
5	1.4599...	−0.084...	−0.125...	−0.0030...	−0.0043...
6	1.3342...	0.041...	0.063...	0.0013...	0.0019...
7	1.3974...	−0.021...	−0.032...	−0.00059...	−0.00085...
8	1.3649...	0.010...	0.016...	0.00026...	0.00039...
9	1.3814...	−0.005...	−0.008...	−0.00012...	−0.00017...
10	1.3730...	0.002...	0.004...	0.000056...	0.00008...

◇

3.2.2 Alternating continued fractions

A continued fraction $b_0 + \mathbf{K}(a_n/b_n)$ with real elements a_n and b_n is called a *real* continued fraction. And if all $b_n \ge 0$ and a_n alternates in sign, we say that $\mathbf{K}(a_n/b_n)$ is an *alternating* continued fraction. We include the following result due to Perron ([Perr57], p 53):

Theorem 3.20. (Alternating continued fraction.) *Let the even part of the alternating continued fraction $\mathbf{K}(a_n/1)$ with $a_2 > -1$ be a positive continued fraction. Then $\mathbf{K}(a_n/1)$ converges if and only if its even part converges.*

The canonical even part of $\mathbf{K}(a_n/1)$ is given by

$$\operatorname*{K}_{n=1}^{\infty} \frac{c_n}{d_n} =: \frac{a_1}{1 + a_2} \underset{+}{} \frac{-a_2 a_3}{1 + a_3 + a_4} \underset{+}{} \frac{-a_4 a_5}{1 + a_5 + a_6} \underset{+}{} \cdots . \tag{2.2.1}$$

Since $\{a_n\}$ is alternating in sign, this is a positive continued fraction if and only if $a_{2n-1} > 0$ and $a_{2n} < 0$ for all n and

$$a_{2n-1} + a_{2n} \ge -1 \quad \text{for } n \ge 2. \tag{2.2.2}$$

Proof of Theorem 3.20: Clearly, $\mathbf{K}(a_n/1)$ diverges if its even part diverges, so let its even part converge. That is, $f_{2n} = (A_{2n}/B_{2n}) \to f$ where $0 \le f < \infty$ by the Seidel-Stern Theorem. Now,

$$\begin{aligned} \left| \frac{f_{2n+1} - f_{2n}}{f_{2n+2} - f_{2n}} \right| &= \left| \frac{A_{2n+1}B_{2n} - A_{2n}B_{2n+1}}{A_{2n+2}B_{2n} - A_{2n}B_{2n+2}} \cdot \frac{B_{2n+2}}{B_{2n+1}} \right| \\ &= \left| \frac{B_{2n+2}}{B_{2n+1}} \right| = \left| \frac{B_{2n+2}}{B_{2n+2} - a_{2n+2}B_{2n}} \right| \end{aligned}$$

by the recurrence relation for $\{A_n\}$ and $\{B_n\}$. Since $a_{2n+2} < 0$ and all $B_{2n} > 0$ ($\{B_{2n}\}$ are the canonical denominators of the positive continued fraction (2.2.1) with $1 + a_2 > 0$), this ratio is < 1. Hence also $f_{2n+1} \to f$. □

Example 5. The alternating continued fraction

$$\frac{1}{1} - \frac{1}{1} + \frac{2}{1} - \frac{3}{1} + \frac{4}{1} - \frac{5}{1} + \frac{6}{1} - \cdots + \frac{2n}{1} - \frac{2n+1}{1} + \cdots \tag{2.2.3}$$

has even part

$$\mathop{\mathrm{K}}_{n=1}^{\infty} \frac{c_n}{d_n} := \frac{1}{0+} \frac{1\cdot 2}{0} + \frac{3\cdot 4}{0} + \frac{5\cdot 6}{0} + \frac{7\cdot 8}{0} + \cdots$$

which diverges by the Seidel-Stern Theorem. Hence the alternating continued fraction (2.2.3) diverges.

The alternating continued fraction

$$\frac{1}{1-} \frac{1/2}{1} + \frac{1}{1-} \frac{3/2}{1} + \frac{2}{1-} \frac{5/2}{1} + \frac{3}{1-} \frac{7/2}{1} + \cdots + \frac{n}{1-} \frac{n+1/2}{1} + \cdots \tag{2.2.4}$$

has even part

$$\frac{1}{1/2+} \frac{\frac{1}{2}\cdot 1}{1/2} + \frac{\frac{3}{2}\cdot 2}{1/2} + \frac{\frac{5}{2}\cdot 3}{1/2} + \frac{\frac{7}{2}\cdot 4}{1/2} + \cdots$$

which is equivalent to

$$\mathop{\mathrm{K}}_{n=1}^{\infty} \frac{c_n}{1} := \frac{2}{1+} \frac{1\cdot 2}{1} + \frac{3\cdot 4}{1} + \frac{5\cdot 6}{1} + \frac{7\cdot 8}{1} + \cdots .$$

Now, $\mathbf{K}(c_n/1)$ converges by the Seidel-Stern Theorem and Theorem 3.2(i) on page 102 since

$$\sum_{n=1}^{\infty} \frac{1}{\sqrt{c_{n+1}}} > \sum_{n=1}^{\infty} \frac{1}{2n} = \infty.$$

Hence the convergence of (2.2.4) follows by Theorem 3.20. ◇

3.2.3 Stieltjes continued fractions

A continued fraction of the form

$$b_0 + \mathop{\mathrm{K}}_{n=1}^{\infty} \frac{a_n z}{1} = b_0 + \frac{a_1 z}{1} \, {}_{+} \, \frac{a_2 z}{1} \, {}_{+} \, \frac{a_3 z}{1} \, {}_{+} \cdots; \quad b_0 \geq 0, \ a_n > 0 \text{ for all } n \tag{2.3.1}$$

is called a *Stieltjes continued fraction*, or an *S-fraction* for short. A large number of continued fraction expansions of useful functions have this form. S-fractions are also essential in the Stieltjes moment theory (Section 1.5.2 on page 39). Luckily they have nice convergence properties ([Stie94]):

Theorem 3.21. (Stieltjes continued fractions.) *The even and odd parts of an S-fraction converge locally uniformly with respect to z in the cut plane $D := \{z \in \mathbb{C}; |\arg(z)| < \pi\}$ to holomorphic functions. The S-fraction itself converges for $z \in D$ if and only if*

$$\sum_{n=1}^{\infty} \prod_{k=1}^{n} a_k^{(-1)^{n-k+1}} = \infty \,. \tag{2.3.2}$$

Proof : Without loss of generality we set $b_0 := 0$ and assume that $\arg z =: 2\alpha \in [0, \pi)$. (Otherwise we just consider the complex conjugate of $\mathbf{K}(a_n z/1)$.) We shall first see that for fixed z, the closed half plane

$$V(z) := e^{i\alpha}\overline{\mathbb{H}} \quad \text{where} \quad \mathbb{H} := \{w \in \mathbb{C}; \operatorname{Re} w > 0\} \tag{2.3.3}$$

is a simple value set for $\mathbf{K}(a_n z/1)$. Let $w \in V(z)$. Then $-\frac{\pi}{2} + \alpha \leq \arg w \leq \frac{\pi}{2} + \alpha$, so $-\frac{\pi}{2} + \alpha < \arg(1 + w) < \frac{\pi}{2} + \alpha$, and thus

$$2\alpha - (\tfrac{\pi}{2} + \alpha) < \arg \frac{a_n z}{1 + w} < 2\alpha + (\tfrac{\pi}{2} - \alpha),$$

i.e., $a_n z/(1 + V(z)) \subseteq V(z)$. This means that the classical approximants $f_n^{(1)}(z)$ of its first tail satisfy

$$f_n^{(1)}(z) := \frac{a_2 z}{1} \, {}_{+} \, \frac{a_3 z}{1} \, {}_{+} \cdots {}_{+} \, \frac{a_{n+1} z}{1} \in V(z).$$

In particular $f_n^{(1)}(z) \neq -1$. Therefore $f_n(z) := S_n(0) = a_1 z/(1 + f_{n-1}^{(1)}(z)) \neq \infty$ for all n. That is, $\{f_n(z)\}$ is a sequence of rational, holomorphic functions in D. Moreover, $f_n(z)$ never takes a negative value for $n \in \mathbb{N}$ and $z \in D$.

For $z > 0$ the result follows from the Seidel-Stern Theorem. Therefore $\{f_{2n}\}$ and $\{f_{2n+1}\}$ converge locally uniformly in D to holomorphic functions by the Stieltjes-Vitali Theorem on page 115, and $\{f_n\}$ converges locally uniformly in D if and only if (2.3.2) holds. □

Remarks.

1. A very nice consequence of Theorem 3.21 is that an S-fraction converges to a holomorphic function in D if and only if it converges at a single point $z \in D$.
2. If an S-fraction diverges for a $z \in D$, then it diverges generally for all $z \in D$. (See Remark 5 on page 102.)

Limit set.

Simple computation shows that for given $z \in D$ with $\arg z =: 2\alpha \in (-\pi, \pi)$, the angular opening (or ray if $\alpha = 0$)

$$V_\alpha := \begin{cases} (-i\, e^{2i\alpha}\overline{\mathbb{H}}) \cap (i\overline{\mathbb{H}}) & \text{if } \alpha \geq 0, \\ (i\, e^{2i\alpha}\overline{\mathbb{H}}) \cap (-i\overline{\mathbb{H}}) & \text{if } \alpha \leq 0 \end{cases} \tag{2.3.4}$$

between the rays $\overline{\mathbb{R}^+}$ and $e^{2i\alpha}\overline{\mathbb{R}^+}$ is also a value set for every S-fraction $\mathbf{K}(a_n z/1)$. If $\alpha = 0$, then $\mathbf{K}(a_n z/1)$ is a positive continued fraction which was studied in Section 3.2.1. Let $\alpha \neq 0$. Our next theorem shows that then every point in the interior of V_α, and no other point, is the value of an S-fraction with $b_0 = 0$ and $\arg z = 2\alpha$:

Theorem 3.22. *For given $\alpha \in \mathbb{R}$ with $0 < |\alpha| < \frac{\pi}{2}$, V_α° is the limit set for the family of continued fractions $\mathbf{K}(a_n/1)$ from the element set $E := e^{2i\alpha}\mathbb{R}^+$.*

Proof : Let $\mathbf{K}(a_n/1)$ be a convergent continued fraction from E with value f. Since V_α is a closed, simple value set for $\mathbf{K}(a_n/1)$, we know that $f \in V_\alpha$ (Corollary 2.10 on page 71). Since also the first tail value $f^{(1)}$ for $\mathbf{K}(a_n/1)$ is the value of a continued fraction from E, also $f^{(1)} \in V_\alpha$. Since $f = a_1/(1 + f^{(1)})$ and $-1 \notin V_\alpha$, it follows that $f \neq \infty$. Similarly, $f^{(1)} \neq \infty$, and thus $f \neq 0$. Clearly $f > 0$ only if $\arg(1 + f^{(1)}) = 2\alpha$ which is impossible. Therefore $f \notin \overline{\mathbb{R}^+}$, and similarly $f^{(1)} \notin \overline{\mathbb{R}^+}$. Finally, $\arg f = 2\alpha$ only if $(1 + f^{(1)}) > 0$ which already is ruled out. This proves that $f \in V_\alpha^\circ$.

Next, let $f := r\, e^{i\theta} \in V_\alpha^\circ$ be arbitrarily chosen. We need to prove that f is the value of a continued fraction from E. We shall actually prove that f is the value of a 2-periodic continued fraction

$$\frac{r_1 e^{2i\alpha}}{1} + \frac{r_2 e^{2i\alpha}}{1} + \frac{r_1 e^{2i\alpha}}{1} + \frac{r_2 e^{2i\alpha}}{1} + \frac{r_1 e^{2i\alpha}}{1} + \cdots \tag{2.3.5}$$

for some $r_1 > 0$ and $r_2 > 0$. Such a continued fraction clearly converges to a value in V_α° by the arguments above. A necessary condition for f to be its value is that

$$f = r\, e^{i\theta} = \frac{r_1 e^{2i\alpha}}{1} + \frac{r_2 e^{2i\alpha}}{1 + r\, e^{i\theta}} = r_1 e^{2i\alpha} \frac{1 + r\, e^{i\theta}}{1 + r_2 e^{2i\alpha} + r\, e^{i\theta}} \, . \tag{2.3.6}$$

Without loss of generality (complex conjugation) we assume that $\alpha > 0$. Then $0 < \theta < 2\alpha$, and $1 + r\,e^{i\theta} = \rho\,e^{i\psi}$ for some $\rho > 0$ and $0 < \psi < \theta$. Hence we need

$$r\,e^{i\theta} = \frac{r_1\rho\,e^{i(2\alpha+\psi)}}{\rho\,e^{i\psi} + r_2e^{2i\alpha}}\,. \tag{2.3.7}$$

Let $r_2 > 0$ be chosen such that $\arg(\rho\,e^{i\psi} + r_2e^{2i\alpha}) = 2\alpha + \psi - \theta$. This can always be done since $\psi < 2\alpha + \psi - \theta < 2\alpha$. Then (2.3.7) holds for the right choice of $r_1 > 0$. With this choice, (2.3.5) converges to $re^{i\theta}$. □

Truncation error bounds.

Henrici and Pfluger ([HePf66]) have proved a posteriori truncation error bounds for classical approximants $f_n(z) := S_n(0)$ of Stieltjes fractions. These bounds are based on the value set V_α given by (2.3.4). What they actually proved though, is in fact a bound for the diameter of $S_n(V_\alpha)$:

Theorem 3.23. (The Henrici-Pfluger Bounds.) *For given $z \neq 0$ with $\alpha := \frac{1}{2}\arg z \in (-\frac{\pi}{2}, \frac{\pi}{2})$, let V_α be given by (2.3.4). Then*

$$\operatorname{diam} S_n(V_\alpha) \leq \begin{cases} |f_n(z) - f_{n-1}(z)| & \text{if } |\alpha| \leq \frac{\pi}{4}\,, \\ \dfrac{|f_n(z) - f_{n-1}(z)|}{|\sin 2\alpha|} & \text{if } |\alpha| > \frac{\pi}{4} \end{cases} \tag{2.3.8}$$

for the S-fraction $\mathbf{K}(a_nz/1)$.

Proof : Again we may without loss of generality assume that $b_0 = 0$. If $\alpha = 0$, then the bound follows from Theorem 3.16 on page 118. Let

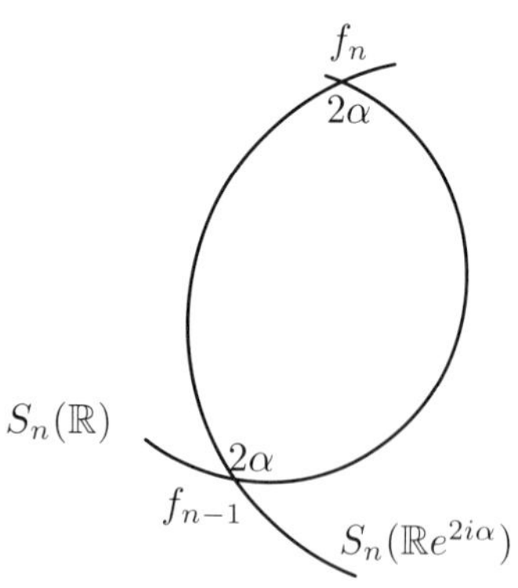

$\alpha \neq 0$. Without loss of generality (complex conjugation) we assume that $\alpha > 0$. The set $K_n := S_n(V_\alpha)$ is bounded by the circular arcs $S_n(\overline{\mathbb{R}^+})$ and $S_n(e^{2i\alpha}\overline{\mathbb{R}^+})$. These two arcs meet at $S_n(0) = f_n = f_n(z)$ and $S_n(\infty) = f_{n-1} = f_{n-1}(z)$ at the angle 2α since S_n is a conformal map. Indeed, 2α is the inner angle since $\zeta_n \notin V_\alpha$ because $S_n(\zeta_n) = \infty$ whereas $S_n(V_\alpha) \subseteq s_1(V_\alpha)$ which is a bounded set. Therefore K_n is a closed, bounded lens-shaped region, for instance as illustrated in the present figure. Now, if K_n is not "too fat", then $\operatorname{diam}(K_n) = |f_n - f_{n-1}|$. Otherwise one measures the diameter along the perpendicular bisector between f_{n-1} and

f_n. The "worst case" occurs if K_n is convex and the two circles $S_n(\mathbb{R})$ and $S_n(\mathbb{R}e^{2i\alpha})$ have the same radius. Therefore

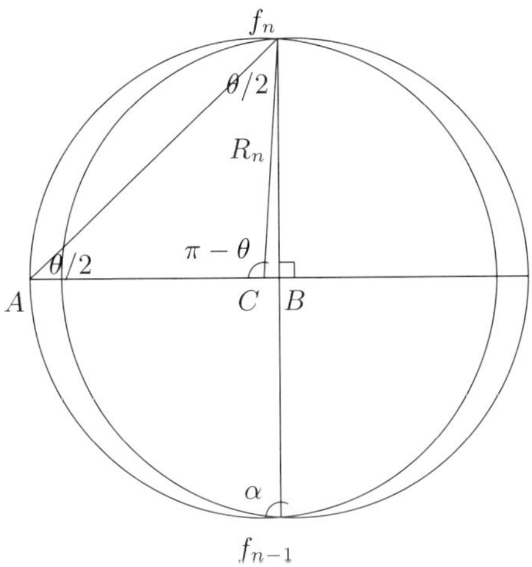

diam(K_n) is less than or equal to the diameter $2R_n$ of two equally large circles which intersect at f_n and f_{n-1} under an angle 2α. The figure illustrates the situation. Here C is the center of the left circle, and B is the midpoint between the two centers.

Let $\frac{\theta}{2}$ denote the angle $\angle CAf_n$. Then also $\angle Af_nC = \frac{\theta}{2}$ and the angle $\angle Af_nB$ is equal to $\frac{\pi}{2} - \frac{\theta}{2}$ which again is less than α; i.e., $\pi - \theta < 2\alpha$. Therefore, since $\frac{\pi}{2} < \pi - \theta < 2\alpha < \pi$, the common radius R_n of these two circles is given by

$$R_n = |f_nC| = \frac{|f_nB|}{\sin\theta} = \frac{|f_n - f_{n-1}|}{2\sin\theta}$$
$$= \frac{|f_n - f_{n-1}|}{2\sin(\pi-\theta)} < \frac{|f_n - f_{n-1}|}{2|\sin 2\alpha|}.$$

□

Of course, this means that if $w_n \in V_\alpha$ for all n, then

$$|S_{n+m}(w_{n+m}) - S_n(w_n)| \le \lambda |f_n(z) - f_{n-1}(z)| \tag{2.3.9}$$

where $\lambda := 1$ if $|\alpha| \le \frac{\pi}{4}$ and $\lambda := 1/|\sin 2\alpha|$ otherwise. If in particular $\mathbf{K}(a_nz/1)$ converges to $f(z)$, then

$$|f(z) - S_n(w_n)| \le \lambda |f_n(z) - f_{n-1}(z)|, \tag{2.3.10}$$

and the choice $w_n := \infty$ gives

$$|f(z) - f_{n-1}(z)| \le \lambda |f_n(z) - f_{n-1}(z)|. \tag{2.3.11}$$

The diameter of $\mathbf{S_n(V(z))}$.

The Henrici-Pfluger Bounds are a posteriori bounds based on the value set V_α in (2.3.4). To prove a priori bounds we follow Thron ([Thron81]) and Gragg and Warner ([GrWa83]), and use the half plane $V := V(z) = e^{i\alpha}\overline{\mathbb{H}}$ which also is a value set for $\mathbf{K}(a_nz/1)$ for given z, even though it is normally quite a lot larger.

Theorem 3.24. (The Thron-Gragg-Warner Bounds.) *For given* $z := re^{i2\alpha}$ *with* $-\frac{\pi}{2} < \alpha < \frac{\pi}{2}$ *and* $r > 0$,

$$\operatorname{diam}(S_n(V)) \le 2\frac{a_1 r}{\cos\alpha}\prod_{k=2}^{n}\frac{\sqrt{1+4a_k r/\cos^2\alpha}-1}{\sqrt{1+4a_k r/\cos^2\alpha}+1} \tag{2.3.12}$$

for the S-fraction $\mathbf{K}(a_n z/1)$ *and* $V := e^{i\alpha}\overline{\mathbb{H}}$.

Since $\infty \in V$, it follows in particular that

$$|f_{n+m} - f_{n-1}| \le 2\frac{a_1 r}{\cos\alpha}\prod_{k=2}^{n}\frac{\sqrt{1+4a_k r/\cos^2\alpha}-1}{\sqrt{1+4a_k r/\cos^2\alpha}+1} \tag{2.3.13}$$

for $m, n \in \mathbb{N}$, which essentially was the original result by Thron and Gragg-Warner. If $\mathbf{K}(a_n z/1)$ converges to $f(z)$, then $|f(z) - f_{n-1}(z)|$ and $|f(z) - S_n(w)|$ for $w \in V$ are bounded by the same expression.

We postpone the proof of Theorem 3.24 to page 158.

Example 6. Let us once again turn to the continued fraction expansion

$$\operatorname*{K}\frac{a_n z}{1} = \frac{z}{1+}\,\frac{z/2}{1}+\frac{z/6}{1}+\frac{2z/6}{1}+\frac{2z/10}{1}+\frac{3z/10}{1}+\cdots$$

of $\operatorname{Ln}(1+z)$. This is an S-fraction, and it converges for $|\arg(z)| < \pi$ (Theorem 3.21). Its classical approximants $f_n(z)$ are rational functions in z with rational coefficients. In Example 3 on page 119 we used this continued fraction to approximate $\operatorname{Ln}(2)$. The bounds we obtained for $|\operatorname{Ln} 2 - f_n|$ are consistent with (2.3.8) and (2.3.12). Let now $z := 1 + i = \sqrt{2}\,e^{i\pi/4}$. Then still

$$|\operatorname{Ln}(1+i) - f_n(1+i)| \le |f_n(1+i) - f_{n+1}(1+i)| \tag{2.3.14}$$

by (2.3.8), and by (2.3.13) we get the a priori bounds

$$\begin{aligned}|\operatorname{Ln}(1+i) - f_n(1+i)| &\le \frac{2\sqrt{2}}{\cos\frac{\pi}{8}}\prod_{k=2}^{n+1}\frac{\sqrt{1+4a_k\sqrt{2}/\cos^2\frac{\pi}{8}}-1}{\sqrt{1+4a_k\sqrt{2}/\cos^2\frac{\pi}{8}}+1}\\ &\le \frac{2\sqrt{2}}{\cos\frac{\pi}{8}}\cdot\frac{\sqrt{1+2w}-1}{\sqrt{1+2w}+1}\cdot\frac{\sqrt{1+\frac{2}{3}w}-1}{\sqrt{1+\frac{2}{3}w}+1}\cdot\frac{\sqrt{1+\frac{4}{3}w}-1}{\sqrt{1+\frac{4}{3}w}+1}\times\\ &\quad\times\frac{\sqrt{1+\frac{4}{5}w}-1}{\sqrt{1+\frac{4}{5}w}+1}\Big(\cdot\frac{\sqrt{1+\frac{6}{5}w}-1}{\sqrt{1+\frac{6}{5}w}+1}\Big)^{n-5}\end{aligned} \tag{2.3.15}$$

where $w := |1+i|/\cos^2\frac{\pi}{8} - 4(\sqrt{2}-1)$. The true value of $\mathrm{Ln}(1+i)$ is $0.804718956217+0.463647609001\,i$, correctly rounded to 12 decimals. The table below shows the order of the true truncation errors compared to the one in (2.3.14) and the two in (2.3.15):

n	$\lvert\mathrm{Ln}(1+i)-f_n\rvert$	$\lvert f_{n+1}-f_n\rvert$	$(2.3.15)_1$	$(2.3.15)_2$
10	$5.7\cdot 10^{-7}$	$7.1\cdot 10^{-7}$	$1.1\cdot 10^{-5}$	$1.7\cdot 10^{-5}$
11	$1.5\cdot 10^{-7}$	$1.8\cdot 10^{-7}$	$2.4\cdot 10^{-6}$	$4.5\cdot 10^{-6}$
12	$3.1\cdot 10^{-8}$	$3.7\cdot 10^{-8}$	$6.0\cdot 10^{-7}$	$1.2\cdot 10^{-6}$
13	$7.8\cdot 10^{-9}$	$9.3\cdot 10^{-9}$	$1.4\cdot 10^{-7}$	$3.2\cdot 10^{-7}$

◇

3.2.4 The Śleszyński-Pringsheim Theorem

A continued fraction $\mathbf{K}(a_n/b_n)$ with

$$|b_n| \geq |a_n| + 1 \quad \text{for } n \geq 1 \tag{2.4.1}$$

is called a *Śleszyński-Pringsheim continued fraction.* The unit disk $\mathbb{D} := \{z \in \mathbb{C};\ |z| < 1\}$ is a simple value set for $\mathbf{K}(a_n/b_n)$ if and only if (2.4.1) holds. Now, $0 \in \mathbb{D}$, so $f_n = S_n(0) \in \mathbb{D}$ for all n. In 1889 Śleszyński ([Śles89]) proved that (2.4.1) was sufficient for the (classical) convergence of $\mathbf{K}(a_n/b_n)$ (although he never used the term "value set"). This result was rediscovered by Pringsheim in 1899 ([Prin99a]). Today it is a simple consequence of Corollary 3.9 on page 114 with $\varphi_n(w) \equiv w$ and $w_n := \infty$. But the classical proof gives some additional information on the convergence:

Theorem 3.25. **(The Śleszyński-Pringsheim Theorem.)** *A Śleszyński-Pringsheim continued fraction converges absolutely to some value f with $0 < |f| \leq 1$. Moreover, $\{|A_n|\}$ and $\{|B_n|\}$ are strictly increasing sequences, and $0 < |S_n(w)| \leq 1$ for all $w \in \overline{\mathbb{D}}$.*

Proof : Solutions $\{X_n\}$ of the recurrence relation $X_n = b_n X_{n-1} + a_n X_{n-2}$ satisfy

$$|X_n| \geq |b_n||X_{n-1}| - |a_n||X_{n-2}| \geq |b_n|X_{n-1}| - (|b_n| - 1)|X_{n-2}|\,, \tag{2.4.2}$$

and hence, if $|X_m| - |X_{m-1}| > 0$, then

$$\begin{aligned} |X_n| - |X_{n-1}| &\geq (|b_n| - 1)(|X_{n-1}| - |X_{n-2}|) \\ &\geq \cdots \geq (|X_m| - |X_{m-1}|) \prod_{k=m+1}^{n} (|b_k| - 1) > 0 \end{aligned} \tag{2.4.3}$$

for $n \geq m$. Now, $\{A_n\}$ and $\{B_n\}$ are such solutions with $|A_1| - |A_0| = |a_1| > 0$ and $|B_0| - |B_{-1}| = 1 > 0$. Therefore $\{|A_n|\}_0^\infty$ and $\{|B_n|\}_0^\infty$ are strictly increasing sequences. Moreover,

$$|B_n| - |B_{n-1}| \geq \prod_{k=1}^{n}(|b_k| - 1) \geq \prod_{k=1}^{n} |a_k| \tag{2.4.4}$$

by (2.4.3). The determinant formula on page 7 therefore gives that

$$\left|\frac{A_n}{B_n} - \frac{A_{n-1}}{B_{n-1}}\right| = \frac{\prod_{k=1}^n |a_k|}{|B_n B_{n-1}|} \leq \frac{|B_n| - |B_{n-1}|}{|B_n B_{n-1}|} = \frac{1}{|B_{n-1}|} - \frac{1}{|B_n|} \tag{2.4.5}$$

where $\sum_{n=1}^\infty (1/|B_{n-1}| - 1/|B_n|) = 1 - 1/\lim|B_n|$. Hence the absolute convergence of $f_n = S_n(0) = A_n/B_n$ is established. That $|S_n(w)| \leq 1$ for $w \in \overline{\mathbb{D}}$ and thus $|f| \leq 1$ follows from the nestedness of $\{S_n(\overline{\mathbb{D}})\}$. This also holds true for the first tail of $\mathbf{K}(a_n/b_n)$; i.e., $|S_{n-1}^{(1)}(w)| \leq 1$ and $|f^{(1)}| \leq 1$. Since $|a_1/(b_1+q)| \geq |a_1|/(|b_1|+1) > 0$ for $|q| \leq 1$, it therefore follows that $|S_n(w)| > 0$ and $|f| > 0$. □

Example 7. We consider the continued fraction

$$b_0 + \mathop{\mathrm{K}}_{n=1}^{\infty} \frac{a_n}{b_n} := 1 + \frac{2z}{2 - z +} \frac{z^2}{6 +} \frac{z^2}{10 +} \cdots \frac{z^2}{4n - 2 +} \cdots . \tag{2.4.6}$$

For given $z \in \mathbb{C}$ there always exists an $n_0 \in \mathbb{N}$ such that $|b_n| \geq |a_n| + 1$ for $n \geq n_0$. Therefore the n_0th tail of $b_0 + \mathbf{K}(a_n/b_n)$ converges, and thus the continued fraction itself converges. Hence (2.4.6) converges for every $z \in \mathbb{C}$. Indeed, its value is e^z ((2.2.1) on page 268). ◇

Remark: If $\mathbf{K}(a_n/b_n)$ is equivalent to a Śleszyński-Pringsheim continued fraction, then also $\mathbf{K}(a_n/b_n)$ converges absolutely, and its classical approximants are still contained in the open unit disk $\mathbb{D}$. Therefore $\mathbf{K}(a_n/b_n)$ converges if

$$|b_1| \geq |a_1| + \sqrt{|a_2|}, \quad |b_n| \geq \sqrt{|a_n|} + \sqrt{|a_{n+1}|} \text{ for } n \geq 2, \tag{2.4.7}$$

since then

$$\mathop{\mathrm{K}}_{n=1}^{\infty} \frac{a_n}{b_n} \sim \frac{a_1/\sqrt{|a_2|}}{b_1/\sqrt{|a_2|}} + \frac{a_2/\sqrt{|a_2 a_3|}}{b_2/\sqrt{|a_3|}} + \frac{a_3/\sqrt{|a_3 a_4|}}{b_3/\sqrt{|a_4|}} + \cdots$$

which is a Śleszyński-Pringsheim continued fraction. Similarly, if

$$|b_1| \geq |a_1| + |a_2|, \quad |b_n| \geq |a_{n+1}| + 1 \quad \text{for } n \geq 2, \tag{2.4.8}$$

then $\mathbf{K}(a_n/b_n)$ converges since

$$\mathop{\mathrm{K}}_{n=1}^{\infty} \frac{a_n}{b_n} \sim \frac{a_1/a_2}{b_1/a_2} + \frac{1/a_3}{b_2/a_3} + \frac{1/a_4}{b_3/a_4} + \cdots$$

is a Śleszyński-Pringsheim continued fraction under condition (2.4.8), ([Prin05], [Prin18]). These are just two examples of infinitely many possibilities.

Limit set.

Every $f \in \overline{\mathbb{D}} \setminus \{0\}$ is actually the value of a Śleszyński-Pringsheim continued fraction:

Theorem 3.26. *$\overline{\mathbb{D}} \setminus \{0\}$ is the limit set for the family of Śleszyński-Pringsheim continued fractions.*

Proof : It follows from Theorem 3.25 that the value of a Śleszyński-Pringsheim continued fraction is contained in $\overline{\mathbb{D}} \setminus \{0\}$. We need to prove that every $f \in \overline{\mathbb{D}} \setminus \{0\}$ is the value of a Śleszyński-Pringsheim continued fraction. For a fixed $b > 2$ let $a := 1 - b$. Then the periodic continued fraction

$$\operatorname*{K}_{n=1}^{\infty} \frac{a}{b} = \frac{a}{b} {+} \frac{a}{b} {+} \frac{a}{b} {+} \cdots$$

is a Śleszyński-Pringsheim continued fraction which converges to a root of the equation $a/(b+x) = x$. Since this value must be $\in \overline{\mathbb{D}}$, it follows that $\mathbf{K}(a/b)$ converges to -1. Let $f \in \overline{\mathbb{D}} \setminus \{0, -1\}$ and $b_1 > 1$ be arbitrarily chosen, and let $a_1 := f \cdot (b_1 - 1)$. Then $|a_1| + 1 \leq b_1 - 1 + 1 = |b_1|$, and

$$\frac{a_1}{b_1} {+} \frac{a}{b} {+} \frac{a}{b} {+} \frac{a}{b} {+} \cdots$$

converges to $a_1/(b_1 - 1) = f$. $\square$

Truncation error bounds.

In [BeLo01] we studied the speed of convergence of Śleszyński-Pringsheim continued fractions. From (2.4.4) it follows immediately that

$$|B_n| \geq \Sigma_n := \sum_{k=0}^{n} P_k \quad \text{where } P_k := \prod_{m=1}^{k} (|b_m| - 1). \tag{2.4.9}$$

Therefore, by (2.4.5)

$$|f - f_n| \leq \sum_{k=n+1}^{\infty} |f_k - f_{k-1}| \leq \sum_{k=n+1}^{\infty} \frac{\prod_{j=1}^{k} |a_j|}{|B_k B_{k-1}|} \leq \sum_{k=n+1}^{\infty} \frac{\prod_{j=1}^{k} |a_j|}{\Sigma_k \Sigma_{k-1}}, \tag{2.4.10}$$

and since $|a_j| \leq |b_j| - 1$ and $P_k = \Sigma_k - \Sigma_{k-1}$,

$$|f - f_n| \leq \sum_{k=n+1}^{\infty} \frac{\Sigma_k - \Sigma_{k-1}}{\Sigma_k \Sigma_{k-1}} = \sum_{k=n+1}^{\infty} \left(\frac{1}{\Sigma_{k-1}} - \frac{1}{\Sigma_k} \right) = \frac{1}{\Sigma_n} - \frac{1}{\Sigma_\infty}. \tag{2.4.11}$$

Alternatively, (2.4.4) and (2.4.9) lead to

$$|f-f_n| \le \frac{1}{\Sigma_n^*} - \frac{1}{\Sigma_\infty^*} \quad \text{where } \Sigma_m^* := \sum_{k=0}^{m} P_k^* \text{ and } P_k^* := \prod_{j=1}^{k} |a_j|\,. \tag{2.4.12}$$

But we can say more:

<u>Case 1:</u> $a_n < 0$, $b_n = 1 - a_n$ for all n.

In this case the constant sequence $\{-1\}$ is a tail sequence for $\mathbf{K}(a_n/b_n)$, so we have full control. From Theorem 2.6 on page 66 with $t_n := -1$ we immediately find that

$$B_n - B_{n-1} = P_n, \quad A_n + B_n = 1 \quad \text{and } 1 + f_n = 1/\Sigma_n$$

where P_k and Σ_n are given by (2.4.9). In particular

$$B_n = \Sigma_n \quad \text{and} \quad f - f_n = \frac{1}{\Sigma_\infty} - \frac{1}{\Sigma_n}\,. \tag{2.4.13}$$

<u>Case 2:</u> $a_n < 0$, $b_n \ge 1 - a_n$ for all n.

It follows by induction that all $B_n > 0$ since $B_0 = B_0 - B_{-1} = 1$ and

$$B_n - B_{n-1} = (b_n - 1)B_{n-1} - |a_n|B_{n-2} \ge (b_n - 1)(B_{n-1} - B_{n-2})$$

if $B_{n-2} \ge 0$. Therefore $B_{n+1} > B_n \ge \Sigma_n$ and

$$f_n - f_{n-1} = -\frac{\prod_{k=1}^{n}(-a_k)}{B_n B_{n-1}} < 0, \tag{2.4.14}$$

so $\{f_n\}$ is a decreasing sequence. Since $f_1 = a_1/b_1 < 0$, this means that

$$-1 < f_n < f_{n-1} < 0 \quad \text{and} \quad B_{n+1} > B_n \ge \Sigma_n. \tag{2.4.15}$$

<u>Case 3:</u> The general case $|a_n| + 1 \le |b_n|$ for all n.

Let $\mathbf{K}(\tilde{a}_n/\tilde{b}_n)$ and $\mathbf{K}(\widehat{a}_n/\widehat{b}_n)$ be the two continued fractions given by

$$\tilde{a}_n := \widehat{a}_n := -|a_n|, \qquad \tilde{b}_n := |b_n| \quad \text{and} \quad \widehat{b}_n := |a_n| + 1. \tag{2.4.16}$$

Then $\mathbf{K}(\widehat{a}_n/\widehat{b}_n)$ is as described in case 1 and $\mathbf{K}(\tilde{a}_n/\tilde{b}_n)$ as described in case 2. We shall prove (with obvious notation):

Theorem 3.27. *Let $\mathbf{K}(a_n/b_n)$ be a Śleszyński-Pringsheim continued fraction and let $\mathbf{K}(\tilde{a}_n/\tilde{b}_n)$ and $\mathbf{K}(\widehat{a}_n/\widehat{b}_n)$ be given by (2.4.16). Then, with Σ_n given by (2.4.9),*

$$|f - f_n| \le \tilde{f}_n - \tilde{f} \le \widehat{f}_n - \widehat{f} = \frac{1}{\Sigma_n} - \frac{1}{\Sigma_\infty}\,. \tag{2.4.17}$$

Proof : The equality follows from (2.4.13) and the second inequality from (2.4.11). It therefore just remains to prove the first inequality. By (2.4.14),

$$\tilde{f}_n - \tilde{f} = -\sum_{k=1}^{n} \Big(\prod_{m=1}^{k} |a_m| \Big) \Big/ \Big(\tilde{B}_k \tilde{B}_{k-1} \Big)$$

with obvious notation. The first inequality in (2.4.17) follows therefore from (2.4.10) if we can prove that $|B_n| \geq \tilde{B}_n$, and thus

$$|B_n| \geq \tilde{B}_n \geq \widehat{B}_n = \Sigma_n \quad \text{for } n \geq 1. \tag{2.4.18}$$

Let $X_n := |B_n| - \tilde{B}_n$. Then $X_{-1} = X_0 = 0$ and

$$X_n \geq |b_n B_{n-1}| - |a_n B_{n-2}| - \tilde{B}_n = |b_n| X_{n-1} \quad |a_n| X_{n-2},$$

so if $X_{n-2} \geq 0$, then

$$X_n - X_{n-1} \geq (|b_n| - 1)(X_{n-1} - X_{n-2}).$$

It follows therefore by induction that $X_n - X_{n-1} \geq 0$ and $X_n \geq 0$ for all n. Therefore (2.4.18) and (2.4.17) follow. □

Remark. The equality $(\widehat{f}_n - \widehat{f}) = (1/\Sigma_n - 1/\Sigma_\infty)$ shows that there are Śleszyński-Pringsheim continued fractions which converge arbitrarily slowly. (Just choose $t_n := -1$ and $b_n > 1$ so that $\Sigma_n = \sum_{k=0}^{n} \prod_{j=1}^{k} (b_j - 1)$ converges as slowly as you want, and set $a_n := -(b_n - 1)$ for all n.) Hence the convergence is not *uniform* with respect to the family of Śleszyński-Pringsheim continued fractions.

The radius of $\mathbf{S_n}(\mathbb{D})$.

The quantity Σ_n is still given by (2.4.9).

Theorem 3.28. *Let* $\mathbf{K}(a_n/b_n)$ *be a Śleszyński-Pringsheim continued fraction. Then*

$$\begin{aligned} \operatorname{rad} S_n(\mathbb{D}) &= \frac{\prod_{k=1}^{n} |a_k|}{(|B_n| - |B_{n-1}|)(|B_n| + |B_{n-1}|)} \\ &\leq \frac{\prod_{k=1}^{n} |a_k|/(|b_k| - 1)}{|B_n| + |B_{n-1}|} \leq \frac{1}{|B_n| + |B_{n-1}|} \leq \frac{1}{\Sigma_n + \Sigma_{n-1}}. \end{aligned} \tag{2.4.19}$$

Let further $\mathbf{K}(\tilde{a}_n/\tilde{b}_n)$ *and* $\mathbf{K}(\widehat{a}_n/\widehat{b}_n)$ *be given by (2.4.16). Then*

$$\operatorname{rad} S_n(\mathbb{D}) \leq \operatorname{rad} \tilde{S}_n(\mathbb{D}) \leq \operatorname{rad} \widehat{S}_n(\mathbb{D}) = \frac{1}{\Sigma_n + \Sigma_{n-1}}, \tag{2.4.20}$$

Proof : The equality in (2.4.19) follows from (1.4.3) on page 110. The first inequality follows from (2.4.4). The second inequality is a consequence of (2.4.1), and the last inequality follows from (2.4.18). The bounds (2.4.20) is a consequence of (2.4.18) and (2.4.19) if we can prove that

$$|B_n| - |B_{n-1}| \geq |\tilde{B}_n| - |\tilde{B}_{n-1}| \geq |\widehat{B}_n| - |\widehat{B}_{n-1}| = \Sigma_n - \Sigma_{n-1}.$$

Of course, $|\widehat{B}_n| = \widehat{B}_n = \Sigma_n$, so the equality is clear. Moreover,

$$\begin{aligned}|\tilde{B}_n| - |\tilde{B}_{n-1}| &= \tilde{B}_n - \tilde{B}_{n-1} = (|b_n| - 1)\tilde{B}_{n-1} - |a_n|\tilde{B}_{n-2} \\ &\geq (|b_n| - 1)(\tilde{B}_{n-1} - \tilde{B}_{n-2}) \geq \prod_{k=1}^{n}(|b_k| - 1) = \widehat{B}_n - \widehat{B}_{n-1},\end{aligned}$$

and finally

$$Q_n := (|B_n| - |B_{n-1}|) - (\tilde{B}_n - \tilde{B}_{n-1}) \geq 0$$

since

$$\begin{aligned}Q_n &\geq (|b_n| - 1)(|B_{n-1}| - \tilde{B}_{n-1}) - |a_n|(|B_{n-2}| - \tilde{B}_{n-2}) \\ &\geq (|b_n| - 1)Q_{n-1} \geq \cdots \geq Q_1 \prod_{k=2}^{n}(|b_k| - 1) \geq 0.\end{aligned}$$

□

Remark. There exist Śleszyński-Pringsheim continued fractions for which the limit point case for $S_n(\overline{\mathbb{D}})$ fails to occur. For instance, for $\mathbf{K}(\widehat{a}_n/\widehat{b}_n)$ the radius $\widehat{R}_n$ converges to a positive value if $\Sigma_\infty < \infty$.

Example 8. For given $q \in \mathbb{C}$ with $|q| = 1$, the continued fraction

$$\mathop{\mathrm{K}}_{n=1}^{\infty} \frac{z}{2q^n} = \frac{z}{2q} + \frac{z}{2q^2} + \frac{z}{2q^3} + \cdots$$

is a Śleszyński-Pringsheim continued fraction for $|z| \leq 1$. So let $|z| \leq 1$. Then $\mathbf{K}(z/2q^n)$ converges to some value $f(z)$ with $0 < |f(z)| \leq 1$. Moreover P_k and Σ_n given by (2.4.9) have values $P_k = 1$ and $\Sigma_n = n + 1$. Therefore, by (2.4.10) and (2.4.18)

$$\begin{aligned}|f(z) - f_n(z)| &\leq \sum_{k=n+1}^{\infty} \frac{|z|^k}{\Sigma_k \Sigma_{k-1}} \leq \sum_{k=n+1}^{\infty} \frac{|z|^k}{(k+1)k} \\ &\leq \sum_{k=n+1}^{\infty} \frac{1}{(k+1)k} = \sum_{k=n+1}^{\infty} \left(\frac{1}{k} - \frac{1}{k+1}\right) = \frac{1}{n+1}.\end{aligned}$$

For $|z| < 1$ we also have the bound

$$|f(z) - f_n(z)| < \sum_{k=n+1}^{\infty} \frac{|z|^k}{(n+2)(n+1)} = \frac{|z|^{n+1}/(1 - |z|)}{(n+2)(n+1)}.$$

We can also majorize the expression by a telescoping series to get

$$\begin{aligned}|f(z) - f_n(z)| &\leq \sum_{k=n+1}^{\infty} \left(\frac{|z|^k}{k} - \frac{|z|^k}{k+1}\right) \\ &< \sum_{k=n+1}^{\infty} \left(\frac{|z|^k}{k} - \frac{|z|^{k+1}}{k+1}\right) = \frac{|z|^{n+1}}{n+1}\,.\end{aligned}$$

◇

3.2.5 Worpitzky's Theorem

Worpitzky was a high school teacher who published deep results in the yearly report from his school. His convergence result from 1865 ([Worp65]) can be stated as follows:

Theorem 3.29. (Worpitzky's Theorem.) *Let*

$$0 < |a_n| \leq \tfrac{1}{4} \quad \textit{for all} \quad n \geq 1\,. \tag{2.5.1}$$

Then $\mathbf{K}(a_n/1)$ *converges absolutely to some value* f *with* $0 < |f| \leq \frac{1}{2}$, *and* $0 < |S_n(w)| \leq \frac{1}{2}$ *for all* $n \in \mathbb{N}$ *and* $|w| \leq \frac{1}{2}$.

This was quite remarkable in 1865. Today we observe that the result follows directly from the Śleszyński-Pringsheim Theorem on page 129 since $2\mathbf{K}(a_n/1)$ is equivalent to the Śleszyński-Pringsheim continued fraction

$$\frac{4a_1}{2} \underset{+}{} \frac{4a_2}{2} \underset{+}{} \frac{4a_3}{2} \underset{+}{} \cdots \underset{+}{} \frac{4a_n}{2} \underset{+}{} \cdots\,. \tag{2.5.2}$$

This equivalence also means that $V := \frac{1}{2}\overline{\mathbb{D}}$ is a simple value set for the continued fractions $\mathbf{K}(a_n/1)$ from $E := \frac{1}{4}\overline{\mathbb{D}}$, and the truncation error bounds for Śleszyński-Pringsheim continued fractions can easily be adjusted to $\mathbf{K}(a_n/1)$ from E. We still chose to include Worpitzky's Theorem since it is so easy to apply and so widely known and used. The following version of Worpitzky's Theorem is more general. It can be derived from the work of Pringsheim ([Prin05]) and Wall ([Wall48], p 45-50):

Theorem 3.30. *For a given sequence $\{g_n\}_{n=0}^{\infty}$ of positive numbers $g_n < 1$, let $\{E_n\}_{n=1}^{\infty}$ be given by*

$$E_n := g_{n-1}(1-g_n)\overline{\mathbb{D}} = \{a \in \mathbb{C};\ |a| \le g_{n-1}(1-g_n)\}, \tag{2.5.3}$$

and let $\mathbf{K}(a_n/1)$ be a continued fraction from $\{E_n\}$. Then $\mathbf{K}(a_n/1)$ converges to some value f with $0 < |f| \le g_0$, and $\{g_n\mathbb{D}\}_{n=0}^{\infty}$ is a sequence of value sets for $\mathbf{K}(a_n/1)$.

This follows since $g_0^{-1}\mathbf{K}(a_n/1)$ is equivalent to the Śleszyński-Pringsheim continued fraction

$$\operatorname*{K} \frac{c_n}{d_n} := \frac{a_1/g_0g_1}{1/g_1} \underset{+}{} \frac{a_2/g_1g_2}{1/g_2} \underset{+}{} \frac{a_3/g_2g_3}{1/g_3} \underset{+\cdots+}{} \frac{a_n/g_{n-1}g_n}{1/g_n} \underset{+\cdots}{} . \tag{2.5.4}$$

The choice $g_n := \frac{1}{2}$ for all n gives back the original Worpitzy Theorem.

The sequence $\{-\mathbf{g_n}\}$.

The sequence $\{-g_n\}_{n=0}^{\infty}$ is a tail sequence for the continued fraction $\mathbf{K}(\widehat{a}_n/1)$ given by

$$\widehat{a}_n := -g_{n-1}(1-g_n) \qquad \text{for all } n. \tag{2.5.5}$$

This continued fraction is special since

- it defines E_n as $\{a \in \mathbb{C};\ |a| \le |\widehat{a}_n|\} = |\widehat{a}_n|\mathbb{D}$,
- it is equivalent to $g_0\mathbf{K}(\widehat{c}_n/\widehat{d}_n)$ where $\widehat{c}_n := -(1-g_n)/g_n$ and $\widehat{d}_n := 1/g_n$, so $\mathbf{K}(\widehat{c}_n/\widehat{d}_n)$ is a Śleszyński-Pringsheim continued fraction as described in case 1 on page 132.

It follows from Theorem 2.6 on page 66 (and also from Property 1 on page 78) that $\mathbf{K}(\widehat{a}_n/1)$ has approximants $\widehat{f}_n$ and denominators $\widehat{B}_n$ given by

$$\begin{gathered} \widehat{B}_n = \widehat{\Sigma}_n,\ \ \widehat{f}_n = \frac{g_0}{\widehat{\Sigma}_n} - g_0,\ \ \widehat{f} = \frac{g_0}{\widehat{\Sigma}_\infty} - g_0,\ \ \widehat{f} - \widehat{f}_n = \frac{g_0}{\widehat{\Sigma}_\infty} - \frac{g_0}{\widehat{\Sigma}_n} \\ \text{where}\ \ \widehat{\Sigma}_n := \frac{\Sigma_n}{\prod_{k=1}^n g_k} = \sum_{k=0}^n \widehat{P}_k \ \ \text{with}\ \ \widehat{P}_k := \prod_{j=1}^k \frac{1-g_j}{g_j}. \end{gathered} \tag{2.5.6}$$

If $\widehat{\Sigma}_\infty = \infty$, then $\{-g_n\}$ is the sequence of tail values for $\mathbf{K}(\widehat{a}_n/1)$ (Corollary 2.7 on page 68). Otherwise $\mathbf{K}(\widehat{a}_n/1)$ still converges, but it has another sequence $\{\widehat{f}^{(n)}\}$ of tail values. Also $\widehat{f}^{(n)} \in V_n$ according to Theorem 3.30, and $\widehat{f}^{(n)} \ne -g_n$. That is, $-g_n < \widehat{f}^{(n)}$, and thus also

$$\widehat{f}^{(n-1)} = \frac{\widehat{a}_n}{1+\widehat{f}^{(n)}} = -\frac{g_{n-1}(1-g_n)}{1+\widehat{f}^{(n)}} < -g_{n-1}(1-g_n) < 0.$$

Therefore we can replace $\{g_n\}$ by $\{-\widehat{f}^{(n)}\}$ without changing $\{E_n\}$:

Theorem 3.31. *For given sequence $\{g_n\}$ of positive numbers < 1, let*

$$\widehat{\Sigma}_\infty := \sum_{k=0}^{\infty} \widehat{P}_k < \infty, \quad \text{where } \widehat{P}_k := \prod_{j=1}^{k} \frac{1-g_j}{g_j}.$$

Then there exists a sequence $\{g_n^*\}$ with $g_n(1-g_{n+1}) < g_n^* < g_n$ for all $n \geq 0$ such that $g_{n-1}^*(1-g_n^*) = g_{n-1}(1-g_n)$ and

$$\widehat{\Sigma}_\infty^* = \infty \quad \text{where } \widehat{\Sigma}_n^* := \sum_{k=0}^{n} \widehat{P}_k^* \text{ and } \widehat{P}_k^* := \prod_{j=1}^{k} \frac{1-g_j^*}{g_j^*}.$$

Remark: Corollary 2.8 on page 69 shows that

$$g_n^* = -\widehat{f}^{(n)} = g_n \frac{(g_0 - g_0^*)\widehat{\Sigma}_n - g_0}{(g_0 - g_0^*)\widehat{\Sigma}_{n-1} - g_0} \quad \text{for } n \geq 1$$

where $g_0^* = -\widehat{f} = -g_0/\widehat{\Sigma}_\infty + g_0$ (Corollary 2.7 on page 68). That is,

$$g_n^* = g_n \frac{\widehat{\Sigma}_\infty - \widehat{\Sigma}_n}{\widehat{\Sigma}_\infty - \widehat{\Sigma}_{n-1}} \quad \text{for } n \geq 1. \tag{2.5.7}$$

Some choices of $\{\mathbf{g_n}\}$.

To make E_n large, we want to choose $\{g_n\}$ such that $|\widehat{a}_n|$ is large, where $\widehat{a}_n$ is given by (2.5.5). If all $g_n = g$, then $g := \frac{1}{2}$ gives the maximal value for $|\widehat{a}_n| = |\widehat{a}|$, and we are back to the original Worpitzky Theorem. If $\{g_n\}$ is allowed to vary, we can actually obtain $|\widehat{a}_n| > \frac{1}{4}$ for all n. For instance, if

$$g_n := \frac{1}{2} + \frac{1}{4n+2} \quad \text{for all } n, \tag{2.5.8}$$

then

$$|\widehat{a}_n| = \left(\frac{1}{2} + \frac{1}{4n-2}\right)\left(\frac{1}{2} - \frac{1}{4n+2}\right) = \frac{1}{4} + \frac{1/4}{4n^2-1}. \tag{2.5.9}$$

This is not an optimal choice, though. Indeed, there exists no optimal choice of this type:

Lemma 3.32. *Let* $\frac{1}{2} < g_n < 1$ *be chosen such that* $|\widehat{a}_n| := g_{n-1}(1-g_n) > \frac{1}{4}$ *for all* n. *Then there exists a sequence* $\{\widehat{g}_n\}$ *with* $g_n < \widehat{g}_n < 1$ *such that* $|\widehat{\widehat{a}}_n| > |\widehat{a}_n|$ *for all* n *for* $\widehat{\widehat{a}}_n := -\widehat{g}_{n-1}(1-\widehat{g}_n)$.

Proof : Let $\widehat{\Sigma}_n$ and $\widehat{P}_n$ be given by (2.5.6), and set

$$\widehat{g}_n := g_n(1+\varepsilon_n) \quad \text{where} \quad \varepsilon_n := \frac{\widehat{P}_n\widehat{P}_{n-1}}{2\widehat{\Sigma}_{n-1}}.$$

Then $g_n < \widehat{g}_n$. Now, $0 < (1-g_n)/g_n < 1$, so $0 < \widehat{P}_n < \widehat{P}_{n-1} < 1$ and $\widehat{P}_{n-1} < \widehat{\Sigma}_{n-1}$ for all $n \geq 2$. Therefore

$$\widehat{g}_n = g_n + g_n\frac{\widehat{P}_n\widehat{P}_{n-1}}{2\widehat{\Sigma}_{n-1}} = g_n + (1-g_n)\frac{\widehat{P}_{n-1}^2}{2\widehat{\Sigma}_{n-1}} < g_n + (1-g_n) = 1.$$

Moreover

$$\begin{aligned}
|\widehat{\widehat{a}}_n| - |\widehat{a}_n| &= g_{n-1}(1-g_n)\left((1+\varepsilon_{n-1})\left(1-\frac{g_n\varepsilon_n}{1-g_n}\right)-1\right)\\
&= |\widehat{a}_n|\left(\varepsilon_{n-1} - \frac{g_n\varepsilon_n}{1-g_n} - \frac{g_n\varepsilon_n\varepsilon_{n-1}}{1-g_n}\right)
\end{aligned}$$

where

$$\begin{aligned}
\varepsilon_{n-1} - \frac{g_n\varepsilon_n}{1-g_n} - \frac{g_n\varepsilon_n\varepsilon_{n-1}}{1-g_n} &= \frac{\widehat{P}_{n-1}\widehat{P}_{n-2}}{2\widehat{\Sigma}_{n-2}} - \frac{\widehat{P}_{n-1}^2}{2\widehat{\Sigma}_{n-1}} - \frac{\widehat{P}_{n-1}^3\widehat{P}_{n-2}}{4\widehat{\Sigma}_{n-1}\widehat{\Sigma}_{n-2}}\\
&= \frac{\widehat{P}_{n-1}}{4\widehat{\Sigma}_{n-1}\widehat{\Sigma}_{n-2}}\left(2\widehat{P}_{n-2}\widehat{\Sigma}_{n-1} - 2\widehat{P}_{n-1}\widehat{\Sigma}_{n-2} - \widehat{P}_{n-1}^2\widehat{P}_{n-2}\right)\\
&> \frac{\widehat{P}_{n-1}}{4\widehat{\Sigma}_{n-1}\widehat{\Sigma}_{n-2}}\left(2\widehat{P}_{n-1}(\widehat{\Sigma}_{n-1} - \widehat{\Sigma}_{n-2}) - \widehat{P}_{n-1}^2\widehat{P}_{n-2}\right)\\
&= \frac{\widehat{P}_{n-1}}{4\widehat{\Sigma}_{n-1}\widehat{\Sigma}_{n-2}}\left(2\widehat{P}_{n-1}\widehat{P}_{n-2} - \widehat{P}_{n-1}^2\widehat{P}_{n-2}\right)\\
&= \frac{\widehat{P}_{n-1}^2\widehat{P}_{n-2}}{4\widehat{\Sigma}_{n-1}\widehat{\Sigma}_{n-2}}\left(2 - \widehat{P}_{n-1}\right) > 0.
\end{aligned}$$

□

Indeed, it was proved in [JaMa90] that there is a hierarchy of sequences $\{g_n^{(k)}\}_{n=0}^{\infty}$ for $k = -1, 0, 1, 2, \ldots$ given by

$$g_{n_k+n}^{(k)} := \frac{1}{2} + \sum_{m=0}^{k} \frac{1}{4L_m(n)}$$

where $L_m(n)$ is defined recursively by

$$L_0(n) := n, \quad L_m(n) := \mathrm{Ln}(L_{m-1}(n)) \quad \text{for } m \geq 1$$

for $n \geq n_m$ sufficiently large. Every continued fraction from $\{E_n\}$ converges when

$$E_n := \{w;\ |w| \leq r_n\} \text{ where } r_n := \frac{1}{4} + \sum_{m=0}^{k} \frac{1}{(4L_m(n))^2} \tag{2.5.10}$$

from some n on, and the continued fraction $\mathbf{K}(a_n/1)$ diverges when

$$a_n := -\frac{1}{4} - \sum_{m=0}^{k-1} \frac{1}{(4L_m(n))^2} - \frac{1+\varepsilon}{(4L_k(n))^2} \tag{2.5.11}$$

for all n sufficiently large, where $k \in \mathbb{N}$ and $\varepsilon > 0$ are arbitrarily chosen. In particular $\mathbf{K}(a_n/1)$ with $a_n := -\frac{1}{4} - \frac{1+\varepsilon}{16n^2}$ diverges for $\varepsilon > 0$.

Limit sets.

If $\widehat{\Sigma}_\infty = \infty$, then every $f \in V_0 \setminus \{0\}$ is the value of a continued fraction $\mathbf{K}(a_n/1)$ from $\{E_n\}$. This follows by the following arguments: $\mathbf{K}(\widehat{a}_n/1)$ with $\widehat{a}_n$ given by (2.5.5) converges to $-g_0$, and thus its first tail converges to $-g_1$. Let $f \in V_0 \setminus \{0, -g_1\}$ be arbitrarily chosen, and set $a := f \cdot (1 - g_1)$. Then $|a| \leq g_0(1 - g_1)$ and

$$\frac{a}{1+}\,\frac{\widehat{a}_2}{1}\,\frac{}{+}\,\frac{\widehat{a}_3}{1}\,\frac{}{+}\,\frac{\widehat{a}_4}{1}\,\frac{}{+}\cdots$$

converges to f. If $\widehat{\Sigma}_\infty < \infty$, this is no longer true. Then V_0 is "too large". Indeed, no value $f \in V_0 \setminus V_0^*$ with V_0^* as in Theorem 3.31 can be the value of a continued fraction from $\{E_n\}$. Therefore:

Theorem 3.33. *Let $\{g_n\}$ and $\{E_n\}$ be as in Theorem 3.30. If $\widehat{\Sigma}_\infty = \infty$, then $\{g_n\mathbb{D} \setminus \{0\}\}_{n=0}^\infty$ is the sequence of limit sets for the family of continued fractions $\mathbf{K}(a_n/1)$ from $\{E_n\}$.*

Truncation error bounds.

We have already (in (2.5.6)) established an expression for the truncation error $\widehat{f} - \widehat{f}_n$ for $\mathbf{K}(\widehat{a}_n/1)$ with $\widehat{a}_n := -g_{n-1}(1 - g_n)$.

$\mathbf{K}(\tilde{a}_n/1)$ with $\widehat{a}_n \leq \tilde{a}_n < 0$ is equivalent to $g_0\mathbf{K}(\tilde{c}_n/\tilde{d}_n)$ where $\tilde{c}_n := \tilde{a}_n/g_n g_{n-1}$ and $\tilde{d}_n := 1/g_n$, and thus $\mathbf{K}(\tilde{c}_n/\tilde{d}_n)$ is a Śleszyński-Pringsheim continued fraction as described in case 2 on page 132. Therefore the denominators $\tilde{B}_n$ and approximants $\tilde{f}_n$ for $\mathbf{K}(\tilde{a}_n/1)$ satisfy

$$\tilde{B}_{n+1} \geq \tilde{B}_n \geq \widehat{\Sigma}_n \quad \text{and} \quad -g_n < \tilde{f}_n < \tilde{f}_{n-1} < 0 \tag{2.5.12}$$

where $\widehat{\Sigma}_n$ still is given by (2.5.6). For the general case we get from Theorem 3.27 (again with obvious notation):

Theorem 3.34. *Let $\{g_n\}$ and $\mathbf{K}(a_n/1)$ be as in Theorem 3.30, and let $\mathbf{K}(\tilde{a}_n/1)$ and $\mathbf{K}(\widehat{a}_n/1)$ be given by $\tilde{a}_n := -|a_n|$ and $\widehat{a}_n := -g_{n-1}(1-g_n)$ for all n. Then*

$$|f - f_n| \le \tilde{f}_n - \tilde{f} \le \widehat{f}_n - \widehat{f} = \frac{g_0}{\widehat{\Sigma}_n} - \frac{g_0}{\widehat{\Sigma}_\infty}. \tag{2.5.13}$$

Corollary 3.35. *For given positive $r \le 1$, let $\kappa := \dfrac{1+\sqrt{1-r}}{1-\sqrt{1-r}}$. If all $|a_n| \le \frac{r}{4}$ for $\mathbf{K}(a_n/1)$, then*

$$|f - f_n| \le \begin{cases} \dfrac{1/2}{n+1} & \text{if } r = 1, \\ \dfrac{\sqrt{1-r}}{\kappa^{n+1}-1} & \text{if } r < 1. \end{cases} \tag{2.5.14}$$

Proof : Let $g_n := g := (1 - \sqrt{1-r})/2$ for all n. Then $|a_n| \le g(1-g) = \frac{r}{4}$ and $\kappa = (1-g)/g$. Therefore $\widehat{P}_k = \kappa^k$ and $\widehat{\Sigma}_n = n+1$ if $\kappa = 1$ and

$$\widehat{\Sigma}_n = \sum_{k=0}^{n} \kappa^n = \frac{\kappa^{n+1}-1}{\kappa - 1}$$

otherwise. In both cases $\widehat{\Sigma}_n \to \widehat{\Sigma}_\infty = \infty$. If $r = 1$, then $\kappa = 1$ and $g = \frac{1}{2}$. If $r < 1$, then $\kappa > 1$ and $g < \frac{1}{2}$ and

$$\frac{g_0}{\widehat{\Sigma}_n} = \frac{(\kappa-1)g}{\kappa^{n+1}-1} = \frac{1-2g}{\kappa^{n+1}-1} = \frac{\sqrt{1-r}}{\kappa^{n+1}-1}.$$

The result follows therefore from Theorem 3.34. □

The radius of $\mathbf{S_n(g_n\mathbb{D})}$.

Theorem 3.36. *Let $\{g_n\}$ and $\{E_n\}$ be as in Theorem 3.30, and let $\mathbf{K}(a_n/1)$ be a continued fraction from $\{E_n\}$. Then*

$$\begin{aligned}\operatorname{rad} S_n(g_n\mathbb{D}) &= \frac{\prod_{k=1}^n |a_k|}{|B_n|^2/g_n - g_n|B_{n-1}|^2} \\ &\le \frac{g_0}{\widehat{\Sigma}_n + \widehat{\Sigma}_{n-1}} \prod_{k=1}^n \frac{|a_k|}{g_{k-1}(1-g_k)} \le \frac{g_0}{\widehat{\Sigma}_n + \widehat{\Sigma}_{n-1}}.\end{aligned} \tag{2.5.15}$$

Let further $\mathbf{K}(\tilde{a}_n/1)$ and $\mathbf{K}(\widehat{a}_n/1)$ be given by $\tilde{a}_n := -|a_n|$ and $\widehat{a}_n := -g_{n-1}(1-g_n)$. Then

$$\operatorname{rad} S_n(g_n\mathbb{D}) \le \operatorname{rad} \tilde{S}_n(g_n\mathbb{D}) \le \operatorname{rad} \widehat{S}_n(g_n\mathbb{D}) = \frac{g_0}{\widehat{\Sigma}_n + \widehat{\Sigma}_{n-1}}. \tag{2.5.16}$$

Proof : The equivalence (2.5.4) shows that the radius R_n of $S_n(g_n\mathbb{D})$ is equal to $g_0 \cdot$ (radius of $T_n(\mathbb{D})$) where

$$T_n(w) = \frac{c_1}{d_1} {+} \frac{c_2}{d_2} {+} \frac{c_3}{d_3} {+} \cdots {+} \frac{c_n}{d_n + w} =: \frac{C_{n-1}w + C_n}{D_{n-1}w + D_n}$$

with $c_n := \dfrac{a_n}{g_n g_{n-1}}$ and $d_n := \dfrac{1}{g_n}$. Here $\mathbf{K}(c_n/d_n)$ is a Śleszyński-Pringsheim continued fraction, and the approximants of $\mathbf{K}(a_n/1)$ satisfy

$$S_n(w) := \frac{A_{n-1}w + A_n}{B_{n-1}w + B_n}; \quad A_n := C_n \prod_{k=0}^n g_k, \quad B_n := D_n \prod_{k=1}^n g_k$$

for $n \ge 2$, so by Theorem 3.28 on page 133

$$R_n = \frac{g_0 \prod_{k=1}^n |c_k|}{|D_n|^2 - |D_{n-1}|^2} \le \frac{g_0}{|D_n| + |D_{n-1}|} \prod_{k=1}^n \frac{|c_k|}{|d_k| - 1} \le \frac{g_0}{|D_n| + |D_{n-1}|}$$

where by (2.4.9)

$$|D_n| \ge \sum_{k=0}^n \prod_{m=1}^k (|d_m| - 1) = \sum_{k=0}^n \prod_{m=1}^k \left(\frac{1}{g_m} - 1\right) = \widehat{\Sigma}_n .$$

Therefore (2.5.15) follows since $|a_k| \le g_{k-1}(1-g_k)$. Similarly we have $g_0\mathbf{K}(\tilde{a}_n/1) \sim \mathbf{K}(\tilde{c}_n/\tilde{d}_n)$ and $g_0\mathbf{K}(\widehat{a}_n/1) \sim \mathbf{K}(\widehat{c}_n/\widehat{d}_n)$ where $\tilde{c}_n := -|a_n|/g_n g_{n-1}$, $\tilde{d}_n := 1/g_n$, $\widehat{c}_n := -(1-g_n)/g_n$ and $\widehat{d}_n := 1/g_n = 1 + |\widehat{c}_n|$. Hence also (2.5.16) follows from Theorem 3.28. □

Remark: The limit point case $S_n(g_n\mathbb{D}) \to \{f\}$ occurs for every continued fraction $\mathbf{K}(a_n/1)$ from $\{E_n\}$ only if $\widehat{\Sigma}_\infty = \infty$. Indeed, if $\widehat{\Sigma}_\infty < \infty$, then $V_n := g_n\mathbb{D}$ contains both $-g_n$ and $-g_n^*$ (as given in Theorem 3.31), and thus $\widehat{S}_n(V_n)$ contains both $-g_0$ and $-g_0^*$ for all n for the continued fraction $\mathbf{K}(\widehat{a}_n/1)$ from $\{E_n\}$ with $\widehat{a}_n := -g_{n-1}(1 - g_n)$, and the limit point case can not occur.

3.2.6 Van Vleck's Theorem

This is a convergence theorem for continued fractions $\mathbf{K}(1/b_n)$ with

$$b_n \in G_\varepsilon := \{w \in \mathbb{C};\ |\arg w| \le \tfrac{\pi}{2} - \varepsilon\} \cup \{0\} \tag{2.6.1}$$

for some arbitrarily given $0 < \varepsilon < \pi/2$. Let

$$V_\varepsilon := G_\varepsilon \cup \{\infty\} \quad \text{for } \varepsilon \ge 0. \tag{2.6.2}$$

Then

$$s_n(V_\varepsilon) = \frac{1}{b_n + V_\varepsilon} \subseteq V_\varepsilon \qquad \text{for} \qquad b_n \in G_\varepsilon. \tag{2.6.3}$$

Therefore V_ε is a simple value set for $\mathbf{K}(1/b_n)$ from G_ε. This even holds for $\varepsilon := 0$. Moreover, V_δ is a value set for G_ε for every $0 \le \delta \le \varepsilon$. We know by Theorem 3.13 on page 117 that a positive continued fraction $\mathbf{K}(1/b_n)$ converges if and only if $\sum b_n = \infty$, where the case $\sum b_n < \infty$ is described by the Stern-Stolz Theorem on page 100. Van Vleck ([VanV01]) showed that this is true, not only for positive continued fractions, but for every continued fraction from G_ε:

Theorem 3.37. (Van Vleck's Theorem.) *Let $b_n \in G_\varepsilon$ for all $n \in \mathbb{N}$ for a given $0 < \varepsilon < \frac{\pi}{2}$ with $b_1 \ne 0$. Then the even and odd approximants of $\mathbf{K}(1/b_n)$ converge to finite values. Moreover, $\mathbf{K}(1/b_n)$ itself converges if and only if $\sum |b_n| = \infty$.*
If $\mathbf{K}(1/b_n)$ diverges, then it diverges generally, and its even and odd classical approximants converge absolutely.

Proof : The nestedness

$$f_n \in S_n(V_\varepsilon) \subseteq S_{n-1}(V_\varepsilon) \subseteq s_1(V_\varepsilon) = \frac{1}{b_1 + V_\varepsilon} \tag{2.6.4}$$

shows that $\{S_n(w)\}$ is uniformly bounded with respect to $n \in \mathbb{N}$ and $w \in V_\varepsilon$ since $0 \notin b_1 + V_\varepsilon$, and thus $1/(b_1 + V_\varepsilon)$ is bounded. In particular all $B_n \ne 0$. If $\sum |b_n| < \infty$, then $\mathbf{K}(1/b_n)$ diverges generally, but its even and odd approximants converge absolutely (the Stern-Stolz Theorem and the subsequent remark on page 101).

Let $\sum |b_n| = \infty$. Following Jones and Thron ([JoTh80], p 89) we shall use the Stieltjes-Vitali Theorem to prove that then $\mathbf{K}(1/b_n)$ converges. Let

$$\beta_n := \arg(b_n) \quad \text{and} \quad d_n(z) := |b_n| e^{i\beta_n z} \quad \text{for all } n\,. \tag{2.6.5}$$

(If $b_n = 0$, we just set $\beta_n := 0$.) Then $|\beta_n| \le \frac{\pi}{2} - \epsilon$ and $d_n(z) \in G_{\epsilon/2}$ if $|\arg(d_n(z))| \le \frac{\pi}{2} - \frac{\epsilon}{2}$; i.e.,

$$d_n(z) \in G_{\epsilon/2} \quad \text{if} \quad z \in D := \left\{ z \in \mathbb{C};\ |\mathrm{Re}(z)| \le \frac{\pi - \epsilon}{\pi - 2\epsilon} \right\}. \tag{2.6.6}$$

Now,

$$d_n(z) \ge 0 \quad \text{if} \quad z \in D^* := \{ z \in D;\ \mathrm{Re}(z) = 0 \}\,. \tag{2.6.7}$$

Since for every fixed $z \in D^*$

$$\sum |d_n(z)| = \sum |b_n| e^{-\beta_n \mathrm{Im}(z)} > e^{-|\mathrm{Im}(z)|\pi/2} \sum |b_n| = \infty\,,$$

it follows from the Seidel-Stern Theorem on page 117 that $\mathbf{K}(1/d_n(z))$ converges to a finite value for every $z \in D^*$. We have proved: the classical approximants $f_n(z)$ of $\mathbf{K}(1/d_n(z))$ form a sequence of uniformly bounded holomorphic functions in D which converges for $z \in D^*$. By the Stieltjes-Vitali Theorem on page 115 it follows therefore that $\mathbf{K}(1/d_n(z))$ converges for all $z \in D^\circ$. In particular it converges for $z = 1$. Hence $\mathbf{K}(1/b_n)$ converges. □

Remark. Equivalence transformations show for instance that the conclusions of Van Vleck's Theorem also hold for $\mathbf{K}(1/(-b_n))$ and $\mathbf{K}(-1/ib_n)$ when all $b_n \in G_\varepsilon$.

Example 9. The *two-periodic* continued fraction

$$\frac{-1}{-1+i} + \frac{-1}{1+i} + \frac{-1}{-1+i} + \frac{-1}{1+i} + \frac{-1}{-1+i} + \frac{-1}{1+i} + \cdots$$

is equivalent to the continued fraction

$$\frac{i}{1+i} + \frac{1}{1-i} + \frac{1}{1+i} + \frac{1}{1-i} + \frac{1}{1+i} + \frac{1}{1-i} + \cdots.$$

(Use $r_n := -i$ in the equivalence transformation.) This new continued fraction has the form $i \cdot \mathbf{K}(1/b_n)$ where all $b_n \in G_\varepsilon$ for $\varepsilon := \frac{\pi}{4}$. Since $\sum |b_n| = \infty$, the original continued fraction converges. It is rather easy to find its value f. It must be $f = ix$ where x satisfies the equation

$$x = \cfrac{1}{1 + i + \cfrac{1}{1 - i + x}}\,.$$

This quadratic equation has the two roots $\frac{1}{2}(-1 \pm \sqrt{3})(1-i)$. Since $x \in V_\varepsilon$, the real part of x can not be negative, so $x = \frac{1}{2}(\sqrt{3}-1)(1-i)$ and thus $f = \frac{1}{2}(\sqrt{3}-1)(1+i)$. ◇

Limit set.

Theorem 3.38. *For given $0 \le \varepsilon \le \pi/2$, $V_\varepsilon \setminus \{0,\infty\}$ is the limit set for the family of continued fractions $\mathbf{K}(1/b_n)$ from $G_\varepsilon \setminus \{0\}$.*

Proof : Let $\mathbf{K}(1/b_n)$ be a convergent continued fraction from G_ε with all $b_n \neq 0$. It is then a consequence of Van Vleck's Theorem that its value f and its first tail value $f^{(1)}$ are finite and belong to V_ε. Hence also $f = 1/(b_1 + f^{(1)}) \neq 0$, and thus $f \in V_\varepsilon \setminus \{0,\infty\}$.

We need to prove that every $f \in V_\varepsilon \setminus \{0,\infty\}$ is the value of a continued fraction $\mathbf{K}(1/b_n)$ from G_ε with all $b_n \neq 0$. Let first f be a point on the boundary; i.e., $f \in \partial V_\varepsilon \setminus \{0,\infty\}$. It suffices to study the case where $\arg f > 0$; i.e., $f = r\,e^{i(\pi/2-\varepsilon)} = i\,r\,e^{-i\varepsilon}$ (complex conjugation). Let $b_1 := r_1\,e^{-i(\pi/2-\varepsilon)} = -i\,r_1\,e^{i\varepsilon}$ and $b_2 := i\,r_2\,e^{-i\varepsilon}$ where $r_1 > 0$, $r_2 > 0$ and $r_1 + 1/(r_2 + r) = 1/r$. Then $b_1, b_2 \in \partial G_\varepsilon$ and

$$\frac{1}{b_2 + f} = \frac{1}{i(r_2 + r)e^{-i\varepsilon}} = \frac{-i}{r_2 + r}\,e^{i\varepsilon} =: \widehat{f} \in \partial V_\varepsilon$$

and

$$\frac{1}{b_1 + \widehat{f}} = \frac{1}{-i(r_1 + \frac{1}{r_2+r})e^{i\varepsilon}} = i\,r\,e^{-i\varepsilon} = f\,.$$

That is, the 2-periodic sequence $f,\ \widehat{f},\ f,\ \widehat{f},\ f,\ \ldots$ is a tail sequence for the 2-periodic continued fraction

$$\mathop{\mathrm{K}}_{n=1}^{\infty} \frac{1}{b_n} := \frac{1}{b_1} \genfrac{}{}{0pt}{}{}{+} \frac{1}{b_2} \genfrac{}{}{0pt}{}{}{+} \frac{1}{b_1} \genfrac{}{}{0pt}{}{}{+} \frac{1}{b_2} \genfrac{}{}{0pt}{}{}{+} \frac{1}{b_1} \genfrac{}{}{0pt}{}{}{+} \frac{1}{b_2} \genfrac{}{}{0pt}{}{}{+\cdots}\,.$$

Now, $\mathbf{K}(1/b_n)$ converges (the Van Vleck Theorem), and its value must belong to V_ε. Moreover, its value must be a solution of the quadratic equation

$$\cfrac{1}{b_1 + \cfrac{1}{b_2 + x}} = x; \quad \text{i.e.,} \quad x = \tfrac{1}{2}\left(-r_1 r_2 \pm \sqrt{r_1^2 r_2^2 + 4 i r_2 e^{-i\varepsilon}}\right).$$

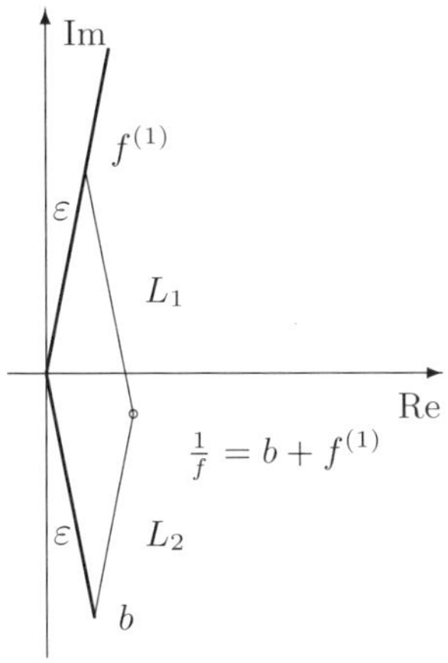

Only one of the two solutions belongs to V_ε. Hence $\mathbf{K}(1/b_n)$ converges to f. This proves that $\partial V_\varepsilon \setminus \{0, \infty\}$ belongs to the limit set. If $\varepsilon = \pi/2$ we are finished, since then $V_\varepsilon^\circ = \emptyset$. Let $\varepsilon < \pi/2$, and let $f := r\,e^{i\alpha} \in V_\varepsilon^\circ$ with $-(\frac{\pi}{2} - \varepsilon) < \alpha < \frac{\pi}{2} - \varepsilon$ and $0 < r < \infty$ be arbitrarily chosen. Then $\frac{1}{f} = \frac{1}{r} e^{-i\alpha} \in V_\varepsilon^\circ$, and it suffices to find a $b \in G_\varepsilon$ and an $f^{(1)} \in \partial V_\varepsilon$ such that $f = r\,e^{i\alpha} = \frac{1}{b + f^{(1)}}$; i.e., $b + f^{(1)} = \frac{1}{f}$. This can for instance be done by drawing lines L_1 and L_2 through $\frac{1}{f}$ parallel to the two boundary lines (see the picture). The points where L_1 and L_2 intersects the boundary of V_ε can then be our points $f^{(1)}$ and b.

□

Remark. Clearly, $\mathbf{K}(1/b_n)$ from G_ε converges to $f = \infty$ only if $b_1 = 0$ and $f^{(1)} = 0$. And $f^{(1)} = 0$ only if $f^{(2)} = \infty$, and so on. Hence, if $\mathbf{K}(1/b_n)$ from G_ε converges to ∞, then all $b_{2n-1} = 0$, and if $\mathbf{K}(1/b_n)$ converges to 0, then all $b_{2n} = 0$. For instance

$$\frac{1}{0} \, \frac{1}{+1} \, \frac{1}{+0} \, \frac{1}{+1} \, \frac{1}{+0} + \cdots \qquad \text{and} \qquad \frac{1}{1} \, \frac{1}{+0} \, \frac{1}{+1} \, \frac{1}{+0} \, \frac{1}{+1} + \cdots$$

converge to ∞ and 0 respectively.

Truncation error bounds.

We have the following a posteriori bounds for $\mathbf{K}(1/b_n)$ from G_ε:

Theorem 3.39. *For given $0 < \varepsilon < \frac{\pi}{2}$, let $\mathbf{K}(1/b_n)$ with $b_1 \neq 0$ be a continued fraction from G_ε. Then*

$$\operatorname{diam} S_n(V_\varepsilon) \leq \begin{cases} |f_n - f_{n-1}| & \text{if } \varepsilon \geq \pi/4, \\ |f_n - f_{n-1}|/\sin 2\varepsilon & \text{if } \varepsilon < \pi/4. \end{cases}$$

Proof : As always, $\mathbb{H}$ denotes the right half plane, so V_ε can be written

$$V_\varepsilon = (e^{i\varepsilon}\overline{\mathbb{H}}) \cap (e^{-i\varepsilon}\overline{\mathbb{H}}).$$

Now, $S_n(e^{i\varepsilon}\overline{\mathbb{H}})$ and $S_n(e^{-i\varepsilon}\overline{\mathbb{H}})$ are two circular disks since $-b_1 \notin e^{\pm i\varepsilon}\overline{\mathbb{H}}$, and thus $S_n(e^{\pm i\varepsilon}\overline{\mathbb{H}})$ is bounded. The boundaries of these two disks intersect at $S_n(0) = f_n$

and $S_n(\infty) = f_{n-1}$ under the inner angle $2\alpha := \pi - 2\varepsilon$. That is, $S_n(V_\varepsilon)$ is a lens-shaped set as illustrated in the figure on page 126. Therefore the result follows just as in the proof of the Henrici-Pfluger Bounds. (We have used that $\sin 2\alpha = \sin(\pi - 2\varepsilon) = \sin 2\varepsilon$.) □

The diameter of $\mathbf{S_n}(\mathbb{H})$.

Our proof of Van Vleck's Theorem was based on the Stieltjes-Vitali Theorem which gives no information on the speed of convergence. We therefore give an alternative proof which also opens up for an extension of Van Vleck's classical result and for truncation error bounds. It is due to Jensen ([Jens09]), and was later found, independently, by Lange ([Lange99a]).

Proof 2 of Van Vleck's Theorem: Also the half plane $V_0 := \overline{\mathbb{H}}$ is a simple value set for $\mathbf{K}(1/b_n)$. Therefore

$$K_n := S_n(V_0) \subseteq S_{n-1}(V_0) \subseteq \cdots \subseteq S_1(V_0) \subseteq V_0. \tag{2.6.8}$$

Now, $\text{Re}(b_1) \neq 0$, so K_1 is a circular disk, and the nestedness (2.6.8) makes all K_n circular disks. Since $0 \in V_0$, we have $f_n \in K_n$, so all $B_n \neq 0$, and thus $\zeta_n := S_n^{-1}(\infty) = -B_n/B_{n-1} \neq \infty$ for all $n \in \mathbb{N}$.

ζ_n is symmetric to $(-\overline{\zeta}_n)$ with respect to $\partial V_0 = i\,\overline{\mathbb{R}^+}$. The center C_n of K_n is symmetric to ∞ with respect to $\partial K_n = \partial S_n(V_0)$. Therefore, by Property 2 on page 109, $C_n = S_n(-\overline{\zeta}_n)$. Hence, the radius of K_n is

$$\begin{aligned} R_n = |S_n(0) - C_n| &= \left| \frac{A_n}{B_n} - \frac{A_n - A_{n-1}\overline{\zeta}_n}{B_n - B_{n-1}\overline{\zeta}_n} \right| \\ &= \frac{|(A_nB_{n-1} - A_{n-1}B_n)\overline{\zeta}_n|}{|B_n(B_n - B_{n-1}\overline{\zeta}_n)|} = \frac{1 \cdot |\overline{B}_n/\overline{B}_{n-1}|}{|B_nB_{n-1}(\frac{B_n}{B_{n-1}} + \frac{\overline{B}_n}{\overline{B}_{n-1}})|} \\ &= \frac{|\overline{B}_n/B_n|}{|B_n\overline{B}_{n-1} + \overline{B}_nB_{n-1}|} = \frac{1}{2|\text{Re}(B_n\overline{B}_{n-1})|}. \end{aligned}$$

Now, set

$$Q_n := \text{Re}(B_n\overline{B}_{n-1}).$$

Then the recurrence relation for $\{B_n\}$ gives

$$Q_n = \text{Re}(b_n|B_{n-1}|^2 + B_{n-2}\overline{B}_{n-1}) = Q_{n-1} + |B_{n-1}|^2\text{Re}(b_n). \tag{2.6.9}$$

Since $Q_1 = \text{Re}(b_1) > 0$ and $\text{Re}(b_n) \geq 0$, this means that $Q_n > 0$ and non-decreasing as n increases. Let $Q := \lim Q_n$. If $Q = \infty$, then $R_n \to 0$, and the convergence is clear. Therefore also $\sum |b_n| = \infty$ (the Stern-Stolz Theorem).

Let $Q < \infty$. Then

$$\sum_{n=2}^{\infty} \left| \frac{1}{Q_{n-1}} - \frac{1}{Q_n} \right| = \sum_{n=2}^{\infty} \left(\frac{1}{Q_{n-1}} - \frac{1}{Q_n} \right) = \frac{1}{Q_1} - \frac{1}{Q} < \infty. \tag{2.6.10}$$

So far we have only used that $\mathrm{Re}(b_1) > 0$ and that V_0 is a value set for $\mathbf{K}(1/b_n)$; i.e., $b_n \in V_0$ for $n \geq 2$. At this point we assume that all $b_n \in G_\varepsilon$ with a given $\varepsilon > 0$. Then $\lambda := \sin \varepsilon > 0$ and $|b_n| \leq \frac{\mathrm{Re}(b_n)}{\cos(\pi/2-\varepsilon)} = \mathrm{Re}(b_n)/\lambda$, so

$$\begin{aligned} |f_n - f_{n-2}| &= \left| \frac{A_n B_{n-2} - B_n A_{n-2}}{B_n B_{n-2}} \right| = \left| \frac{b_n}{B_n B_{n-2}} \right| \leq \frac{\mathrm{Re}(b_n)}{\lambda |B_n B_{n-2}|} \\ &= \frac{|B_{n-1}|^2 \mathrm{Re}(b_n)}{\lambda \cdot |B_n B_{n-1}| \cdot |B_{n-1} B_{n-2}|} = \frac{Q_n - Q_{n-1}}{\lambda |B_n B_{n-1}| \cdot |B_{n-1} B_{n-2}|} \\ &\leq \frac{1}{\lambda} \cdot \frac{Q_n - Q_{n-1}}{Q_n Q_{n-1}} = \frac{1}{\lambda} \cdot \left(\frac{1}{Q_{n-1}} - \frac{1}{Q_n} \right) \end{aligned} \tag{2.6.11}$$

where we have used the equality (2.6.9). Hence $\sum |f_n - f_{n-2}| < \infty$, and the even and odd approximants of $\mathbf{K}(1/b_n)$ converge absolutely. Van Vleck's Theorem follows therefore from the Lane-Wall Characterization on page 103. □

Actually, the limit point case always occurs when $\mathbf{K}(1/b_n)$ converges:

Corollary 3.40. *Let $\mathbf{K}(1/b_n)$ be as in Van Vleck's Theorem. Then $\mathrm{diam}\, S_n(\overline{\mathbb{H}}) \to 0$ if and only if $\sum |b_n| = \infty$.*

Proof : If $\sum |b_n| < \infty$, then $\mathbf{K}(1/b_n)$ diverges, and $\mathrm{diam}\, S_n(\overline{\mathbb{H}})$ can not vanish as $n \to \infty$. Let $\sum |b_n| = \infty$. We want to prove that now $R_n \to 0$; i.e., $Q_n \to \infty$. From (2.6.9)

$$\begin{aligned} Q_n &= Q_{n-1} + \mathrm{Re}(b_n) \cdot |\zeta_{n-1}| \cdot |B_{n-1} B_{n-2}| \\ &\geq Q_{n-1}(1 + \mathrm{Re}(b_n)|\zeta_{n-1}|) \geq Q_1 \prod_{k=2}^{n} (1 + \mathrm{Re}(b_k)|\zeta_{k-1}|) \end{aligned} \tag{2.6.12}$$

where $Q_1 = \mathrm{Re}(b_1) > 0$ and $\mathrm{Re}(b_k) \geq \lambda |b_k|$. Therefore it suffices to prove that $\sum |b_k \zeta_{k-1}| = \infty$. For convenience we set $\delta_{k-1} := -b_k \zeta_{k-1} = b_k B_{k-1}/B_{k-2}$. Then $\{\delta_k\}$ is exactly as in (1.2.3) on page 104, and thus, if $\sum |\delta_n| < \infty$, then $\sum |b_n| < \infty$ as in the proof the Lane-Wall Characterization. Therefore $\sum |\delta_n| = \sum |b_k \zeta_{k-1}| = \infty$. □

Corollary 3.41. *For given* $\mathbf{K}(1/b_n)$ *with* $\text{Re}(b_1) > 0$ *and* $\text{Re}(b_n) \geq 0$ *for* $n > 2$,

$$\text{diam}\, S_n(\overline{\mathbb{H}}) = \frac{1}{\text{Re}(B_n \overline{B}_{n-1})} \leq \frac{1/\text{Re}(b_1)}{\prod_{k=2}^{n}(1 + |\zeta_{k-1}|\text{Re}(b_k))}$$

$$\leq \frac{1/\text{Re}(b_1)}{\prod_{k=2}^{n}(1 + \text{Re}(b_{k-1})\text{Re}(b_k))} \quad \textit{for } n \in \mathbb{N}. \qquad (2.6.13)$$

Proof : The first inequality follows from (2.6.12). The second one follows since $-\zeta_k \in b_k + \overline{\mathbb{H}}$ (Theorem 2.3 on page 63). □

Remark. This proves that if $\text{Re}(b_1) > 0$ and $\text{Re}(b_n) \geq 0$ for $n \geq 2$, then $\mathbf{K}(1/b_n)$ converges if $\text{Re}(B_n \overline{B}_{n-1}) \to \infty$, which happens if $\sum |\zeta_{n-1}| \cdot \text{Re}(b_n) = \infty$. Although this extends Van Vleck's Theorem, the criterion is more difficult to check. Beardon and Short ([BeSh07]) have proved that the limit point case may fail to occur for a convergent continued fraction in this case.

3.2.7 The Thron-Lange Theorem

The Thron-Lange Theorem concerns a family of continued fractions $\mathbf{K}(1/b_n)$ for which $\{b_n\}$ stays away from a circular disk $\mathfrak{B}(-2\gamma, 2r)^\circ$ symmetric about the real axis. The unifying factor for this family is that the disk $\mathfrak{B}(\gamma, r)$ is a simple value set for all its continued fractions.

Theorem 3.42. (The Thron-Lange Theorem). *For given* $\gamma \in \mathbb{R}$, *let* $r := \sqrt{1+\gamma^2}$, $V := \mathfrak{B}(\gamma, r)$ *and* $G := \mathfrak{B}(-2\gamma, -2r)$. *Then* G *is the element set for continued fractions* $\mathbf{K}(1/b_n)$ *corresponding to the value set* V, *and every continued fraction* $\mathbf{K}(1/b_n)$ *from* G *converges with*

$$\text{diam}\, S_n(V) \leq 2r \prod_{k=2}^{n} \frac{2k-\sigma}{2k+\sigma} \leq 2r\left(\frac{4+\sigma}{2n+2+\sigma}\right)^{\sigma}; \quad \sigma := 1 - \frac{|\gamma|}{r}. \qquad (2.7.1)$$

The convergence of $\mathbf{K}(1/b_n)$ was proved by Thron. Actually, it is a corollary of a much more general result from [Thron49]. Lange ([Lange99b]) has proved a number of corollaries of Thron's general theorem, but our particular version can be found in Thron's paper. However, Lange is responsible for the truncation error bounds ([Lange99b]).

To prove the convergence in Theorem 3.42 is not difficult. It follows from Corollary 3.9 on page 114 with $\varphi_n(w) := \varphi(w) := \gamma + rw$ and $w_n := \infty$. However, it is also a consequence of Lange's truncation error bound (2.7.1) since this vanishes as $n \to \infty$, and thus the limit point case occurs. We shall prove this bound:

Proof : We shall first prove that the inclusion $1/(b+V) \subseteq V$ holds if and only if $b \in G$. Evidently, $1/(b+V) \subseteq V$ if and only if

$$b + V = \mathfrak{B}(b+\gamma, r) \subseteq \frac{1}{V} = \mathfrak{B}\Big(\frac{\gamma}{\gamma^2 - r^2}, \frac{r}{\gamma^2 - r^2}\Big) = \mathfrak{B}(-\gamma, -r)$$

(Lemma 3.6 on page 110); that is, if and only if

$$|b + \gamma - (-\gamma)| \geq r + r$$

which proves the assertion.

In the following we assume that $\gamma \geq 0$. We can do so without loss of generality since $\mathbf{K}(1/(-b_n))$ is equivalent to $-\mathbf{K}(1/b_n)$. Let $\mathbf{K}(1/b_n)$ be a continued fraction from G. Since $0 \in V$, we know that $f_n \in V$, and thus $B_n \neq 0$ since V is bounded. Hence it follows from Lemma 3.6 on page 110 that the radius R_n of $S_n(V)$ is given by

$$R_n = \frac{r}{|B_{n-1}|^2(|\gamma - \zeta_n|^2 - r^2)},$$

and thus

$$\frac{R_{n+1}}{R_n} = \frac{1}{|\zeta_n|^2} \cdot \frac{|\gamma - \zeta_n|^2 - r^2}{|\gamma - \zeta_{n+1}|^2 - r^2} \tag{2.7.2}$$

where

$$\zeta_{n+1} = -b_{n+1} + \frac{1}{\zeta_n}. \tag{2.7.3}$$

Now, $b_{n+1} \in \mathfrak{B}(-2\gamma, -2r)$, and we shall first prove that

$$\zeta_n \in \mathfrak{B}\Big(\frac{n+1}{n}\gamma, -\frac{n+1}{n}r\Big) \quad \text{for } n \geq 1 \tag{2.7.4}$$

which by Lemma 3.6 is equivalent to

$$\frac{1}{\zeta_n} \in \mathfrak{B}\Big(-\frac{n\gamma}{n+1}, \frac{nr}{n+1}\Big) \quad \text{for } n \geq 1. \tag{2.7.5}$$

(2.7.4) holds trivially for $n = 1$ since $\zeta_1 = -b_1 \in (-G) = \mathfrak{B}(2\gamma, -2r)$. Assume it holds for a given $n \in \mathbb{N}$. Then we can write

$$\frac{1}{\zeta_n} = -\frac{n\gamma}{n+1} + \frac{nr}{n+1}(1-\mu)e^{i\theta}, \;\; b_{n+1} = -2\gamma + 2(1+\lambda)r\, e^{i\varphi} \tag{2.7.6}$$

for some $0 \leq \mu \leq 1$, $\lambda \geq 0$ and $\theta, \varphi \in \mathbb{R}$, and thus

$$\zeta_{n+1} = 2\gamma - 2(1+\lambda)r\, e^{i\varphi} - \frac{n\gamma}{n+1} + \frac{nr}{n+1}\, e^{i\theta} = \Big(2 - \frac{n}{n+1}\Big)\gamma + \rho\, e^{i\psi}$$

where $\rho \geq 2(1+\lambda)r - \dfrac{nr}{n+1} = \dfrac{r(n+2)}{n+1}$. That is, $\zeta_{n+1} \in \mathfrak{B}\Big(\dfrac{n+2}{n+1}\gamma, -\dfrac{n+2}{n+1}r\Big)$. Hence (2.7.4) holds for all n by induction.

With the notation (2.7.6), the ratio (2.7.2) takes the form

$$\begin{aligned}\frac{R_{n+1}}{R_n} &= \frac{1}{|\zeta_n|^2}\cdot\frac{|\gamma-\zeta_n|^2-r^2}{|\gamma+b_{n+1}-\frac{1}{\zeta_n}|^2-r^2} = \frac{|\frac{\gamma}{\zeta_n}-1|^2-\frac{r^2}{|\zeta_n|^2}}{|\gamma+b_{n+1}-\frac{1}{\zeta_n}|^2-r^2}\\ &= \frac{\big|-\frac{n\gamma^2}{n+1}+\frac{n\gamma r}{n+1}(1-\mu)e^{i\theta}-1\big|^2-r^2\big|-\frac{n\gamma}{n+1}+\frac{nr}{n+1}(1-\mu)e^{i\theta}\big|^2}{\big|-\gamma+2(1+\lambda)r\,e^{i\varphi}+\frac{n\gamma}{n+1}-\frac{nr}{n+1}(1-\mu)e^{i\theta}\big|^2-r^2}.\end{aligned} \tag{2.7.7}$$

Let

$$u_n := \gamma - \frac{n\gamma}{n+1} + \frac{nr}{n+1}(1-\mu)e^{i\theta} = \frac{\gamma}{n+1} + \frac{nr}{n+1}(1-\mu)e^{i\theta}.$$

Then

$$|u_n| \leq \frac{\gamma}{n+1} + \frac{nr}{n+1} < \frac{r+nr}{n+1} = r,$$

so the denominator of (2.7.7) is bounded below by

$$(2r-|u_n|)^2 - r^2 = (3r-|u_n|)(r-|u_n|) > 0.$$

The numerator of (2.7.7) can be written

$$\begin{aligned}&\Big|\gamma\Big(-\frac{n\gamma}{n+1}+u_n-\frac{\gamma}{n+1}\Big)-1\Big|^2 - \Big|-\frac{n\gamma}{n+1}+u_n-\frac{\gamma}{n+1}\Big|^2 r^2\\ &= |1+\gamma^2-\gamma u_n|^2 - |u_n-\gamma|^2r^2 = |r^2-\gamma\,u_n|^2 - |u_n-\gamma|^2r^2\\ &= r^4+\gamma^2|u_n|^2 - 2r^2\gamma\,\mathrm{Re}(u_n) - (\gamma^2+|u_n|^2-2\gamma\,\mathrm{Re}(u_n))r^2\\ &= r^4+\gamma^2|u_n|^2 - r^2\gamma^2 - r^2|u_n|^2 = (r^2-|u_n|^2)(r^2-\gamma^2)\end{aligned}$$

where $r^2-\gamma^2=1$. Therefore

$$\frac{R_{n+1}}{R_n} \leq \frac{r^2-|u_n|^2}{(3r-|u_n|)(r-|u_n)} = \frac{r+|u_n|}{3r-|u_n|} \leq \frac{r+\max|u_n|}{3r-\max|u_n|}$$

where $|u_n| \leq \gamma/(n+1)+nr/(n+1)$. That is,

$$\frac{R_{n+1}}{R_n} \leq \frac{(n+1)r+\gamma+nr}{3(n+1)r-\gamma-nr} = \frac{(2n+1)r+\gamma}{(2n+3)r-\gamma} = \frac{2n+2-\sigma}{2n+2+\sigma}.$$

Therefore

$$R_{n+1} = R_1\prod_{k=1}^{n}\frac{R_{k+1}}{R_k} \leq r\prod_{k=1}^{n}\frac{2k+2-\sigma}{2k+2+\sigma} = r\prod_{k=1}^{n}\frac{(1+\sigma/2)+k-\sigma}{(1+\sigma/2)+k}$$

which proves the first inequality in (2.7.1). The second one follows from Lemma 3.11 on page 115. □

3.2.8 The parabola theorems

The parabola theorems are convergence theorems for continued fractions $\mathbf{K}(a_n/1)$ with half planes as value sets. The version with the simple value set

$$V_\alpha := -\tfrac{1}{2} + e^{i\alpha}\overline{\mathbb{H}} = \{w \in \mathbb{C};\ \mathrm{Re}(we^{-i\alpha}) \geq -\tfrac{1}{2}\cos\alpha\} \cup \{\infty\} \tag{2.8.1}$$

for some fixed $\alpha \in \mathbb{R}$ with $|\alpha| < \frac{\pi}{2}$, has got the prominent name *the Parabola Theorem.* The convergence theorem for S-fractions on page 124 implies that $\mathbf{K}(a_n/1)$ from the ray $\arg a_n = 2\alpha$ converges if and only if its Stern-Stolz Series

$$\mathcal{S} := \sum_{n=1}^{\infty} \left| \frac{a_1 a_3 \cdots a_{2n-1}}{a_2 a_4 \cdots a_{2n}} \right| + \sum_{n=1}^{\infty} \left| \frac{a_2 a_4 \cdots a_{2n}}{a_1 a_3 \cdots a_{2n+1}} \right| \tag{2.8.2}$$

diverges to ∞. The Parabola Theorem extends this to continued fractions $\mathbf{K}(a_n/1)$ from a closed parabolic neighborhood

$$E_\alpha := \{a \in \mathbb{C};\ |a| - \mathrm{Re}(ae^{-i2\alpha}) \leq \tfrac{1}{2}\cos^2\alpha\} \tag{2.8.3}$$

of this ray. More precisely, with the notation above we have:

Theorem 3.43. (The Parabola Theorem.) *For fixed $\alpha \in \mathbb{R}$ with $|\alpha| < \frac{\pi}{2}$, the set E_α is the element set for continued fractions $\mathbf{K}(a_n/1)$ corresponding to the value set V_α.*
Let $\mathbf{K}(a_n/1)$ be a continued fraction from E_α. If $\mathcal{S} = \infty$, then $\mathbf{K}(a_n/1)$ converges to a finite value. If $\mathcal{S} < \infty$, then $\{f_{2n}\}$ and $\{f_{2n+1}\}$ converge absolutely to distinct finite values, and $\{S_{2n}\}$ and $\{S_{2n+1}\}$ converge generally to these values.

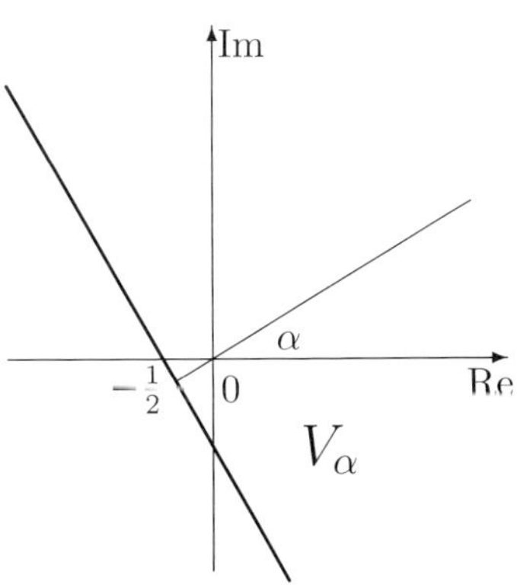

Both V_α° and E_α° contain the origin. The boundary of the half plane V_α is a line passing through $-\frac{1}{2}$. The boundary of E_α is a parabola with axis along the ray $\arg z = 2\alpha$, focus at the origin and vertex at the point $-\frac{1}{4}e^{i2\alpha}\cos^2\alpha$. It intersects the real axis at $z = -\frac{1}{4}$. The half plane V_α is illustrated to the left and the parabolic region E_α on the next page. Since 0 is an interior point in V_α, it follows that the approximants $S_n(0) = A_n/B_n$ of a continued fraction $\mathbf{K}(a_n/1)$ from E_α are interior points in V_α. In particular $S_n(0) \neq \infty$, and thus $B_n \neq 0$ and $\zeta_n \neq \infty$ for $n \geq 1$.

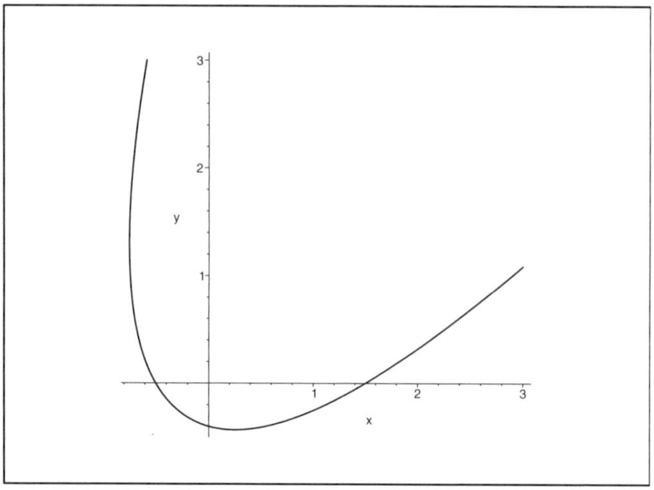

For $\alpha := 0$, the set E_0 contains the Worpitzky disk $E := \{a \in \mathbb{C};\, |a| \leq \frac{1}{4}\}$. The Parabola Theorem therefore generalizes the classical Worpitzky Theorem considerably. It also generalizes the Seidel-Stern Theorem on page 117 which says that a positive continued fraction of the form $\mathbf{K}(a_n/1)$ converges if and only if $\mathcal{S} = \infty$. The Parabola Theorem implies that this holds for real continued fractions $\mathbf{K}(a_n/1)$ with $a_n \geq -\frac{1}{4}$.

Proof of the Parabola Theorem: We shall first prove that $s(V_\alpha) := a/(1 + V_\alpha) \subseteq V_\alpha$ if and only if $a \in E_\alpha$. The inclusion is clear for $a = 0$, so let $a \neq 0$. Then s maps V_α onto the closed circular disk with center at $\gamma_a := (ae^{-i\alpha})/\cos\alpha$ and radius $\rho_a := |a|/\cos\alpha$ (Theorem 3.6 on page 110). This disk is contained in V_α if and only if $\gamma_a \in V_\alpha$ and γ_a has a distance δ_a to ∂V_α such that $\delta_a \geq \rho_a$. Now

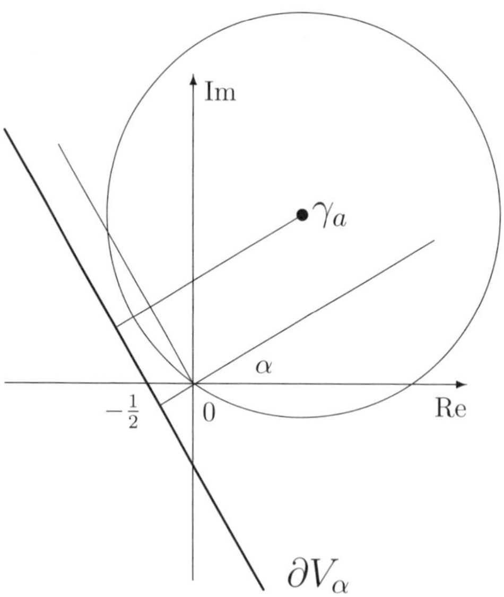

$$\delta_a = \frac{1}{2}\cos\alpha + \operatorname{Re}(\gamma_a e^{-i\alpha}),$$

where $\delta_a > 0$ if and only if $\gamma_a \in V_\alpha$. (See the figure.)

So $s(V_\alpha) \subseteq V_\alpha$ if and only if

$$\frac{1}{2}\cos\alpha + \operatorname{Re}\left(\frac{ae^{-i\alpha}}{\cos\alpha}e^{-i\alpha}\right) \geq \frac{|a|}{\cos\alpha},$$

i.e. if and only if $a \in E_\alpha$.

Let $\mathbf{K}(a_n/1)$ be a continued fraction from E_α. Since $\infty \notin S_1(V_\alpha)$, the nested sets $K_n := S_n(V_\alpha)$ are bounded disks. If $\operatorname{diam}(K_n) \to 0$, the limit point case, then the convergence is clear, and it follows from the Stern-Stolz Theorem that $\mathcal{S} = \infty$. Assume that $K_n \to K$ where $\operatorname{diam}(K) > 0$. The linear fractional transformation

$$\varphi(w) := \frac{-1 + e^{i\alpha}\cos\alpha - w}{1 + w} \tag{2.8.4}$$

maps the closed unit disk $\overline{\mathbb{D}}$ onto V_α with $\varphi(\infty) = -1$ and $\varphi(-1) = \infty$. Let

$$\tau_n := \varphi^{-1} \circ s_{2n-1} \circ s_{2n} \circ \varphi \quad \text{for } n = 1, 2, 3, \dots . \tag{2.8.5}$$

Then, $\tau_n(\overline{\mathbb{D}}) = \varphi^{-1} \circ s_{2n-1} \circ s_{2n}(V_\alpha) \subseteq \varphi^{-1} \circ s_{2n-1}(V_\alpha) \subseteq \varphi^{-1}(V_\alpha) = \overline{\mathbb{D}}$, and $\tau_n(\infty) = \varphi^{-1} \circ s_{2n-1} \circ s_{2n}(-1) = \varphi^{-1} \circ s_{2n-1}(\infty) = \varphi^{-1}(0) = -1 + e^{i\alpha}\cos\alpha =: k$ where $|k\,e^{-i\alpha}| = |i \sin\alpha| < 1$. It follows therefore by Lemma 3.8 on page 113 with $w_n := \infty$ that the sequence $\{\mathcal{T}_n\}$ given by $\mathcal{T}_n := \tau_1 \circ \tau_2 \circ \cdots \circ \tau_n$ converges generally to some value $\gamma \in \overline{\mathbb{D}}$ and that $\sum |\mathcal{T}_n(\infty) - \mathcal{T}_{n-1}(\infty)| < \infty$. Since $\mathcal{T}_n = \varphi^{-1} \circ S_{2n} \circ \varphi$, this means that $\mathcal{T}_n(\infty) = \varphi^{-1}(S_{2n}(-1)) = \varphi^{-1}(S_{2n-2}(0)) = \varphi^{-1}(f_{2n-2})$ where φ^{-1} is a fixed linear fractional transformation with pole at -1. Since all $f_n \in s_1(V_\alpha)$, and $\{f_n\}$ thus is bounded away from ∞ and -1, it follows that $\sum |f_{2n} - f_{2n-2}| < \infty$. Similarly, also $\sum |f_{2n+1} - f_{2n-1}| < \infty$. (Just use $\tau_n := \varphi^{-1} \circ s_{2n} \circ s_{2n+1} \circ \varphi$.) It is therefore a consequence of the Lane-Wall Characterization on page 103 that $\mathbf{K}(a_n/1)$ converges if and only if $\mathcal{S} = \infty$. That $\{S_{2n}\}$ and $\{S_{2n+1}\}$ converge generally, also when $\mathbf{K}(a_n/1)$ diverges, follows since $S_{n+2}(-1) = S_n(0)$. □

The Parabola Theorem is in many ways the queen among the convergence theorems for continued fractions $\mathbf{K}(a_n/1)$. It is best in several respects. For instance, one can not enlarge the set E_α, not even by adding just one point, without destroying the property that every continued fraction $\mathbf{K}(a_n/1)$ from E_α with divergent Stern-Stolz Series converges, ([Lore92]).

The Parabola Theorem has an extension of type similar to the extension of Worpitzky's Theorem on page 136. It is due to Thron, ([Thron58]). This time we have a sequence $\{V_{\alpha,n}\}_{n=0}^\infty$ of half planes given by

$$V_{\alpha,n} := -g_n + e^{i\alpha}\overline{\mathbb{H}} = \{w \in \mathbb{C};\ \operatorname{Re}(we^{-i\alpha}) \geq -g_n \cos\alpha\} \cup \{\infty\} \tag{2.8.6}$$

where

$$-\tfrac{\pi}{2} < \alpha < \tfrac{\pi}{2}, \quad g_0 > 0 \text{ and } 0 < \varepsilon \leq g_n \leq 1 - \varepsilon \text{ for } n \geq 1 \tag{2.8.7}$$

are given constants.

Theorem 3.44. (The Parabola Sequence Theorem.) *For given constants (2.8.7), the sequence $\{E_{\alpha,n}\}_{n=1}^{\infty}$ given by*

$$E_{\alpha,n} := \{a \in \mathbb{C};\ |a| - \mathrm{Re}(ae^{-i2\alpha}) \leq 2g_{n-1}(1-g_n)\cos^2\alpha\} \qquad (2.8.8)$$

is the sequence of element sets corresponding to the value sets $\{V_{\alpha,n}\}$ given by (2.8.6).
Let $\mathbf{K}(a_n/1)$ be a continued fraction from $\{E_{\alpha,n}\}$, and let $\mathcal{S}$ be the sum of its Stern-Stolz Series (2.8.2). If $\mathcal{S} = \infty$, then $\mathbf{K}(a_n/1)$ converges to a finite value. If $\mathcal{S} < \infty$, then $\{f_{2n}\}$ and $\{f_{2n+1}\}$ converge absolutely to distinct finite values, and $\{S_{2n}\}$ and $\{S_{2n+1}\}$ converge generally to these values.

Proof : The proof is similar to the proof of the Parabola Theorem. This time the disk $s_n(V_{\alpha,n})$ has center at $\gamma_{\alpha,n} := (a_n\, e^{-i\alpha})/(2(1-g_n)\cos\alpha)$ and radius $\rho_{\alpha,n} := |a_n|/(2(1-g_n)\cos\alpha)$. Hence $s_n(V_{\alpha,n}) \subseteq V_{\alpha,n-1}$ if and only if $\gamma_{\alpha,n} \in V_{\alpha,n-1}$ and $\gamma_{\alpha,n}$ has a distance $\delta_{\alpha,n} \geq \rho_{\alpha,n}$ to $\partial V_{\alpha,n-1}$; i.e.,

$$\delta_{\alpha,n} = g_{n-1}\cos\alpha + \mathrm{Re}\left(\frac{a_n e^{-i\alpha}}{2(1-g_n)\cos\alpha}\, e^{-i\alpha}\right) \geq \frac{|a_n|}{2(1-g_n)\cos\alpha};$$

i.e., if and only if $a_n \in E_{\alpha,n}$. If the limit point case occurs for $S_n(V_{\alpha,n})$, then $\mathbf{K}(a_n/1)$ converges and $\mathcal{S} = \infty$.

Assume that the limit circle case occurs. The linear fractional transformation

$$\varphi_n(w) := \frac{-1 + 2(1-g_n)e^{i\alpha}\cos\alpha - w}{1+w}$$

maps $\overline{\mathbb{D}}$ onto $V_{\alpha,n}$ with $\varphi_n(\infty) = -1$ and $\varphi_n(-1) = \infty$. Hence $\tau_{n+1} := \varphi_{2n}^{-1} \circ s_{2n+1} \circ s_{2n+2} \circ \varphi_{2n+2}$ maps $\mathbb{D}$ into $\mathbb{D}$ with $\tau_{n+1}(\infty) = \varphi_{2n}^{-1}(0) = -1 + 2(1-g_{2n})e^{i\alpha}\cos\alpha =: k_n$.

Now, $|k_n|$ is bounded away from 1 since $k_n e^{-i\alpha} = (1-2g_{2n})\cos\alpha + i\sin\alpha$ where $|1-2g_{2n}| \leq 1-2\varepsilon$, and thus $|k_n e^{-i\alpha}|^2 \leq (1-2\varepsilon)^2\cos^2\alpha + \sin^2\alpha = 1-4\varepsilon(1-\varepsilon)\cos^2\alpha < 1$. Therefore we also now get $\sum |\mathcal{T}_n(\infty) - \mathcal{T}_{n-1}(\infty)| < \infty$. Since $\mathcal{T}_n = \varphi_0^{-1} \circ S_{2n} \circ \varphi_{2n}$ where φ_0^{-1} is a fixed linear fractional transformation with pole at -1 and $\{\varphi_{2n}\}$ is totally non-restrained, we still have $\sum |f_{2n} - f_{2n-2}| < \infty$ in the limit circle case. Similarly, also $\sum |f_{2n+1} - f_{2n-1}| < \infty$. Finally, $S_{n+2}(-1) = S_n(0)$ as before. □

The diameter of $\mathbf{S_n}(\mathbf{V}_{\alpha,\mathbf{n}})$.

As in the Worpitzky situation, the quantities

$$\widehat{\Sigma}_n := \sum_{m=0}^{n} \widehat{P}_m \quad \text{where} \quad \widehat{P}_m := \prod_{k=1}^{m} \frac{1-g_k}{g_k} \qquad (2.8.9)$$

are important. The following truncation error bound is due to Thron ([Thron58]):

Theorem 3.45. *Let* α, $\{g_n\}$ *and* $\mathbf{K}(a_n/1)$ *be as in the Parabola Sequence Theorem, and let* $\widehat{P}_m$ *and* $\widehat{\Sigma}_n$ *be given by (2.8.9). Then*

$$\operatorname{diam} S_n(V_{\alpha,n}) \leq \frac{|a_1|/((1-g_1)\cos\alpha)}{\prod_{j=2}^{n}\left(1+\frac{\widehat{P}_{j-1}g_{j-1}(1-g_j)\cos^2\alpha}{|a_j|\widehat{\Sigma}_{j-2}}\right)}. \tag{2.8.10}$$

Proof : Since $a_1/(1+V_{\alpha,1})$ is bounded, we know that $S_n(V_{\alpha,n})$ is a sequence of nested circular disks. Therefore $f_{n-1} = A_{n-1}/B_{n-1} \neq \infty$ which means that $B_n \neq 0$ and $\zeta_n = -B_n/B_{n-1} \neq \infty$ for $n \geq 1$. We can therefore write

$$S_n(w) = f_{n-1} + \frac{A_nB_{n-1} - B_nA_{n-1}}{B_{n-1}^2(w-\zeta_n)},$$

and thus the radius R_n of $S_n(V_{\alpha,n})$ is given by

$$R_n = \frac{1}{2}\sup_{w\in\partial V_{\alpha,n}}|S_n(w) - S_n(\infty)| = \frac{1}{2}\sup_{w\in\partial V_{\alpha,n}}\left|\frac{A_nB_{n-1} - B_nA_{n-1}}{B_{n-1}^2(w-\zeta_n)}\right|$$

where $\zeta_n \notin V_{\alpha,n}$ since $\zeta_1 = -1 \notin V_{\alpha,1}$. Now, points on $\partial V_{\alpha,n}$ can be written $g_ne^{i\alpha}(-1+i\,v_n)\cos\alpha$ with $v_n \in \mathbb{R}$. Therefore ζ_n can be written

$$\zeta_n = g_ne^{i\alpha}(-1-u_n+iv_n)\cos\alpha \text{ where } u_n > 0,\ \ v_n \in \mathbb{R}, \tag{2.8.11}$$

and so

$$R_n = \frac{1}{2}\left|\frac{A_nB_{n-1} - B_nA_{n-1}}{B_{n-1}^2g_nu_n\cos\alpha}\right|.$$

By means of the determinant formula on page 7 we therefore get that

$$\frac{R_n}{R_{n-1}} = \frac{|a_n|}{|\zeta_{n-1}|^2}\,\frac{g_{n-1}u_{n-1}}{g_nu_n}. \tag{2.8.12}$$

Since $a_n \in E_{\alpha,n}$, we may also write

$$a_n = g_{n-1}(1-g_n)e^{2i\alpha}(x_n+iy_n)\cos^2\alpha \text{ where } y_n^2 \leq 4x_n + 4.$$

Then $g_n(1+u_n)\cos\alpha = -\operatorname{Re}(\zeta_ne^{-i\alpha})$ where

$$\begin{aligned}-\operatorname{Re}(\zeta_ne^{-i\alpha}) &= -\operatorname{Re}\left(\left(-1+\frac{a_n}{\zeta_{n-1}}\right)e^{-i\alpha}\right)\\ &= -\operatorname{Re}\left(-e^{-i\alpha}+\frac{a_ne^{-2i\alpha}}{\zeta_{n-1}e^{-i\alpha}}\right)\\ &= \cos\alpha - \operatorname{Re}\left(\frac{g_{n-1}(1-g_n)(x_n+iy_n)\cos^2\alpha}{g_{n-1}(-1-u_{n-1}+i\,v_{n-1})\cos\alpha}\right).\end{aligned}$$

This means that

$$\begin{aligned} g_n u_n &= g_n(1+u_n) - g_n = (1-g_n)\Big(1 + \operatorname{Re}\frac{x_n + iy_n}{1+u_{n-1} - iv_{n-1}}\Big) \\ &= (1-g_n)\Big(1 + \frac{x_n(1+u_{n-1}) - y_n v_{n-1}}{(1+u_{n-1})^2 + v_{n-1}^2}\Big). \end{aligned} \tag{2.8.13}$$

Therefore

$$\begin{aligned} \frac{R_n}{R_{n-1}} &= \frac{g_{n-1}(1-g_n)\sqrt{x_n^2+y_n^2}}{g_{n-1}^2\{(1+u_{n-1})^2+v_{n-1}^2\}} \cdot \frac{g_{n-1}u_{n-1}/(1-g_n)}{1+\dfrac{x_n(1+u_{n-1})-y_n v_{n-1}}{(1+u_{n-1})^2+v_{n-1}^2}} \\ &= \frac{\sqrt{x_n^2+y_n^2}\,u_{n-1}}{(1+u_{n-1})^2+v_{n-1}^2+x_n(1+u_{n-1})-y_n v_{n-1}}. \end{aligned} \tag{2.8.14}$$

Since $y_n^2 \le 4x_n + 4$, the last denominator can be written

$$\begin{aligned} &(1+u_{n-1})^2 + (v_{n-1} - y_n/2)^2 - y_n^2/4 + x_n(1+u_{n-1}) \\ &\ge (1+u_{n-1})^2 - \tfrac{1}{4}(4x_n+4) + x_n(1+u_{n-1}) = (2+u_{n-1}+x_n)u_{n-1}. \end{aligned}$$

Therefore, since $x_n^2 + y_n^2 \le x_n^2 + 4x_n + 4 = (x_n+2)^2$,

$$\frac{R_n}{R_{n-1}} \le \frac{\sqrt{x_n^2+y_n^2}}{x_n+2+u_{n-1}} \le \frac{\sqrt{x_n^2+y_n^2}}{\sqrt{x_n^2+y_n^2}+u_{n-1}} = \frac{1}{1+\frac{u_{n-1}}{\sqrt{x_n^2+y_n^2}}}. \tag{2.8.15}$$

Equality (2.8.13) also gives a lower bound for u_n: we first observe that $|y_n| \le 2\sqrt{x_n+1}$ implies that

$$\frac{x_n(1+u_{n-1}) - y_n v_{n-1}}{(1+u_{n-1})^2+v_{n-1}^2} \ge \frac{x_n(1+u_{n-1}) - 2\sqrt{x_n+1}\cdot|v_{n-1}|}{(1+u_{n-1})^2+v_{n-1}^2},$$

where a standard minimalization procedure shows that the right hand side attains its minimum for $v_{n-1} := (1+u_{n-1})\sqrt{x_n+1}$; that is,

$$\begin{aligned} &\frac{x_n(1+u_{n-1}) - 2\sqrt{x_n+1}\cdot|v_{n-1}|}{(1+u_{n-1})^2+v_{n-1}^2} \\ &\qquad \ge \frac{x_n(1+u_{n-1}) - 2(1+u_{n-1})(x_n+1)}{(1+u_{n-1})^2+(1+u_{n-1})^2(x_n+1)} = \frac{-1}{1+u_{n-1}}. \end{aligned}$$

With this inequality we get from (2.8.13) that

$$g_n u_n \ge (1-g_n)\left(1 - \frac{1}{1+u_{n-1}}\right) = \frac{(1-g_n)u_{n-1}}{1+u_{n-1}},$$

i.e.,

$$\frac{1}{u_n} \le \frac{g_n}{1-g_n}\left(1 + \frac{1}{u_{n-1}}\right). \tag{2.8.16}$$

Now, $\zeta_1 = -1$, and thus, by (2.8.11), $\dfrac{1}{u_1} = \dfrac{g_1}{1-g_1}$. Hence it follows by induction, using (2.8.16), that

$$\frac{1}{u_n} \leq \frac{g_n}{1-g_n} + \frac{g_n}{1-g_n}\,\frac{g_{n-1}}{1-g_{n-1}} + \cdots + \prod_{k=1}^{n} \frac{g_k}{1-g_k}$$
$$= (\widehat{P}_{n-1} + \widehat{P}_{n-2} + \cdots + 1)/\widehat{P}_n = \widehat{\Sigma}_{n-1}/\widehat{P}_n\,.$$

Therefore the bound (2.8.15) is again bounded by

$$\frac{R_n}{R_{n-1}} \leq \frac{1}{1+\dfrac{\widehat{P}_{n-1}/\widehat{\Sigma}_{n-2}}{\sqrt{x_n^2+y_n^2}}} = \frac{1}{1+\dfrac{\widehat{P}_{n-1} g_{n-1}(1-g_n)\cos^2\alpha}{|a_n|\widehat{\Sigma}_{n-2}}}\,.$$

Now, $\operatorname{diam} S_n(V_{\alpha,n}) = 2R_n = 2R_1 \prod_{k=2}^{n} R_k/R_{k-1}$ where R_1 is the radius of $a_1/(1+V_{\alpha,1})$; i.e., $2R_1 = |a_1|/\delta_1$ where δ_1 is the euclidean distance between -1 and $\partial V_{\alpha,1}$; i.e., $\delta_1 = (1-g_1)\cos\alpha$. This proves (2.8.10). □

The choice $g_n := \frac{1}{2}$ for all n in the Parabola Sequence Theorem gives back the Parabola Theorem. The truncation error bound (2.8.10) then takes the form:

Corollary 3.46. *Let α and $\mathbf{K}(a_n/1)$ be as in the Parabola Theorem. Then*

$$\operatorname{diam} S_n(V_\alpha) \leq \frac{2|a_1|/\cos\alpha}{\displaystyle\prod_{j=2}^{n}\left(1+\frac{\cos^2\alpha}{4(j-1)|a_j|}\right)}\,. \qquad (2.8.17)$$

Proof : For all $g_n := \frac{1}{2}$ we get $\widehat{P}_k = 1$ and $\widehat{\Sigma}_n = n+1$. □

Remarks.

1. The bound (2.8.17) depends on $\{a_n\}$. If also $|a_n| \leq M$ for all n, then this bound is again bounded by

$$\frac{2M/\cos\alpha}{\prod_{j=1}^{n-1}(1+(\cos^2\alpha)/(4Mj))} \to 0 \text{ as } n \to \infty. \qquad (2.8.18)$$

   That is, every continued fraction $\mathbf{K}(a_n/1)$ from $E_\alpha \cap \mathfrak{B}(0,M)$ converges, and the convergence is uniform with respect to $\mathbf{K}(a_n/1)$ from $E_\alpha \cap \mathfrak{B}(0,M)$. Moreover, the limit point case for $S_n(V_\alpha)$ occurs for all these continued fractions.

2. Also the truncation error bound (2.8.10) depends on $\mathbf{K}(a_n/1)$. If we add the condition that $|a_n| \leq M_n$ for all n, then

$$\operatorname{diam} S_n(V_{\alpha,n}) \leq \frac{M_1/((1-g_1)\cos\alpha)}{\prod_{j=1}^{n-1}\left(1 + \frac{\widehat{P}_j g_j(1-g_{j+1})\cos^2\alpha}{\widehat{\Sigma}_{j-1}M_{j+1}}\right)}. \tag{2.8.19}$$

If $\sum \widehat{P}_j/(\widehat{\Sigma}_{j-1}M_{j+1}) = \infty$, then every continued fraction $\mathbf{K}(a_n/1)$ from $\{E_{\alpha,n} \cap \mathfrak{B}(0, M_n)\}$ converges to some finite value $f \in V_{\alpha,0}$, uniformly with respect to $\mathbf{K}(a_n/1)$ from $\{E_{\alpha,n} \cap \mathfrak{B}(0, M_n)\}$, and $S_n(V_{\alpha,n})$ approaches a limit point.

3. Let $\mathbf{K}(a_n/1)$ be from $\{E_{\alpha,n} \cap \mathfrak{B}(0, M)\}$ for some $M > 0$. Then the bound (2.8.19) vanishes as $n \to \infty$ if and only if $\sum \widehat{P}_j/\widehat{\Sigma}_{j-1} = \infty$, which happens if and only if $\widehat{\Sigma}_\infty = \infty$ (standard property of positive series).

4. If $\widehat{\Sigma}_n \to \widehat{\Sigma}_\infty < \infty$, then the limit point case fails to occur for the continued fraction $\mathbf{K}(\widehat{a}_n/1)$ with $\widehat{a}_n := -g_{n-1}(1-g_n)$ from $\{E_{\alpha,n}\}$, since the limit set must contain both $-g_0$ and $f \neq -g_0$. As in Theorem 3.31 on page 137, we can replace $\{g_n\}$ by some smaller constants $\{g_n^*\}$ to get smaller value sets

$$V_{\alpha,n}^* := -g_n^* + e^{i\alpha}\overline{\mathbb{H}}$$

for $\{E_{\alpha,n}\}$ where $\widehat{\Sigma}_\infty^* := \sum_{n=0}^\infty \prod_{k=1}^n (1+g_k^*)/g_k^* = \infty$. If $0 < \varepsilon \leq g_n \leq 1-\varepsilon$ for all n, then $0 < \varepsilon^2 \leq g_n^* \leq 1-\varepsilon < 1-\varepsilon^2$ for all n.

Proof of Theorem 3.24 on page 128: For fixed $z := r\,e^{2i\alpha}$ with $|\alpha| < \pi/2$, the S-fraction $\mathbf{K}(a_n z/1)$ is a continued fraction from E_α given by (2.8.3), and the half plane V_α in (2.8.1) is a simple value set for $\mathbf{K}(a_n z/1)$. By (2.8.14) (with all $g_n = \frac{1}{2}$) it follows therefore that the radius R_n of $S_n(V_\alpha)$ satisfies

$$\frac{R_n}{R_{n-1}} = \frac{\sqrt{x_n^2 + y_n^2}\,u_{n-1}}{(1+u_{n-1})^2 + v_{n-1}^2 + x_n(1+u_{n-1}) - y_n v_{n-1}}.$$

Now, with this notation, $a_n z = a_n r\,e^{2i\alpha} = \frac{1}{4}\,e^{2i\alpha} x_n \cos^2\alpha$ and $\zeta_n = \frac{1}{2}\,e^{i\alpha}(-1 - u_n + i\,v_n)$ with $u_n > 0$ by (2.8.11). That is, $x_n = 4a_n r/\cos^2\alpha > 0$, $y_n = 0$ and

$$\frac{R_n}{R_{n-1}} \leq \frac{x_n u_{n-1}}{(1+u_{n-1})^2 + x_n(1+u_{n-1})}$$

which obtains its maximum with respect to u_{n-1} for $u_{n-1} := \sqrt{1+x_n}$. That is,

$$\begin{aligned}
\frac{R_n}{R_{n-1}} &\leq \frac{x_n\sqrt{1+x_n}}{(1+\sqrt{1+x_n})^2 + x_n(1+\sqrt{1+x_n})} \\
&= \frac{x_n\sqrt{1+x_n}}{2(1+x_n) + (2+x_n)\sqrt{1+x_n}} \\
&= \frac{x_n}{2\sqrt{1+x_n} + 2 + x_n} = \frac{x_n}{(1+\sqrt{1+x_n})^2} = \frac{1}{\left(\frac{1}{\sqrt{x_n}} + \sqrt{1+\frac{1}{x_n}}\right)^2}
\end{aligned}$$

where $x_n = 4a_n r/\cos^2\alpha$. Since $R_1 = a_1 r/(\frac{1}{2}\cos\alpha)$, this proves that

$$\operatorname{diam} S_n(V_\alpha) = 2R_n = 2R_1 \prod_{k=2}^{n} \frac{R_k}{R_{k-1}} \le \frac{4a_1 r}{\cos\alpha} \prod_{k=2}^{n} \left(\frac{1}{\sqrt{x_k}} + \sqrt{1 + \frac{1}{x_k}} \right)^{-2}$$

which actually was the way Thron presented his error bound. To get our formulation (which agrees with the formulation by Gragg and Warner), we just observe that

$$\frac{\sqrt{1+x_n}-1}{\sqrt{1+x_n}+1} = \frac{x_n}{(\sqrt{1+x_n}+1)^2} = \frac{1}{\left(\frac{1}{\sqrt{x_n}} + \sqrt{1+\frac{1}{x_n}}\right)^2}.$$

□

Limit sets.

Also for the Parabola Theorem it turns out that every point in $V_\alpha \setminus \{0, \infty\}$ is the value of a continued fraction $\mathbf{K}(a_n/1)$ from E_α:

Theorem 3.47. *For fixed $\alpha \in \mathbb{R}$ with $|\alpha| < \frac{\pi}{2}$ let E_α and V_α be given by (2.8.3) and (2.8.1). Then $V_\alpha \setminus \{0, \infty\}$ is the limit set for the family of continued fractions $\mathbf{K}(a_n/1)$ from E_α.*

Proof : Every convergent continued fraction $\mathbf{K}(a_n/1)$ from E_α has its value in $V_\alpha \setminus \{0, \infty\}$ (the Parabola Theorem). Let $w_0 \in V_\alpha \setminus \{0, \infty\}$ be arbitrarily chosen, and let $\widehat{w} \in \partial V_\alpha$ and $\mu \ge 0$ be the (uniquely determined) numbers which make $w_0 = \widehat{w} + \mu e^{i\alpha}$. Let further

$$w_1 := -1 - \widehat{w} + \mu\, e^{i\alpha}, \quad a_1 := -\widehat{w}^2 + \mu^2 e^{2i\alpha}, \quad a_2 := -(1+\widehat{w})^2 + \mu^2 e^{2i\alpha}.$$

Then also $w_1 \in V_\alpha \setminus \{0, \infty\}$. Moreover, $\widehat{w}^2 \subset \partial E_\alpha$ since $\widehat{w}$ can be written $\widehat{w} = -\frac{1}{2} + it\, e^{i\alpha}$ (see Problem 19 on page 169). Similarly, $-(1+\widehat{w})^2 \in \partial E_\alpha$. Therefore $a_1, a_2 \in E_\alpha$ and the 2-periodic continued fraction

$$\operatorname*{K}_{n=1}^{\infty} \frac{a_n}{1} := \frac{a_1}{1} + \frac{a_2}{1} + \frac{a_1}{1} + \frac{a_2}{1} + \frac{a_1}{1} + \cdots$$

is a continued fraction from E_α. Therefore $\mathbf{K}(a_n/1)$ converges to some $f \in V_\alpha$. Indeed, by Remark 1 on page 157 $S_n(w_n) \to f$ whenever all $w_n \in V_\alpha$. Now,

$$\frac{a_1}{1+w_1} = \frac{-\widehat{w}^2 + \mu^2 e^{2i\alpha}}{1 - 1 - \widehat{w} + \mu\, e^{i\alpha}} = \frac{\widehat{w}^2 - \mu^2 e^{2i\alpha}}{\widehat{w} - \mu\, e^{i\alpha}} = \widehat{w} + \mu\, e^{i\alpha} = w_0,$$

$$\frac{a_2}{1+w_0} = \frac{-(1+\widehat{w})^2 + \mu^2 e^{2i\alpha}}{1 + \widehat{w} + \mu\, e^{i\alpha}} = -(1+\widehat{w}) + \mu\, e^{i\alpha} = w_1.$$

That is, $S_{2n}(w_0) = S_{2n+1}(w_1) = w_0$ for all n, and therefore $\mathbf{K}(a_n/1)$ converges to w_0. □

3.3 Additional convergence theorems

3.3.1 Simple bounded circular value sets

The Śleszyński - Pringsheim Theorem required that $V = \overline{\mathbb{D}}$. This easily generalizes to cases where V is a disk $\mathfrak{B}(\Gamma, \rho)$ with $0 \in V^\circ$; i.e., $\rho > |\Gamma|$.

Theorem 3.48. *For given $\Gamma \in \mathbb{C}$ and $\rho > 0$ with $\rho > |\Gamma|$, let*

$$\Omega := \{(a,b) \in \mathbb{C}^2;\ |a(\overline{b}+\overline{\Gamma}) - \Gamma d| + \rho|a| \leq \rho d \text{ for } d := |b+\Gamma|^2 - \rho^2\}. \quad (3.1.1)$$

Then every continued fraction $\mathbf{K}(a_n/b_n)$ from Ω converges to some value $f \in V := \mathfrak{B}(\Gamma, \rho)$.

Proof : Let $a, b \in \mathbb{C}$ with $a \neq 0$. Then $s(V) := a/(b+V) \subseteq V$ only if $0 \notin (b+V)$; i.e., $|b+\Gamma| > \rho$. So assume that $d > 0$. Then $s(V) = \mathfrak{B}(\Gamma_1, \rho_1)$ given by

$$\Gamma_1 := \frac{a}{d}(\overline{b}+\overline{\Gamma}) \quad \text{and} \quad \rho_1 := \frac{|a|}{d}\rho \quad \text{(Lemma 3.6 on page 110.)} \qquad (3.1.2)$$

Hence $s(V) \subseteq V$ if and only if $|\Gamma - \Gamma_1| + \rho_1 \leq \rho$; i.e., if and only if $(a,b) \in \Omega$. In other words, Ω is the element set corresponding to the value set V. This was first proved by Lane ([Lane45]). Since moreover $s(\infty) = 0 \in V^\circ$ for all $(a,b) \in \Omega$ with $a \neq 0$, the convergence follows from Corollary 3.9 on page 114. (See the remark on page 113.) □

If we set $\Gamma := 0$ and $\rho := 1$, we get back the Śleszyński-Pringsheim criterion on page 135.

If $0 \notin V^\circ$, we still get general convergence under mild conditions:

Theorem 3.49. *For given $\Gamma \in \mathbb{C}$, and $0 < \rho \neq |\Gamma|$ and $0 < \varepsilon < 1$, let*

$$\Omega^* := \{(a,b) \in \Omega;\ |a| \leq (1-\varepsilon)d\} \qquad (3.1.3)$$

where Ω and d are given by (3.1.1). Then every continued fraction $\mathbf{K}(a_n/b_n)$ from Ω^* converges generally to some value $f \in V := \mathfrak{B}(\Gamma, \rho)$ with exceptional sequence $\{w_n^\dagger\}$ from $\widehat{\mathbb{C}} \setminus V$.

Proof : With Γ_1 and ρ_1 as in the previous proof, the extra condition $|a| \leq (1-\varepsilon)d$ makes $\rho_1 = |a|\rho/d \leq (1-\varepsilon)\rho$, and thus (1.4.11) on page 114 holds with $\varphi_n(w) := \Gamma + \rho w$ for $\mathbf{K}(a_n/b_n)$ from Ω^*.

If $0 \in V^\circ$, the convergence follows from Theorem 3.48. If $0 \notin V$, the general convergence follows from Corollary 3.9 on page 114 with $w_n := \infty$. Since V contains more than two points, the value f of $\mathbf{K}(a_n/1)$ belongs to V. In particular $f \neq \infty$, so $\{S_n^{-1}(\infty)\}$ is an exceptional sequence. Finally, $w_n^\dagger := S_n^{-1}(\infty) \in \widehat{\mathbb{C}} \setminus V$ since $S_n(w_n^\dagger) = \infty \notin V$ whereas $S_n(V) \subseteq V$. □

For continued fractions of the form $\mathbf{K}(a_n/1)$, the element set Ω simplifies to

$$E := \{a \in \mathbb{C};\ |a(1+\overline{\Gamma}) - \Gamma d| + \rho|a| \leq \rho d \text{ for } d := |1+\Gamma|^2 - \rho^2\}. \tag{3.1.4}$$

If $\Gamma = 0$, then E is the circular disk $|a| \leq \rho(1-\rho)$ where $0 < \rho < 1$, a situation we recognize from the generalized version of the Worpitzky Theorem. If $\Gamma \neq 0$, then

$$\begin{aligned} E = a^* E^* \quad &\text{where } a^* := \Gamma(1+\Gamma)\Big(1 - \frac{\rho^2}{|1+\Gamma|^2}\Big) = \frac{\Gamma}{1+\overline{\Gamma}}\, d \\ &\text{and } E^* := \Big\{\xi \in \mathbb{C};\ |\xi - 1| + \frac{\rho}{|1+\Gamma|}|\xi| \leq \frac{\rho}{|\Gamma|}\Big\}. \end{aligned} \tag{3.1.5}$$

Clearly $E^* \neq \emptyset$ if and only if $1 \in E^*$; i.e., if and only if $|\Gamma| \leq |1+\Gamma|$. And $E^\circ \neq \emptyset$ if and only if $|\Gamma| < |1+\Gamma|$. The boundary ∂E of E is then called a *cartesian oval.*

Lemma 3.50. *The cartesian oval $\mathcal{O} := \{\xi \in \mathbb{C};\ |\xi - 1| + |\xi|k = \ell\}$ with $0 < k < 1$ and $k < \ell$ is a closed convex curve, symmetric about the real axis with $\mathcal{O} \cap \mathbb{R} = \{\mu, \nu\}$ given by*

$$\mu := \frac{1-\ell}{1-k} \text{ if } \ell \leq 1,\quad \mu := \frac{1-\ell}{1+k} \text{ if } \ell \geq 1 \text{ and } \nu := \frac{\ell+1}{k+1}. \tag{3.1.6}$$

Moreover, $\mu \leq |\xi| \leq \nu$ for all $\xi \in \mathcal{O}$.

Proof : Let $t \in (0,1)$ and $\xi_1, \xi_2 \in \mathcal{O}$. Then $\xi := t\xi_1 + (1-t)\xi_2$ satisfies $\xi - 1 = t(\xi_1 - 1) + (1-t)(\xi_2 - 1)$, so

$$\begin{aligned} |\xi - 1| + |\xi|k &\leq t|\xi_1 - 1| + (1-t)|\xi_2 - 1| + t|\xi_1|k + (1-t)|\xi_2|k \\ &= t\ell + (1-t)\ell = \ell, \end{aligned}$$

and thus $\mathcal{O}$ is convex. The symmetry follows since $\xi \in \mathcal{O}$ implies that its complex conjugate $\overline{\xi}$ also is an element in $\mathcal{O}$. For $\xi \in \mathcal{O} \cap \mathbb{R}$,

$$\begin{aligned} \xi \geq 1 &\implies \xi - 1 + \xi k = \ell \implies \xi = \frac{\ell+1}{k+1}, \\ 0 < \xi < 1 &\implies 1 - \xi + \xi k = \ell \implies \xi = \frac{1-\ell}{1-k} \implies \ell < 1, \\ \xi \leq 0 &\implies 1 - \xi - \xi k = \ell \implies \xi = \frac{1-\ell}{1+k} \implies \ell \geq 1. \end{aligned}$$

That is, $\mathcal{O} \cap \mathbb{R} = \{\mu, \nu\}$. Finally, if $\xi \in \mathcal{O}$, then

$$\begin{aligned} \ell = |\xi - 1| + |\xi|k \leq |\xi| + 1 + |\xi|k &\implies |\xi| \geq (\ell-1)/(k+1), \\ \ell = |\xi - 1| + |\xi|k \geq |\xi| - 1 + |\xi|k &\implies |\xi| \leq (\ell+1)/(k+1), \\ \ell = |\xi - 1| + |\xi|k \geq 1 - |\xi| + |\xi|k &\implies |\xi| \geq (1-\ell)/(1-k) \end{aligned}$$

which proves that $\mu \leq |\xi| \leq \nu$. □

Hence E is a convex set, symmetric about the line $a^*\mathbb{R}$. For more information on the shape of E we refer to Section 5.2.4 on page 244 in Chapter 5.

Theorem 3.51. (The Oval Theorem.) *Let E be given by (3.1.4) with $|\Gamma| < |1+\Gamma|$ and $|1+\Gamma| > \rho$. Then every continued fraction $\mathbf{K}(a_n/1)$ from E converges to some value $f \in V := \mathfrak{B}(\Gamma, \rho)$ with exceptional sequence from $-1-V$, and V is a simple value set for $\mathbf{K}(a_n/1)$.*

Proof : Let $\mathbf{K}(a_n/1)$ be a continued fraction from E. Assume first that $|a| = d$. Then E consists of the single point $a = a^*$. But this is impossible since we have assumed that $|\Gamma| \neq |1+\Gamma|$. Since E is a compact set, there therefore exists an $\varepsilon > 0$ such that $|a| \leq (1-\varepsilon)d$ for all $a \in E$. Hence it follows from Theorem 3.49 that $\mathbf{K}(a_n/1)$ converges generally to some value $f \in V := \mathfrak{B}(\Gamma, \rho)$. It has an exceptional sequence $\{S_n^{-1}(-1-w)\}$ for some $w \in V$ with $-1-w \neq f$, where

$$S_n^{-1}(-1-w) = -1 - \frac{a_n}{1} \, + \, \frac{a_{n-1}}{1} \, + \cdot + \, \frac{a_1}{1+w} \in -1 - V \quad \text{for all } n.$$

Since V is bounded, this means that also $S_n(\infty) = S_{n-1}(0) \to f$. □

Truncation error bounds.

As a corollary to Theorem 5.9 which we prove on page 241, we get the following truncation error bound:

Theorem 3.52. *Let Ω be given by (3.1.1) with $\Gamma \in \mathbb{C}$ and $0 < \rho \neq |\Gamma|$. Then*

$$|f - S_n(\Gamma)| \leq \rho \frac{|\Gamma| + \rho}{|b_n + \Gamma| - \rho} \prod_{k=1}^{n-1} M_k \tag{3.1.7}$$

for every convergent continued fraction $\mathbf{K}(a_n/b_n)$ from Ω, where $M_k := \max\{|w/(b_k + w)|;\ w \in V\}$.

For continued fractions $\mathbf{K}(a_n/1)$ from E the error bound simplifies:

Theorem 3.53. *Let E be given by (3.1.4) with $|1 + \Gamma| > |\Gamma| \neq \rho$. Then*

$$|f - S_n(\Gamma)| < \rho \frac{|\Gamma| + \rho}{|1 + \Gamma| - \rho} M^{n-1} \tag{3.1.8}$$

for every continued fraction $\mathbf{K}(a_n/1)$ from E, where $M := \max\{|w/(1 + w)|;\ w \in V\}$.

Remarks.

1. Since $s(w) := w/(1 + w)$ maps $\mathfrak{B}(\Gamma, \rho)$ onto $\mathfrak{B}(\Gamma_1, \rho_1)$ given by

$$\Gamma_1 := 1 - \frac{1 + \overline{\Gamma}}{|1 + \Gamma|^2 - \rho^2} = \frac{\Gamma(1 + \overline{\Gamma}) - \rho^2}{|1 + \Gamma|^2 - \rho^2}, \qquad \rho_1 := \frac{\rho}{|1 + \Gamma|^2 - \rho^2} \tag{3.1.9}$$

(Lemma 3.6 on page 110), it follows that

$$M = |\Gamma_1| + \rho_1 = \frac{|\Gamma(1 + \overline{\Gamma}) - \rho^2| + \rho}{|1 + \Gamma|^2 - \rho^2}. \tag{3.1.10}$$

2. The convergence of $\mathbf{K}(a_n/1)$ in the Oval Theorem follows from the Parabola Theorem since Lange ([Lange95]) proved that E is a subset of the parabolic region E_α with $\alpha := \arg(\Gamma + \frac{1}{2})$. The importance of the Oval Theorem lies therefore with its better truncation error bound for this smaller family of continued fractions $\mathbf{K}(a_n/1)$. We shall return to this idea in Chapter 5.

3.3.2 Simple unbounded circular value sets

An unbounded, closed circular set V is either a half plane or the exterior of a disk. In particular $\infty \in V$, so if V is a simple value set for $\mathbf{K}(a_n/b_n)$, then $a_n/(b_n + \infty) =$

$0 \in V$. Hence we are back to the classical type of value sets which contain the classical approximants.

If we stick to continued fractions of the form $\mathbf{K}(a_n/1)$ or $\mathbf{K}(1/b_n)$, we have the following interesting observation:

Theorem 3.54.
A. Let V be a simple value set for $\mathbf{K}(a_n/1)$. Then $W := \widehat{\mathbb{C}} \setminus (-1-V)$ and $W \cap V$ are also simple value sets for $\mathbf{K}(a_n/1)$.
B. Let V be a simple value set for $\mathbf{K}(1/b_n)$. Then $W := \widehat{\mathbb{C}} \setminus (-1/V)$ and $W \cap V$ are also simple value sets for $\mathbf{K}(1/b_n)$.

Proof : A. Since $s_n(V) := a_n/(1+V) \subseteq V$, we have $V \subseteq s_n^{-1}(V) = -1 + a_n/V$; i.e., $(-1-V) \subseteq a_n/(-V) = s_n(-1-V)$, and thus $W \supseteq s_n(W)$. Clearly, $s_n(V) \subseteq V$ and $s_n(W) \subseteq W$ implies that $s_n(V \cap W) \subseteq V \cap W$.

B. Since $s_n(V) := 1/(b_n + V) \subseteq V$, we have $(b_n + V) \subseteq 1/V$, and thus $-V \subseteq b_n - 1/V$, which means that

$$-\frac{1}{V} \subseteq \frac{1}{b_n - 1/V} = s_n(-1/V)$$

and the result follows. □

From this follows immediately:

1. If $V := \mathfrak{B}(\Gamma, \mu)$ with $\mu < 0$ is a simple value set for $\mathbf{K}(a_n/1)$, then also the disk $\mathfrak{B}(-1-\Gamma, |\mu|)$ is a simple value set for $\mathbf{K}(a_n/1)$, and $\mathbf{K}(a_n/1)$ converges by virtue of the Oval Theorem.

2. If $V := \mathfrak{B}(\Gamma, \mu)$ with $\mu < 0$ is a simple value set for $\mathbf{K}(1/b_n)$, then also $-1/V$ is the exterior of a disk since $\infty \in V^\circ$ implies that $0 \in V^\circ$. Therefore $W := \widehat{\mathbb{C}} \setminus (-1/V)$ is a circular disk, and we can apply Theorem 3.49.

3. If $V := w_0 + e^{i\alpha}\overline{\mathbb{H}}$ is a simple value set for $\mathbf{K}(a_n/1)$, then $\infty \in \partial V$, and thus $0 \in V$ is necessary. With the extra condition $0 \in V^\circ$ and $-1 \notin V$, we are back in the Parabola Theorem.

4. If $V := w_0 + e^{i\alpha}\overline{\mathbb{H}}$ is a simple value set for $\mathbf{K}(1/b_n)$, then $0 \in V$ is still necessary. With the extra condition $0 \in V^\circ$, it follows that $(-1/V)$ is the exterior of a disk, and thus $W := \widehat{\mathbb{C}} \setminus (-1/V)$ is a disk, and we can again apply Theorem 3.49.

3.4 Remarks

1. **Lemma 3.7.** This crucial lemma is inspired by the work of W.J.Thron. In 1965 he and his student K.L.Hillam ([HiTh65]) published the result that $\{\mathcal{T}_n(0)\}$ converges if there exist two points, $w_1 \in \mathbb{D}$ and $w_2 \in \widehat{\mathbb{C}} \setminus \overline{\mathbb{D}}$, such that $\tau_n(w_2) = w_1$ for all n. The present version dates from 1994 ([Lore94a]) where it also was proved that the conclusions still hold if we replace (1.4.6) by $\liminf \operatorname{rad}(\tau_n^{-1}(\mathbb{D})) > 1$. (If $\tau_n^{-1}(\mathbb{D})$ is unbounded, we set $\operatorname{rad} \tau_n^{-1}(\mathbb{D}) := \infty$.) Also Lemma 3.8 still holds under this condition. Lemma 3.7 has later been further generalized in [Lore07].

2. **Truncation error bounds.** The demand for truncation error estimates became more prominent in the 1960s with the growing use of computers. W. J. Thron ([Thron58]) realized that value sets could also be used for the purpose of deriving such estimates. This started a series of useful publications in this area, such as [HePf66], [JoTh76], [CrJT94], [BaJo85], [FiJo72] and [CJPVW7]. For further references we refer to ([JoTh76], p 298).

3. **The parabola theorems.** The first parabola theorem was published by Scott and Wall in 1940, ([ScWa40]). It was valid for the special case $\alpha = 0$. It was proved by exploiting what they called the fundamental inequalities. The result was generalized almost immediately ([PaWa42] and [LeTh42]). The most general parabola theorem is due to Jones and Thron, ([JoTh68]) who also proved convergence for cases of continued fractions $\mathbf{K}(a_n/b_n)$ with half planes as value sets.

4. **Classical convergence theorems.** The Worpitzky Theorem was proved already in 1865, but remained unknown to workers in the field until Pringsheim rediscovered it more than 30 years later. It was not until 1905, through Van Vleck, that Worpitzky got the credit he deserved. Part of the reason may be the way it was published, (in an annual report from the school where Worpitzky was teaching, [Worp65]), but there may be other more significant reasons, see [JaTW89]. Beardon ([Bear01a]) has given a generalization of Worpitzky's Theorem.

   Theorem 3.25 on page 129 usually carries the name of Pringsheim only. However, as pointed out to us by W. J. Thron, J. Śleszyński is the right one to credit, since he already proved the theorem in 1888, see [Śles89].

5. **Boundary versions of convergence theorems.** The limit regions in Worpitzky's Theorem and the Parabola Theorem are created by using the whole element set in both cases. A natural question to ask is: what happens to the limit regions if we restrict the elements to the boundary of the element sets? The answer is given in [Waad89]: the Worpitzky limit set is reduced to the annulus $\frac{1}{6} \leq |f| \leq \frac{1}{2}$ whereas the half plane minus $\{0, \infty\}$ remains unchanged in the parabola case.

   In [Waad92] an investigation of the Oval Theorem is carried out. The process and the result are somewhat more complicated, but the limit region is a circular disk with a hole (not centered). Proper limits of parameters make the oval region approach the parabolic region of the Parabola Theorem and the corresponding limit sets approach the half plane limit set in the parabola case.

   A probabilistic aspect also turned up in the Worpitzky case. When we computed values (or rather high order approximants) of continued fractions from the boundary of the Worpitzky disk, they all seemed to lie in a more narrow annulus that expected from the theoretic results. The reason for this turned out to be that the probability density for the values is very small close to the boundary of the annulus, ([Waad98]).

6. **More convergence theorems.** It is clear that with the help of simple value sets one can produce large numbers of convergence theorems. This has been done by a number of authors, in particular by Jones and Thron ([Thron48], [Thron74], [JoTh68]). Let us also mention Lange's strip convergence regions ([Lange94]) and Overholts polygons ([Over82]) which are based on non-circular value sets, and the general work by Córdova Yevenes ([Cord92]).
7. **Twin convergence theorems.** The situation where $\{V_n\}$ is a periodic sequence of value sets has also been studied in detail. In particular the 2-periodic case is very interesting. Here Thron has given a large number of convergence results based on circular value sets, in later years in collaboration with Jones and Lange, ([Thron43], [Thron49], [Thron59], [LaTh60], [Lange66], [JoTh68], [JoTh70]). In [Lore08a] the cases where $0 \notin V_n$ are also included.

3.5 Problems

1. **Convergence to ∞.** Prove that if the continued fraction $\mathbf{K}(a_n/b_n)$ with $b_{2n-1} = 0$ for all n converges, then it converges to ∞. (Broman [Brom77].)

2. **Convergence of continued fractions.** Determine whether $\mathbf{K}(a_n/b_n)$ converges or diverges and whether its even and odd parts converge or diverge in each of the following cases.

   (a) All $a_n = 1$ and $b_n = z/n^2$ for $z \in \mathbb{C}$.

   (b) All $a_n = z/n^2$ and $b_n = 1$ for $z \in \mathbb{C} \setminus \{0\}$.

   (c) All $a_n = zn^2$ and $b_n = 1$ for $z \in \mathbb{C} \setminus \mathbb{R}$.

3. **Convergence of positive continued fractions.** Prove that the two positive continued fractions $\mathbf{K}(c_n/1)$ and $\mathbf{K}(d_n/1)$ given by

$$c_1 := \frac{1}{3}, \quad c_2 := 2, \quad c_n := \frac{(n-1)^2(2n-1)}{2n-3} \quad \text{for } n \geq 3,$$
$$d_1 := \frac{2}{3}, \quad d_n := \frac{n^2(2n-1)}{2n+1} \quad \text{for } n \geq 2$$

   converge to finite values. (Hint: Theorem 3.2(i) on page 102 and the Seidel-Stern Theorem on page 117 may be of help.)

4. **Convergence of continued fractions.** Determine whether $\mathbf{K}(1/b_n)$ converges or diverges when

   (a) $b_n := (-1)^n n$. (b) $b_n := 1 + (-1)^n$. (c) $b_n := i^n$.
   (d) $b_n := 1/n$. (e) $b_n := (2i)^n$. (f) $b_n = (\sin n)/n^2$.

5. **Convergence of continued fractions.** Determine whether $\mathbf{K}(a_n/1)$ converges or diverges when

   (a) $a_n := (-1)^n n$. (b) $a_n := 2 + (-1)^n$. (c) $a_n := (-2)^n$.
   (d) $a_n := 1/n$. (e) $a_n := 1/n^2$. (f) $a_n = (\sin n)/n^2$.

6. **Mapping of disks.** Find the center and radius of the disk

   (a) $3/(1+2\mathbb{D})$.

   (b) $\tau(\mathfrak{B}(1,-3))$ where $\tau(w) := \dfrac{w+2}{w-2}$.

7. **Positive continued fraction.** Given the continued fraction

$$b + \mathop{\mathrm{K}}_{n=1}^{\infty} \frac{a}{2b} = b + \frac{a}{2b+} \frac{a}{2b+} \frac{a}{2b+\cdots}$$

Prove that if $a > 0$ and $b > 0$ then $b + \mathbf{K}(a/2b)$converges to $\sqrt{a+b^2}$. Use this to find a rational approximation to $\sqrt{13}$ with an error less than 10^{-4}.

8. **Positive continued fractions.** Let $b > 0$ and $p > 0$. Prove that $\mathbf{K}(n^p/b)$ converges if and only if $p \leq 2$. (Perron [Perr57], p 48.)

9. **Oscillations of approximants.** The continued fraction

$$\mathop{\mathrm{K}}_{n=1}^{\infty} \frac{a_n}{1} = \frac{1}{1+} \frac{1/2}{1+} \frac{1/(2\cdot 3)}{1} + \frac{2/(2\cdot 3)}{1} + \frac{2/(2\cdot 5)}{1} + \frac{3/(2\cdot 5)}{1} + \cdots + \frac{n/(2(2n-1))}{1} + \frac{n/(2(2n+1))}{1} + \cdots$$

converges to $\mathrm{Ln}(2)$. Suggest three sequences $\{S_n(w_n)\}$ of approximants converging to $\mathrm{Ln}(2)$. Does any of these sequences have an oscillating character which can be used to obtain upper bounds for the truncation error $|\mathrm{Ln}(2) - S_n(w_n)|$?

10. **Truncation error bounds.** Let $\mathbf{K}(1/b_n)$ be the continued fraction where $b_n = 4 + (0.9)^n$ for all n.

    (a) Find a connected value set V for $\mathbf{K}(1/b_n)$. (Try to make V small.)

    (b) Does $\mathbf{K}(1/b_n)$ converge to a finite value f?

    (c) Are the classical approximants f_n of $\mathbf{K}(1/b_n)$ all distinct; i.e. $f_n \neq f_m$ if $n \neq m$?

    (d) Use the value set V found in (a) to derive upper bounds for the truncation error $|f - S_n(w)|$ for suitably chosen $w \in \mathbb{C}$.

11. **Truncation error bounds.** Use Theorem 3.46 on page 157 to derive a priori truncation error bounds for

$$\mathrm{Ln}(1+i) = \mathop{\mathrm{K}}_{n=1}^{\infty} \frac{a_n i}{1} = \frac{i}{1+} \frac{i/2}{1+} \frac{i/(2\cdot 3)}{1} + \frac{2i/(2\cdot 3)}{1} + \frac{2i/(2\cdot 5)}{1} + \cdots$$

where $a_{2n} = n/(2(2n-1))$ and $a_{2n+1} = n/(2(2n+1))$ for all $n \geq 1$. Compare these with the Gragg-Warner truncation error bounds in Theorem 3.24 on page 128.

12. ♠ **Alternating continued fraction.** Let $\mathbf{K}(a_n/1)$ have real elements a_n such that $(-1)^n a_n > 0$ and

$$|a_{2n-1}| < 1 + a_{2n}, \quad |a_{2n+1}| < 1 + a_{2n} \quad \text{for all } n\,.$$

Prove that $\{S_{4n+p}(0)\}_{n=1}^{\infty}$ converges for $p = 1, 2, 3$ and 4.

13. **Oscillating approximants.** Suggest expressions for w_n such that the sequence $S_n(w_n)$ of approximants for

$$\mathop{\mathrm{K}}_{n=1}^{\infty} \frac{a_n}{1} = \frac{1^2}{1} + \frac{5 \cdot 2^2}{1} + \frac{3^2}{1} + \frac{5 \cdot 4^2}{1} + \frac{5^2}{1} + \frac{5 \cdot 6^2}{1} + \frac{7^2}{1} + \cdots$$

(hopefully) converges faster to the value of $\mathbf{K}(a_n/1)$ than $S_n(0)$. Compute the first 6 approximants of $S_n(0)$ and $S_n(w_n)$. Use the oscillating character to determine an error bound for $S_6(0)$, and if possible for $S_6(w_6)$.

14. **Convergence of continued fractions.** Let α be a positive number. For which values of α does the continued fraction $\mathbf{K}_{n=1}^{\infty}(1/n^{-\alpha})$ converge, and for which values does it diverge? (Hint: Use Van Vleck's Theorem.)

15. ♠ **Convergence neighborhoods.** Use the Parabola Theorem to prove:

    (a) For given $a \in \mathbb{C}$ with $|a| < \frac{1}{4}$, every continued fraction $\mathbf{K}(a_n/1)$ from $\mathfrak{B}(a, |a + \frac{1}{4}|)$ converges.

    (b) For given $a := r\,e^{2i\alpha}$ with $r \geq \frac{1}{4}$ and $-\frac{\pi}{2} < \alpha < \frac{\pi}{2}$, every continued fraction $\mathbf{K}(a_n/1)$ from $\mathfrak{B}(a, \sqrt{r/2}\,\cos\alpha)$ converges. (Jones and Thron [JoTh80], p 108.)

16. **From Van Vleck to Stieltjes.** Prove the following implications:

    (a) Van Vleck's Theorem on page 142 $\Longrightarrow$ Theorem 3.21 on page 124.

    (b) Theorem 3.39 on page 145 $\Longrightarrow$ the Henrici-Pfluger Bounds on page 126.

17. **Convergence of even and odd parts.** Prove that if the even and odd parts of $\mathbf{K}(a_n/1)$ with $a_n \in \mathbb{C} \setminus \{0\}$ converge to distinct values, then $a_n \to \infty$. (Bowman and McLaughlin [BoML04].)

18. **The Cardioid Theorem.** Use the Parabola Sequence Theorem to prove that if

$$|a_n| - \mathrm{Re}(a_n) \leq k\, g_{n-1}(1 - g_n) \quad \text{for } n = 1, 2, \ldots$$

and

$$\sum_{n=1}^{\infty} \prod_{k=1}^{n} |a_k|^{n-k+1} = \infty,$$

then $\mathbf{K}(a_n z/1)$ converges locally uniformly to a holomorphic function in the cardioid domain

$$D := \{z = r\,e^{2i\theta} \in \mathbb{C};\ r < \tfrac{1}{k}\cos\theta\}.$$

(Jones and Thron [JoTh80], p 134.)

19. ♠ **The Parabola Theorem.** Let V_α and E_α be as in the Parabola Theorem. Prove that

   (a) $a \in E_\alpha$ if and only if $a = c^2$ for some $c \in \mathbb{C}$ with $|\mathrm{Im}(c\,e^{-i\alpha})| \leq \frac{1}{2}\cos\alpha$.

   (b) $a \in E_\alpha$ if and only if $a = -c^2$ for some $c \in \mathbb{C}$ with $|\mathrm{Re}(c\,e^{-i\alpha})| \leq \frac{1}{2}\cos\alpha$.

20. ♠ **A convergence theorem.** Let $\{\rho_n\}$ be a sequence of positive numbers. Prove that

   (a) $\{G_n\}_{n=1}^\infty$ given by

   $$G_n := \mathfrak{B}(0, -\rho_n - 1/\rho_{n-1}) = \{b \in \mathbb{C};\ |b| \geq \rho_n + 1/\rho_{n-1}\}$$

   is the sequence of element sets for continued fractions $\mathbf{K}(1/b_n)$ corresponding to the value sets $V_n := \mathfrak{B}(0, \rho_n)$ for $n \geq 0$. (Jones and Thron [JoTh80], p 75.)

   (b) $\{-\rho_n\}$ is a tail sequence for $\mathbf{K}(-1/b_n)$ with $b_n := \rho_n + 1/\rho_{n-1}$.

   (c) $\mathbf{K}(-1/b_n)$ in (b) converges. What is its value?

   (d) $\mathbf{K}(-1/b_n)$ with

   $$b_n := \sqrt{\frac{b+n+1}{b+n}} + \sqrt{\frac{b+n-1}{b+n}}$$

   for a fixed constant $0 \leq b \leq 1$ converges to $-\sqrt{1+1/b}$.

   (e) Prove that

   $$\mathrm{diam} S_n(V_n) \leq 2\rho_0 \prod_{j=2}^{n} \frac{\rho_j\rho_{j-1}(1+3\rho_{j-1}\rho_{j-2})}{2\rho_j\rho_{j-1}^2\rho_{j-2} + 2\rho_j\rho_{j-1} - \rho_{j-1}\rho_{j-2} + 1}$$

   for every $\mathbf{K}(1/b_n)$ from $\{G_n\}$. (Craviotto et al [CrJT94].)

21. ♠ **Comment to the Stern-Stolz Theorem.** Let $\mathbf{K}(1/b_n)$ be given by

   $$b_1 := 1, \quad b_{2n+1} := 2(-1)^n/\sqrt{n+1} \quad \text{and} \quad b_{2n} := (-1)^{n-1}/(\sqrt{n+1}+\sqrt{n}).$$

   (a) Prove that $\sum b_n$ converges to a finite value.

   (b) Prove that $\mathbf{K}(1/b_n)$ converges (and thus that convergence of $\sum b_n$ to a finite value is not sufficient for divergence in general).

22. ♠ **An extension of Thron-Lange's Theorem.** For given $\gamma \in \mathbb{C}$, let $r := \sqrt{1+|\gamma|^2}$, $V := \mathfrak{B}(\gamma, r)$ and $G := \mathfrak{B}(-2\gamma, -2r)$. Let $\mathbf{K}(1/b_n)$ be a continued fraction from G.

   (a) Let $\varphi(w) := (w-\gamma)/r$. Prove that $\tau_n := \varphi \circ s_n \circ \varphi^{-1}$ maps $\mathbb{D}$ into $\mathbb{D}$ with $\tau_n(\infty) = \gamma/r \in \mathbb{D}$.

   (b) Use Lemma 3.8 on page 113 to prove that $\{\mathcal{T}_n\}_{n=1}^\infty$ given by $\mathcal{T}_n := \tau_1 \circ \tau_2 \circ \cdots \circ \tau_n$ converges generally.

   (c) Prove that $\mathbf{K}(1/b_n)$ converges (in the classical sense).

23. ♠ **Element sets.** Show that the sequence $\{E_n\}_{n=1}^{\infty}$ of element sets (possibly empty) for continued fractions $\mathbf{K}(a_n/1)$ corresponding to a given sequence $\{V_n\}_{n=0}^{\infty}$ of value sets, is given by

$$E_n = \cap\{(1+w)V_{n-1};\ w \in V_n \setminus \{-1, \infty\}\}.$$

(Jones and Thron, [JoTh80] p 77.)

24. ♠ **Element sets.** Show that the sequence $\{G_n\}_{n=1}^{\infty}$ of element sets (possibly empty) for continued fractions $\mathbf{K}(1/b_n)$ corresponding to a given sequence $\{V_n\}_{n=0}^{\infty}$ of value sets, is given by

$$G_n = \cap\{-w + 1/V_{n-1};\ w \in V_n \setminus \{\infty\}\}.$$

(Jones and Thron, [JoTh80] p 78.)

25. **Location of the value of a continued fraction.** Use Worpitzky's Theorem to prove the following:

   (a) The value of any continued fraction

$$\frac{1/4}{1} + \frac{-1/4}{1} + \frac{a_3}{1} + \frac{a_4}{1} + \cdots,$$

   with $|a_n| \leq 1/4$ for all n must lie in the disk

$$\left| w - \frac{2}{5} \right| \leq \frac{1}{10}.$$

   (b) The value of any continued fraction

$$\frac{i/4}{1} + \frac{a_2}{1} + \frac{a_3}{1} + \cdots,$$

   with $|a_n| \leq 1/4$ for all n must lie in the disk

$$\left| w - \frac{i}{3} \right| \leq \frac{1}{6}.$$

   (c) Show that the disk in (b) is a limit set for the family of continued fractions from (a).

26. ♠ **The Parabola Theorem.** Prove that E_α in the Parabola Theorem on page 151 also is given by

$$E_\alpha := \left\{ r\, e^{i(\theta+2\alpha)};\ 0 \leq r \leq \frac{\frac{1}{2}\cos^2\alpha}{1-\cos\theta} \right\}.$$

Chapter 4

Periodic and limit periodic continued fractions

Periodic continued fractions are well understood. We can determine when they converge, when they diverge, their values if they converge and the asymptotic behavior of their tail sequences. Indeed, it is all a matter of iterations of linear fractional transformations, and we have closed expressions for their approximants $S_n(w)$.

Only few continued fraction expansions of interesting functions are periodic, but large numbers are limit periodic, where the tail of $\mathbf{K}(a_n/b_n)$ looks more and more like a periodic continued fraction the further out it starts. Then $\mathbf{K}(a_n/b_n)$ essentially inherits the asymptotic properties from the periodic one in most cases. This means that also limit periodic continued fractions are reasonably well understood.

Equivalence transformations are sometimes useful to bring a continued fraction to a wanted form. Naturally, there is no point in transforming a given continued fraction to some $\mathbf{K}(a_n/b_n)$ with $\lim a_n = \lim b_n = 0$ or with $\lim a_n = \lim b_n = \infty$. But sometimes one has to live with outcomes like $\lim a_n = \infty$, $\lim b_n = b \in \mathbb{C} \setminus \{0\}$ or $\lim a_n = a \in \mathbb{C} \setminus \{0\}$, $\lim b_n = 0$. Also some of these continued fractions are analyzed in this chapter.

4.1 Periodic continued fractions

4.1.1 Introduction

A continued fraction $\mathbf{K}(a_n/b_n)$ is called *strictly periodic* with *period length* $p \in \mathbb{N}$, or just *p-periodic* or *periodic* for short, if the sequences $\{a_n\}_{n=1}^{\infty}$ and $\{b_n\}_{n=1}^{\infty}$ are p-periodic; i.e. if

$$a_{n+p} = a_n, \quad b_{n+p} = b_n \quad \text{for all } n \in \mathbb{N}, \tag{1.1.1}$$

and p is the smallest positive integer for which this holds. For such continued fractions the approximants can be written

$$S_{np+m}(w) = S_p^{[n]} \circ S_m(w) = S_m \circ (S_p^{(m)})^{[n]}(w) \tag{1.1.2}$$

where $F^{[n]} := F \circ F \circ \cdots \circ F$ is the nth *iterate* of a function F, and

$$S_p^{(m)}(w) := S_m^{-1} \circ S_p \circ S_m(w) = \frac{a_{m+1}}{b_{m+1}} + \frac{a_{m+2}}{b_{m+2}} + \cdots + \frac{a_{m+p}}{b_{m+p} + w}. \tag{1.1.3}$$

The convergence behavior of $\mathbf{K}(a_n/b_n)$ therefore depends on how sequences of iterates of linear fractional transformations $S_p^{(m)}$ behave asymptotically. For convenience we also use the notation

$$S_0(w) := w \quad \text{and} \quad S_m^{(0)} := S_m(w). \tag{1.1.4}$$

4.1.2 Iterations of linear fractional transformations

We consider linear fractional transformations $\tau \in \mathcal{M}$; i.e.,

$$\tau(w) = \frac{aw + b}{cw + d}, \qquad a,\ b,\ c,\ d \in \mathbb{C} \text{ with } \Delta := ad - bc \neq 0. \tag{1.2.1}$$

If $\{\tau^{[n]}\}$ converges generally to some $x \in \widehat{\mathbb{C}}$, then x must be a *fixed point* for τ; i.e. $\tau(x) = x$. Unless τ is the *identity transformation* $I(w) \equiv w$; i.e. $a = d \neq 0, b = c = 0$, τ has only two (possibly coinciding) fixed points x and y. From (1.2.1) we see that ∞ is a fixed point for τ if and only if $c = 0$. For $c \neq 0$ and $a + d \neq 0$, the fixed points are

$$\begin{aligned} x, y &= \frac{a - d \pm \sqrt{(a-d)^2 + 4bc}}{2c} = \frac{a - d \pm \sqrt{(a+d)^2 - 4\Delta}}{2c} \\ &= \frac{a - d \pm (a+d)u}{2c} \quad \text{where} \quad u := \sqrt{1 - 4\Delta/(a+d)^2} \end{aligned} \tag{1.2.2}$$

(with $\operatorname{Re}\sqrt{\ldots} \geq 0$ as always), and for $c \neq 0$ and $a = -d$

$$x, y = \begin{cases} \frac{a}{c}(1 \pm \sqrt{-\Delta/a^2}) & \text{if } a \neq 0, \\ \pm\sqrt{b/c} & \text{if } a = 0. \end{cases} \tag{1.2.3}$$

Thereby we obtain the standard identity

$$(cx+d)(cy+d) = \Delta \quad \text{if } c \neq 0. \tag{1.2.4}$$

If $c = 0$ and $\tau \neq I$, then the fixed points are at ∞ and $b/(d-a)$.

Case 1: τ has only one fixed point.

Let first $c \neq 0$. Then it follows from (1.2.2) that

$$(a+d)^2 = 4\Delta \quad \text{and} \quad x = y = (a-d)/2c.$$

In particular $a + d \neq 0$. In this case

$$\begin{aligned} \frac{1}{\tau(w)-x} &= \frac{1}{\tau(w)-\tau(x)} = \frac{(cx+d)(cw+d)}{(ad-bc)(w-x)} \\ &= \frac{cx+d}{ad-bc}\left(c+\frac{cx+d}{w-x}\right) = \frac{2c}{a+d} + \frac{1}{w-x}. \end{aligned} \tag{1.2.5}$$

That is, $\varphi \circ \tau(w) = q + \varphi(w)$ where $q := 2c/(a+d)$ and $\varphi(w) := \frac{1}{w-x}$, and so

$$\tilde{\tau}(w) := \varphi \circ \tau \circ \varphi^{-1}(w) = q + w. \tag{1.2.6}$$

Next, let $c = 0$. Then $x = y = \infty$ and $\tilde{\tau}(w) := \tau(w) = w + b/d$ already has the form $q + w$, this time with $q := b/d$. The advantage of this form is that iterations of $\tilde{\tau}$ just gives $\tilde{\tau}^{[n]}(w) = w + nq$, and thus we know how iterations of $\tau = \varphi^{-1} \circ \tilde{\tau} \circ \varphi$ behave. It is also clear that $\tau^{[n]} = \varphi^{-1} \circ \tilde{\tau}^{[n]} \circ \varphi$, so the following result can be stated:

Theorem 4.1. *Let τ given by (1.2.1) have only one fixed point x. Then*

$$\tau^{[n]}(w) = \begin{cases} x + \frac{w-x}{nq(w-x)+1} \; \text{where } x = \frac{a-d}{2c}, \; q := \frac{2c}{a+d} & \text{if } c \neq 0 \\ nq + w \; \text{where } q := b/d & \text{if } c = 0. \end{cases}$$

Case 2: τ has exactly two distinct fixed points.

Let first $c \neq 0$. Then x and y are finite, and

$$\frac{\tau(w)-x}{\tau(w)-y} = \frac{\tau(w)-\tau(x)}{\tau(w)-\tau(y)} = \frac{cy+d}{cx+d} \cdot \frac{w-x}{w-y} \tag{1.2.7}$$

where $cx + d \neq 0$ and $cy + d \neq 0$ by (1.2.4). That is, $\varphi \circ \tau(w) = \Re \cdot \varphi(w)$ where $\varphi(w) := \dfrac{w-x}{w-y}$ and $\Re := \frac{cy+d}{cx+d}$, so

$$\tilde{\tau}(w) := \varphi \circ \tau \circ \varphi^{-1}(w) = \Re w. \tag{1.2.8}$$

Next, let $c = 0$. Then $ad \neq 0$ since $\Delta \neq 0$. Moreover, τ has the *two* distinct fixed points ∞ and $b/(d-a)$, where $a \neq d$. Since $\tau(w) = (a/d)w + (b/d)$, we have

$$\begin{aligned} \tau(w) - \frac{b}{d-a} &= \frac{a}{d}\,w + \frac{b}{d} - \frac{b}{d-a} = \frac{a}{d}\Big(w - \frac{b}{d-a}\Big) \\ \varphi \circ \tau &= \Re \cdot \varphi(w) \quad \text{for } \Re := \frac{a}{d} \text{ and } \varphi(w) := w - \frac{b}{d-a}\,, \end{aligned} \tag{1.2.9}$$

so $\tilde{\tau}(w) := \varphi \circ \tau \circ \varphi^{-1}(w) = \Re w$, and thus $\tilde{\tau}^{[n]}(w) = \Re^n w$, which tells the whole story about iterations of $\tau = \varphi^{-1} \circ \tilde{\tau} \circ \varphi$:

Theorem 4.2. *Let τ given by (1.2.1) have exactly two fixed points $x \neq y$. Then*

$$\tau^{[n]}(w) = \begin{cases} \dfrac{(x - \Re^n y)w - xy(1-\Re^n)}{(1-\Re^n)w + \Re^n x - y} & \text{for } \Re := \dfrac{cy+d}{cx+d} \quad \text{if } c \neq 0, \\ (\frac{a}{d})^n w + (1 - (\frac{a}{d})^n)\frac{b}{d-a} & \text{if } c = 0. \end{cases}$$

4.1.3 Classification of linear fractional transformations

We classify $\tau \in \mathcal{M}$ according to the behavior of iterations $\tau^{[n]}$ as $n \to \infty$. A linear fractional transformation $\widehat{\tau}$ is *conjugate* (or *similar*) to τ if there exists a linear fractional transformation φ such that $\widehat{\tau} = \varphi \circ \tau \circ \varphi^{-1}$. As seen in the previous section, we may therefore look at the conjugates $\tilde{\tau}(w) = \Re w$ or $\tilde{\tau}(w) = w + q$ of τ, or equivalently, use Theorem 4.2 or Theorem 4.1.

Definition 4.1. *For τ given by (1.2.1) with fixed points x and y (coinciding if necessary), the ratio $\Re := \Re(\tau)$ of τ is a complex number with $0 < |\Re| \leq 1$ given by*

$$\Re := \begin{cases} (cy+d)/(cx+d) & \text{if } c \neq 0, \\ a/d & \text{if } c = 0 \text{ and } |a| \leq |d|, \\ d/a & \text{if } c = 0 \text{ and } |a| > |d|. \end{cases} \tag{1.3.1}$$

(If $|cy + d| > |cx + d|$, we just interchange the names of the two fixed points.) Then $\Re$ is uniquely defined for given τ, except if $\Re \neq 1$ with $|\Re| = 1$. This lack of uniqueness will not be a problem for our applications. With u as given by (1.2.2),

$\Re$ can be written

$$\Re = \begin{cases} \dfrac{1-u}{1+u} & \text{if } c \neq 0, \quad a+d \neq 0, \\ -1 & \text{if } c \neq 0, \quad a+d = 0. \end{cases} \tag{1.3.2}$$

This ratio determines to a large degree the asymptotic behavior of $\{\tau^{[n]}\}$:

- $\tau \in \mathcal{M}$ is a *loxodromic* transformation if $|\Re| < 1$. Then τ has exactly two distinct fixed points $x, y \in \widehat{\mathbb{C}}$, and τ is conjugate to $\tilde{\tau}(w) := \Re w$ which has fixed points at 0 and ∞. Since $\tilde{\tau}^{[n]}(w) = \Re^n w \to 0$ for $w \neq \infty$, the sequence $\{\tau^{[n]}\}$ converges generally to x with exceptional sequence $\{y\}_{n=1}^{\infty}$. We say that x is the *attracting fixed point* and y is the *repelling fixed point* for τ.

- $\tau \in \mathcal{M}$ is an *elliptic* transformation if $|\Re| = 1$ with $\Re \neq 1$. Also now τ has exactly two distinct fixed points $x, y \in \widehat{\mathbb{C}}$, and τ is conjugate to $\tilde{\tau}(w) = \Re w$, but $\{\tilde{\tau}^{[n]}\}$ is totally non-restrained and diverges for every $w \neq 0, \infty$. Therefore $\{\tau^{[n]}\}$ is totally non-restrained, and $\{\tau^{[n]}(w)\}$ diverges for every $w \neq x, y$. We say that x and y are *indifferent fixed points* for τ.

- $\tau \in \mathcal{M}$ is *the identity transformation* if $\tau(w) = I(w) \equiv w$. Then every point in $\widehat{\mathbb{C}}$ is a fixed point for τ, and $\Re = 1$. $\tau = I$ is only conjugate to itself, and $\tau^{[n]}(w) = w$.

- $\tau \in \mathcal{M}$ is a *parabolic* transformation if $\tau \neq I$ and $\Re = 1$. Then τ has only one fixed point $x = y \in \widehat{\mathbb{C}}$, and τ is conjugate to $\tilde{\tau}(w) := w + q$ with $q \neq 0$. Since $\tilde{\tau}^{[n]}(w) = w + nq \to \infty$ for every fixed $w \in \widehat{\mathbb{C}}$, it follows that $\tau^{[n]}(w) \to x$ for every $w \in \widehat{\mathbb{C}}$. Also in this case we say that x is an attracting fixed point.

Remarks:

1. This classification is invariant under conjugation.

2. This classification is invariant under inversion since w is a fixed point for τ if and only if w is a fixed point for τ^{-1}, and straight forward computation shows that $\Re$ is invariant under inversion. The roles of x and y must be interchanged, though. (See Remark 4 below.)

3. This classification is essentially invariant under iteration. This follows since for τ with ratio $\Re$ and fixed points x, y, $\tau^{[n]}$ has ratio $\Re^n$ and fixed points x, y. The only exception occurs when τ is elliptic with ratio $\Re$, where $\Re^N = 1$ for some $N \in \mathbb{N}$. Then $\tau^{[N]}$ is the identity transformation. Indeed, $\tau^{[nN]} = I$ for every $n \in \mathbb{N}$.

4. If τ is loxodromic with attracting fixed point x and repelling fixed point y, then τ^{-1} is loxodromic with attracting fixed point y and repelling fixed point x.

5. If $w \neq \infty$ is a fixed point for τ, then the derivative $\tau'(w) = \mathfrak{R}$ if w is the attracting fixed point and $\tau'(w) = 1/\mathfrak{R}$ if w is the repelling fixed point for τ (Problem 2 on page 212).

Example 1. The linear fractional transformation

$$\tau(w) := \frac{4w+3}{2w+5}$$

has fixed points 1 and $-3/2$. Since $|2 \cdot 1 + 5| = 7$ and $|2 \cdot (-3/2) + 5| = 2$, we set $x := 1$, $y := -\frac{3}{2}$, and the ratio for τ is $\mathfrak{R} = 2/7$. Hence $\{\tau^{[n]}\}$ converges generally to 1 with exceptional sequence $\{-\frac{3}{2}\}$. Indeed, by Theorem 4.2,

$$\tau^{[n]}(w) = \frac{(1 + \frac{3}{2}(\frac{2}{7})^n)w + \frac{3}{2}(1 - (\frac{2}{7})^n)}{(1 - (\frac{2}{7})^n)w + (\frac{2}{7})^n + \frac{3}{2}} \to 1 \text{ for } w \neq -\frac{3}{2} = y.$$

◇

Even though $\tau^{[n]}$ has ratio $\mathfrak{R}^n$ when τ has ratio $\mathfrak{R}$, the ratio is not multiplicative in general:

Example 2. The transformation $s_1(w) := (-1/4)/(1+w)$ is parabolic with fixed point $x = -\frac{1}{2}$ and ratio 1, and the transformation $s_2(w) := 2/(1+w)$ is loxodromic with fixed points 1 and -2 and ratio $-1/2$. Their composition is

$$s_1 \circ s_2(w) = \frac{-1/4}{1 + 2/(1+w)} = -\frac{1}{4} \cdot \frac{w+1}{w+3}$$

which is loxodromic with fixed points $x, y = \frac{1}{8}(-13 \pm \sqrt{153})$ and ratio $\mathfrak{R} = (3 + \frac{1}{8}(-13 + \sqrt{153}))/(3 + \frac{1}{8}(-13 - \sqrt{153}))$. ◇

4.1.4 General convergence of periodic continued fractions

We say that a p-periodic continued fraction $\mathbf{K}(a_n/b_n)$ is of *parabolic* or *loxodromic* or *elliptic* or *identity type* according to the classification of S_p. We further say that $\mathfrak{R}$ is the ratio for the p-periodic continued fraction if $\mathfrak{R}$ is the ratio for S_p. The convergence properties for iterations of S_p are described in the previous section. By (1.1.3) we see that $S_p^{(m)}$ is conjugate to S_p, and that x (or y) is a fixed point for S_p if and only if

$$x^{(m)} := S_m^{-1}(x) \quad (\text{or } y^{(m)} := S_m^{-1}(y)) \tag{1.4.1}$$

is a fixed point for $S_p^{(m)}$.

If S_p is loxodromic, then x and $x^{(m)}$ shall always denote the *attracting* fixed point of S_p and $S_p^{(m)}$ respectively, and y and $y^{(m)}$ the repelling ones. By (1.1.2) we then have:

Theorem 4.3. *Let* $\mathbf{K}(a_n/b_n)$ *be a p-periodic continued fraction.*

A. *If* S_p *is parabolic or loxodromic, then* $\mathbf{K}(a_n/b_n)$ *converges generally to the attracting fixed point of* S_p.

B. *If* S_p *is elliptic or the identity transformation, then* $\{S_n\}$ *is totally non-restrained, and* $\mathbf{K}(a_n/b_n)$ *diverges generally.*

Example 3. Straight forward checking shows that the 1-periodic continued fraction $\mathbf{K}(a/b)$ with $a, b \in \mathbb{C}$, $a \neq 0$ is of

- elliptic type if $b = 0$ or $\frac{a}{b^2} < -\frac{1}{4}$,
- parabolic type if $\frac{a}{b^2} = -\frac{1}{4}$,
- loxodromic type otherwise.

Hence $\mathbf{K}(a/b)$ converges generally if and only if $b \neq 0$ and (a/b^2) lies in the cut plane $\mathbb{C} \setminus (-\infty, -\frac{1}{4})$. Its value is then

$$x = \tfrac{b}{2}(-1 + u) \quad \text{where } u := \sqrt{1 + 4a/b^2} \text{ (with } \operatorname{Re} u \geq 0). \tag{1.4.2}$$

If $b = 0$, then $\Re = -1$, and $\{S_n\}$ is the 2-periodic sequence $\frac{a}{w}$, w, $\frac{a}{w}$, $w, \dots$ of linear fractional transformations. ◇

The picture does not change much if we consider tail sequences $\{S_n^{-1}(t_0)\}$ instead of approximants $\{S_n(w)\}$. Our classification is independent of inversion, and

$$t_{np+m} = S_{np+m}^{-1}(t_0) = (S_p^{(m)^{-1}})^{[n]}(S_m^{-1}(t_0)) = (S_p^{(m)^{-1}})^{[n]}(t_m). \tag{1.4.3}$$

Hence $\{t_{np+m}\}_n$ is the sequence of iterates of $S_p^{(m)^{-1}}$ evaluated at t_m.

Theorem 4.4. *Let* $\mathbf{K}(a_n/b_n)$ *be a* p*-periodic continued fraction, and let* $\{t_n\} \subseteq \widehat{\mathbb{C}}$ *be a tail sequence for* $\mathbf{K}(a_n/b_n)$.

A. *If* S_p *is parabolic with fixed point* x*, then*

$$\lim_{n\to\infty} t_{pn+m} = S_m^{-1}(x) =: x^{(m)} \quad \text{for } 0 \le m \le p-1. \tag{1.4.4}$$

B. *If* S_p *is loxodromic with attracting fixed point* x *and repelling fixed point* y*, and* $t_0 \neq x$*, then*

$$\lim_{n\to\infty} t_{pn+m} = S_m^{-1}(y) =: y^{(m)} \quad \text{for } 0 \le m \le p-1. \tag{1.4.5}$$

Of course, if $t_0 = x$, then the p-periodic sequence $\{x^{(m)}\}_{m=0}^{\infty}$ is the sequence of tail values for $\mathbf{K}(a_n/b_n)$.

Example 4. We consider the 1-periodic continued fraction $\mathbf{K}(a/b)$ from Example 3. In the parabolic case, $\mathbf{K}(a/b)$ has one and only one periodic tail sequence $\{x\}_{n=0}^{\infty}$, where $x = -\frac{b}{2}$ is the value of $\mathbf{K}(a/b)$. Indeed, $\{-\frac{b}{2}\}_{n=0}^{\infty}$ is its sequence of tail values, and every tail sequence for $\mathbf{K}(a/b)$ converges to $-\frac{b}{2}$.

In the loxodromic case $\mathbf{K}(a/b)$ has two periodic tail sequences, $\{x\}_{n=0}^{\infty}$ and $\{y\}_{n=0}^{\infty}$ where x and y are the attracting and the repelling fixed point for $S_1(w) = s_1(w) = a/(b+w)$; that is, $x = b(-1+u)/2$ and $y = b(-1-u)/2$, where u is given by (1.4.2). Hence $\{x\}_{n=0}^{\infty}$ is the sequence of tail values for $\mathbf{K}(a/b)$, and $t_n \to y$ for every other tail sequence $\{t_n\}$ for $\mathbf{K}(a/b)$. In particular $\{y\}_{n=0}^{\infty}$ is an exceptional sequence. ◇

Example 5. The 4-periodic continued fraction

$$\mathop{\mathrm{K}}_{n=1}^{\infty} \frac{a_n}{1} := \frac{1}{1} + \frac{1}{1} - \frac{3}{1} - \frac{2}{1} + \frac{1}{1} + \frac{1}{1} - \frac{3}{1} - \frac{2}{1} + \frac{1}{1} + \cdots \tag{1.4.6}$$

converges generally to the attracting fixed point $x = 1$ of the loxodromic transformation

$$S_4(w) = \frac{1}{1} + \frac{1}{1} - \frac{3}{1} - \frac{2}{1+w} = \frac{4+2w}{5+w}\,.$$

The repelling fixed point of S_4 is $y = -4$, so

$$y^{(0)} = -4, \qquad y^{(1)} = s_1^{-1}(-4) = -\tfrac{5}{4}\,,$$
$$y^{(2)} = s_2^{-1}(-\tfrac{5}{4}) = -\tfrac{9}{5}, \qquad y^{(3)} = s_3^{-1}(-\tfrac{9}{5}) = \tfrac{2}{3}\,,$$

and $y^{(4)} = s_4^{-1}(y^{(3)}) = -4 = y^{(0)}$, and so on. Therefore, every tail sequence $\{t_n\}$ with $t_0 \neq 1$ satisfies

$$\lim_{n\to\infty} t_{4n+m} = \begin{cases} -4 & \text{if } m = 0\,, \\ -\frac{5}{4} & \text{if } m = 1\,, \\ -\frac{9}{5} & \text{if } m = 2\,, \\ \frac{2}{3} & \text{if } m = 3\,. \end{cases}$$

◇

4.1.5 Convergence in the classical sense

The parabolic case.

Let $\mathbf{K}(a_n/b_n)$ be p-periodic of parabolic type. Iterations $\{\tau^{[n]}\}$ of a parabolic transformation τ converge to the fixed point for τ for *every* $w \in \mathbb{C}$. Therefore $\lim_{n\to\infty} S_{np}^{(m)}(w) = x^{(m)}$ for *every* $w \in \widehat{\mathbb{C}}$ and $m \in \{0, 1, \ldots, p-1\}$, and thus $\lim S_n(w) = x$ for every $w \in \widehat{\mathbb{C}}$. In particular $S_n(0) \to x$. Therefore:

Theorem 4.5. *A periodic continued fraction of parabolic type converges to x in the classical sense and $S_n(w) \to x$ for every $w \in \widehat{\mathbb{C}}$.*

Still, $\{S_n(w)\}$ does not converge *uniformly* to x in $\widehat{\mathbb{C}}$, not even with respect to the chordal metric. We always have exceptional sequences when $\{S_n\}$ converges generally. For example, every tail sequence $\{t_n\}$ with $t_0 \neq x$ is an exceptional sequence.

The loxodromic case.

This is a case where $\mathbf{K}(a_n/1)$ converges generally, but may diverge in the classical sense:

Example 6. We have seen in Example 2 on page 59 that the three-periodic continued fraction

$$\mathop{\mathrm{K}}_{n=1}^{\infty} \frac{a_n}{1} := \frac{2}{1} + \frac{1}{1} - \frac{1}{1} + \frac{2}{1} + \frac{1}{1} - \frac{1}{1} + \frac{2}{1} + \cdots$$

converges generally to $\frac{1}{2}$, but diverges in the classical sense. The linear fractional transformation

$$S_3(w) := \frac{2}{1} \underset{+}{} \frac{1}{1} \underset{-}{} \frac{1}{1+w} = \frac{2w}{1+2w}$$

is loxodromic with attracting fixed point $1/2$. But $S_3(0) = 0$, and thus $S_{3n}(0) = 0$ for all n whereas $\lim_{n\to\infty} S_{3n+m}(0) = 1/2$ for $m := 1, 2$. ◇

Theorem 4.6. *Let* $\mathbf{K}(a_n/b_n)$ *be a* p*-periodic continued fraction of loxodromic type. If* $S_m(0) = y$ *for some* $m \in \{1, 2, \dots, p\}$*, then* $\mathbf{K}(a_n/b_n)$ *diverges in the classical sense. Otherwise,* $\mathbf{K}(a_n/b_n)$ *converges to* x *in the classical sense.*

Proof : $\mathbf{K}(a_n/b_n)$ converges generally to x (Theorem 4.3). Hence $\mathbf{K}(a_n/b_n)$ converges to x in the classical sense if its exceptional sequences have no limit point at 0. Now, $\{S_n^{-1}(y)\}$ is an exceptional sequence. It is p-periodic, so if $S_m^{-1}(y) \neq 0$; i.e., $S_m(0) \neq y$ for $m = 1, 2, \dots, p$, then 0 is no limit point for this sequence.

Let $S_m(y) = 0$ for some $m \in \{1, 2, \dots, p\}$. Evidently $S_m(0) = y$ can not happen for all $m \in \{1, 2, \dots, p\}$ since $S_p(0) = y$ implies that $y = 0$, whereas $S_1(0) = a_1/b_1 \neq 0$. Therefore $S_{np+m}(0) = y$ for all n, whereas $S_{np+k}(0) \to x$ as $n \to \infty$ for all k for which $S_k(0) \neq y$. Hence $\mathbf{K}(a_n/b_n)$ diverges in this case. □

Remark: This type of divergence of periodic continued fractions was first described by Thiele ([Thie79]), and is therefore called *Thiele oscillation.* If Thiele oscillation occurs for a p-periodic continued fraction of loxodromic type, then

$$\lim_{n\to\infty} S_{np+m}(0) = \begin{cases} x \text{ whenever } S_m(0) \neq y, \\ y \text{ whenever } S_m(0) = y. \end{cases}$$

Since all $y^{(m)} \neq 0$ in Example 5, the continued fraction in this example converges in the classical sense.

Corollary 4.7. *A 1- or 2-periodic continued fraction* $\mathbf{K}(a_n/b_n)$ *of loxodromic type with* $b_1 b_2 \neq 0$*, converges to* x *in the classical sense.*

Proof : Let first $\mathbf{K}(a_n/b_n)$ be 1-periodic. Since $S_1(w) = a_1/(b_1+w)$ is loxodromic, we know from Example 3 that

$$\begin{aligned} &a_1/b_1^2 \in \mathbb{C} \setminus (-\infty, -\tfrac{1}{4}], \\ &x = \frac{b_1}{2}(-1+u_1), \quad y = \frac{b_1}{2}(-1-u_1) \quad \text{where } u_1 := \sqrt{1+4a_1/b_1^2}. \end{aligned} \tag{1.5.1}$$

Hence $y \neq 0$.

Next, let $\mathbf{K}(a_n/b_n)$ be 2-periodic with $b_1 b_2 \neq 0$. The fixed points of $S_2(w)$ are

$$\begin{aligned} &x = \frac{2a_1 - \lambda + \lambda u_2}{2b_1}, \quad y = \frac{2a_1 - \lambda - \lambda u_2}{2b_1} \\ &\text{where } \lambda := a_1 + a_2 + b_1 b_2 \quad \text{and } u_2 := \sqrt{1 - 4a_1 a_2/\lambda^2}. \end{aligned} \tag{1.5.2}$$

Since $x \neq y$, this means that $u_2 \neq 0$ and $\lambda \neq 0$. Therefore $y = 0$ only if $(2a_1 - \lambda)^2 = \lambda^2 u_2^2$; i.e., if $b_1 b_2 = 0$ which we have excluded. Moreover, also $y^{(1)} := a_2/(b_2+y) \neq 0$, and the result follows. □

Example 7. The continued fraction

$$\frac{z}{2} \underset{+}{} \frac{z}{2} \underset{+}{} \frac{z}{2} \underset{+}{} \frac{z}{2} \underset{+}{} \cdots$$

converges to $f(z) = \sqrt{1+z} - 1$ for all z in the cut plane $D := \{z \in \mathbb{C};\ |\arg(1+z)| < \pi\}$ since $S_1(w) = z/(2+w)$ is loxodromic with attracting fixed point $x = f(z)$ for $z \in D$. Every tail sequence $\{t_n\}$ for $\mathbf{K}(z/2)$ with $t_0 \neq x$ converges to $y = -\sqrt{1+z} - 1$. The continued fraction also converges to $f(z)$ for $z := -1$, since then S_1 is parabolic. But for z on the cut $z < -1$, S_1 is elliptic and the continued fraction diverges generally (Theorem 4.3B). ◇

4.1.6 Approximants on closed form

The expressions for $\tau^{[n]}(w)$ in Section 4.1.2 can be used to find closed expressions for $S_n(w) = (A_{n-1}w + A_n)/(B_{n-1}w + B_n)$ when $\mathbf{K}(a_n/b_n)$ is p-periodic. But a periodic tail sequence also simplifies the formulas in Theorem 2.6 on page 66. For instance:

Corollary 4.8. *Let $\mathbf{K}(a/b)$ be 1-periodic, and let x and y be the two possibly coinciding fixed points for $s(w) := a/(b+w)$, where x is attracting or indifferent. Then $x \neq 0, \infty, -b$, $\Re = (b+y)/(b+x) = x/y$, $y = -b - x$ and*

$$B_n + B_{n-1}x = (b+x)^n, \qquad A_n - B_n x = (-x)^{n+1},$$

$$A_n = x(-x)^n \sum_{k=1}^{n} \Re^{-k}, \ B_n = (-x)^n \sum_{k=0}^{n} \Re^{-k}, \ x - \frac{A_n}{B_n} = x \Big/ \sum_{k=0}^{n} \Re^{-k}.$$

Proof : We know that $a \neq 0$. If $b \neq 0$, it follows from (1.5.1) that $y = -b - x$ where $x \neq 0, \infty, -b$. If $b = 0$, then S_1 has the two fixed points $\pm\sqrt{a} \neq 0, \infty$. The expression for $\Re$ follows since $a/(b+x) = x$ and $a/(b+y) = y$. □

Remarks:

1. If s is parabolic, then $a = -b^2/4$, $x = -b/2$, $\Re = 1$ and the formulas in Corollary 4.8 imply that

$$A_n = -n(\tfrac{b}{2})^{n+1}, \qquad B_n = (n+1)(\tfrac{b}{2})^n,$$
$$x - \frac{A_n}{B_n} = -\frac{b/2}{n+1}, \qquad \frac{A_n}{B_n} = -\frac{nb/2}{n+1}.$$

   If $|b| < 2$, then $\{A_n\}$ and $\{B_n\}$ therefore converge to 0, and if $|b| \geq 2$, they converge to ∞.

2. For $\Re \neq 1$ we can sum the geometric sums in the formulas in Corollary 4.8 to get

$$A_n = x(-x)^n \frac{\Re^{-n} - 1}{1 - \Re} = xy\frac{(-y)^n - (-x)^n}{y - x},$$
$$B_n = (-x)^n \frac{\Re^{-n} - \Re}{1 - \Re} = \frac{(-x)^{n+1} - (-y)^{n+1}}{y - x},$$
$$x - \frac{A_n}{B_n} = \frac{x(1-\Re)}{\Re^{-n} - \Re}, \qquad \frac{A_n}{B_n} = \frac{x(\Re^{-n} - 1)}{\Re^{-n} - \Re}.$$

   Hence, for $|\Re| < 1$, the sequences $\{A_n\}$ and $\{B_n\}$ both converge to ∞ if $|y| > 1$, and to 0 if $|y| < 1$.

Our next result gives an alternative expression for the ratio $\Re$ of a transformation S_p:

Theorem 4.9. *Let $\mathbf{K}(a_n/b_n)$ be p-periodic with a p-periodic tail sequence $\underline{t} := \{t_n\}$ for which all $t_n \neq \infty$, and let*

$$X := X(\underline{t}) := \prod_{m=0}^{p-1} (-t_m) \quad \text{and} \quad Y := Y(\underline{t}) := \prod_{m=1}^{p} (b_m + t_m).$$

Then either $X/Y = \Re$ or $X/Y = \Re^{-1}$. If $\widehat{\underline{t}} := \{\widehat{t}_n\}$ is a second p-periodic tail sequence for $\mathbf{K}(a_n/b_n)$ with all $\widehat{t}_n \neq \infty$, then

$$X(\underline{t}) = Y(\widehat{\underline{t}}) \quad \text{and} \quad X(\widehat{\underline{t}}) = Y(\underline{t}). \tag{1.6.1}$$

Remark. We recognize the expression

$$P_p := \frac{Y}{X} = \prod_{m=1}^{p} \frac{b_m + t_m}{-t_m} \quad \text{(since } t_0 = t_p\text{)}$$

from Theorem 2.6 on page 66.

Proof of Theorem 4.9: Evidently t_{np+m} is a fixed point for $S_p^{(m)}$, say $t_{np+m} = x^{(m)}$. The derivative of S_p at $x^{(p)}$ is (using the chain rule)

$$S_p'(x^{(0)}) = S_p'(x^{(p)}) = s_1'(x^{(1)}) \cdot s_2'(x^{(2)}) \cdots s_p'(x^{(p)}) \tag{1.6.2}$$

where

$$s_m'(x^{(m)}) = \frac{-a_m}{(b_m + x^{(m)})^2} = \frac{-x^{(m-1)}}{b_m + x^{(m)}}\,. \tag{1.6.3}$$

This proves that $S_p'(x^{(0)}) = X/Y$. That $X/Y = \Re$ follows now from Remark 5 on page 176. Next, assume that $\{\widehat{t}_n\}$ is a second p-periodic tail sequence with all $\widehat{t}_n \neq \infty$. Then $\widehat{t}_{np+m}$ must be a second fixed point for $S_p^{(m)}$; i.e., $\widehat{t}_{np+m} = y^{(m)}$. Writing $\widehat{X} := X(\widehat{\underline{t}})$ and $\widehat{Y} := Y(\widehat{\underline{t}})$ we similarly find that $\widehat{X}/\widehat{Y} = 1/\Re$. That is, $\Re = X/Y = \widehat{Y}/\widehat{X}$. Moreover,

$$B_p + B_{p-1}t_p = \prod_{m=1}^{p} (b_m + t_m) \tag{1.6.4}$$

by Theorem 2.6 on page 66. Hence by Definition 4.1 of $\Re$ on page 174

$$\Re = \frac{B_p + B_{p-1}y^{(p)}}{B_p + B_{p-1}x^{(p)}} = \prod_{m=1}^{p} \frac{b_m + y^{(m)}}{b_m + x^{(m)}} = \frac{\widehat{Y}}{Y}.$$

This proves that $X = \widehat{Y}$ and $Y = \widehat{X}$. □

4.1.7 A connection to the Parabola Theorem

In the Parabola Theorem on page 151, the half plane $V_\alpha := -\frac{1}{2} + e^{i\alpha}\overline{\mathbb{H}}$ is a simple value set for every continued fraction $\mathbf{K}(a_n/1)$ from

$$E_\alpha := \{a \in \mathbb{C};\ |a| - \text{Re}(a\, e^{-2i\alpha}) \leq \tfrac{1}{2}\cos^2\alpha\} \tag{1.7.1}$$

for given $\alpha \in \mathbb{R}$ with $|\alpha| < \pi/2$. It follows from this Parabola Theorem that every periodic continued fraction from E_α converges. Hence:

- there are no periodic continued fractions of elliptic or identity type from E_α.
- every periodic continued fraction from E_α converges also in the classical sense.

With this notation we have:

Theorem 4.10. *Let $x \in \partial V_\alpha \setminus \{\infty\}$. Then $a_1 := -x^2 \in \partial E_\alpha$ and $a_2 := -(1+x)^2 \in \partial E_\alpha$, and*

$$S_2 := s_1 \circ s_2 \text{ for } s_1(w) := \frac{a_1}{1+w}, \quad s_2(w) := \frac{a_2}{1+w} \tag{1.7.2}$$

is a parabolic transformation.

Proof : That S_2 is parabolic follows since

$$S_2(w) = \frac{-x^2}{1} \underset{-}{} \frac{(1+x)^2}{1+w} = \frac{-x^2(1+w)}{w-2x-x^2}$$

has only one fixed point, namely x. Since $\infty \neq x \in \partial V_\alpha$, it can be written $x = -\frac{1}{2} + it\,e^{i\alpha}$ for some $t \in \mathbb{R}$, and thus

$$\begin{aligned}|a_1| - \mathrm{Re}(a_1 e^{-2i\alpha}) &= |x|^2 - \mathrm{Re}[(-\tfrac{1}{4} + it\,e^{i\alpha} + t^2 e^{2i\alpha})e^{-2i\alpha}] \\ &= \tfrac{1}{4} + t^2 + t\sin\alpha + \tfrac{1}{4}\cos 2\alpha - t\sin\alpha - t^2 = \tfrac{1}{2}\cos^2\alpha\end{aligned}$$

which proves that $a_1 \in \partial E_\alpha$. Since $-(1+x) = -\frac{1}{2} - it\,e^{i\alpha}$, also $a_2 \in \partial E_\alpha$. □

Remarks.

1. This means that
$$\partial E_\alpha = -(\partial V_\alpha)^2, \tag{1.7.3}$$
and that ∂E_α is made up of parabolic pairs (a_1, a_2) in the sense that S_2 in (1.7.2) is parabolic with fixed point x. Similarly, $S_2^{(1)} := s_2 \circ s_1$ is parabolic with fixed point $-(1+x)$.

2. Theorem 4.10 implies that the boundary of E_α is a kind of natural boundary: for every $a_1 = -x^2 \in \partial E_\alpha$ there exist points $a_2^\dagger$ arbitrarily close to $a_2 := -(1+x)^2 \in \partial E_\alpha$ such that $s_1 \circ s_2^\dagger$ is elliptic.

3. To every $0 \neq a_1 = -x^2 \in \mathbb{C} \setminus (-\infty, -\frac{1}{4}]$ there exist exactly two points a_2 such that S_2 in (1.7.2) is parabolic. They are given by $-(1+x)^2$ and $-(1-x)^2$.

4. To every $0 \neq a_1 \in \mathbb{C} \setminus (-\infty, -\frac{1}{4}]$ there exist exactly two parabolas of the type ∂E_α through a_1. They are the parabola through $a_1 = -x^2$ and $a_2 := -(1+x)^2$ and the parabola through $a_1 = -x^2$ and $a_2 := -(1-x)^2$.

Theorem 4.11. $\mathbf{K}(a_n/1)$ *is a p-periodic continued fraction of parabolic type from* E_α *if and only if* $p = 1$ *or* $p = 2$ *and*

$$\mathop{K}_{n=1}^{\infty} \frac{a_n}{1} = \frac{-x^2}{1} - \frac{(1+x)^2}{1} - \frac{x^2}{1} - \cdots \quad \text{where } x = S_p(x) \in \partial V_\alpha. \qquad (1.7.4)$$

Proof : That (1.7.4) is a parabolic continued fraction from E_α follows from Theorem 4.10. Let $\mathbf{K}(a_n/1)$ be a p-periodic continued fraction of parabolic type from E_α. It is clear that $a_m \in \partial E_\alpha$ for all m since there are no p-periodic continued fractions of elliptic type from E_α. Since $\mathbf{K}(a_n/1)$ is from E_α, it converges to a finite value, and so do also all its tails (the Parabola Theorem). Hence, the attracting fixed point $x^{(m)}$ of $S_p^{(m)}$ is $\neq \infty$ for all m. Moreover, all $x^{(m)} \in \partial V_\alpha$ since all $a_m = x^{(m-1)}(1 + x^{(m)}) \in \partial E_\alpha$. That is,

$$\begin{aligned} x^{(m)} &= -\tfrac{1}{2} + i\mu_m e^{i\alpha} \quad \text{with } \mu_m \in \mathbb{R} \text{ for } m = 1, 2, \ldots, p \\ a_m &= -(-\tfrac{1}{2} + i\lambda_m e^{i\alpha})^2 \quad \text{with } \lambda_m \in \mathbb{R}, \end{aligned}$$

and since $a_m = x^{(m-1)}(1 + x^{(m)})$,

$$\begin{aligned} -(\tfrac{1}{4} - i\lambda_m e^{i\alpha} - \lambda_m^2 e^{2i\alpha}) &= (-\tfrac{1}{2} + i\mu_{m-1}e^{i\alpha})(\tfrac{1}{2} + i\mu_m e^{i\alpha}) \\ i\lambda_m e^{i\alpha} + \lambda_m^2(\cos\alpha + i\sin\alpha)e^{i\alpha} &= \tfrac{i}{2}(\mu_{m-1} - \mu_m)e^{i\alpha} - \mu_{m-1}\mu_m e^{2i\alpha}, \end{aligned}$$

which gives the two equations

$$\lambda_m + \lambda_m^2 \sin\alpha = \tfrac{1}{2}(\mu_{m-1} - \mu_m) - \mu_{m-1}\mu_m \sin\alpha\,, \qquad \lambda_m^2 \cos\alpha = -\mu_{m-1}\mu_m \cos\alpha$$

where $\cos\alpha \neq 0$. That is,

$$\lambda_m = \tfrac{1}{2}(\mu_{m-1} - \mu_m) \text{ and } \lambda_m^2 = \tfrac{1}{4}(\mu_{m-1} - \mu_m)^2 = -\mu_m\mu_{m-1}$$

which only holds for $\mu_{m-1} = -\mu_m$. Hence $\{\mu_m\}$ is either the constant sequence $\{0\}$ or a 2-periodic sequence $\mu_0, -\mu_0, \mu_0, -\mu_0, \ldots$. Therefore $x^{(2n)} = x^{(0)} =: x$ and $x^{(2n+1)} = -\frac{1}{2} - i\mu_0 e^{i\alpha} = -(1 + x)$, and $\mathbf{K}(a_n/1)$ has the form (1.7.4). □

For later reference we note:

Lemma 4.12. *The tangent of* ∂E_α *at a finite point* $a = -x^2 \in \partial E_\alpha$ *is the line* $a + xie^{i\alpha}\mathbb{R}$.

Proof : The boundary of V_α is the line

$$\partial V_\alpha : \qquad w = -\tfrac{1}{2} + ti\, e^{i\alpha} \quad \text{for } t \in \mathbb{R} \cup \{\infty\},$$

and thus the curve $\partial E_\alpha = -(\partial V_\alpha)^2$ is given by

$$w = -(-\tfrac{1}{2} + ti\, e^{i\alpha})^2. \tag{1.7.5}$$

The result follows therefore since

$$\frac{\partial w}{\partial t} = -2\left(-\tfrac{1}{2} + tie^{i\alpha}\right) \cdot ie^{i\alpha} = -2xie^{i\alpha} \text{ for } x := -\tfrac{1}{2} + tie^{i\alpha}.$$

□

4.2 Limit periodic continued fractions

4.2.1 Definition

A continued fraction $\mathbf{K}(a_n/b_n)$ is *limit periodic with period length* p, or just *limit* p*-periodic* or *limit periodic* for short, if the limits

$$\lim_{n\to\infty} a_{np+m} =: \tilde{a}_m\,, \qquad \lim_{n\to\infty} b_{np+m} =: \tilde{b}_m \quad \text{for} \quad m = 1, 2, \ldots, p \tag{2.1.1}$$

exist in $\widehat{\mathbb{C}}$, and p is the smallest positive integer for which (2.1.1) holds (although we sometimes find it convenient to relax this last condition). An arbitrary continued fraction $\mathbf{K}(a_n/b_n)$ can always be made limit periodic by an equivalence transformation. But if the result is $\mathbf{K}(c_n/d_n)$ where $c_n \to 0$, $d_n \to 0$ or where $c_n \to \infty$, $d_n \to \infty$, this just hides the structure of $\mathbf{K}(a_n/b_n)$. We are interested in continued fractions for which $\lim_{n\to\infty} a_{np+m}/(b_{np+m-1}b_{np+m})$ exists in $\widehat{\mathbb{C}}$ for $m = 1, 2, \ldots, p$.

Let us assume that the limits in (2.1.1) are finite with all $\tilde{a}_m \neq 0$. (Other cases will be treated later on.) Then the (np)th tail of $\mathbf{K}(a_n/b_n)$ looks more and more like the *corresponding* p*-periodic* continued fraction

$$\operatorname*{K}_{n=1}^{\infty} \frac{\tilde{a}_n}{\tilde{b}_n} = \frac{\tilde{a}_1}{\tilde{b}_1} \, \frac{}{+} \, \frac{\tilde{a}_2}{\tilde{b}_2} \, \frac{}{+} \cdots \frac{}{+} \, \frac{\tilde{a}_p}{\tilde{b}_p} \, \frac{}{+} \, \frac{\tilde{a}_1}{\tilde{b}_1} \, \frac{}{+} \, \frac{\tilde{a}_2}{\tilde{b}_2} \, \frac{}{+} \cdots \frac{}{+} \, \frac{\tilde{a}_p}{\tilde{b}_p} \, \frac{}{+} \, \frac{\tilde{a}_1}{\tilde{b}_1} \, \frac{}{+} \, \frac{\tilde{a}_2}{\tilde{b}_2} \, \frac{}{+} \cdots \tag{2.1.2}$$

as n increases. The convergence of $\mathbf{K}(\tilde{a}_n/\tilde{b}_n)$ is determined by the classification of the linear fractional transformation

$$\tilde{S}_p(w) := \frac{\tilde{a}_1}{\tilde{b}_1} \, \frac{}{+} \, \frac{\tilde{a}_2}{\tilde{b}_2} \, \frac{}{+} \cdots \frac{}{+} \, \frac{\tilde{a}_p}{\tilde{b}_p + w}\,. \tag{2.1.3}$$

So we expect the convergence properties of $\mathbf{K}(a_n/b_n)$ to depend on this classification too. We say that $\mathbf{K}(a_n/b_n)$ is a limit periodic continued fraction of *parabolic, loxodromic, elliptic* or *identity* type depending on the classification of $\tilde{S}_p$. We further say that $\Re$ is the ratio for $\mathbf{K}(a_n/b_n)$ if $\Re$ is the ratio for $\tilde{S}_p$. The important point in our investigations is actually that

$$\lim_{n\to\infty} S_p^{(np+m)} = \tilde{S}_p^{(m)} \quad \text{for } m = 1, 2, \ldots, p, \tag{2.1.4}$$

but this follows in general from (2.1.1).

Notation. We let $x^{(m)}$ be an attracting or indifferent fixed point of

$$\tilde{S}_p^{(m)}(w) = \frac{\tilde{a}_{m+1}}{\tilde{b}_{m+1}} + \frac{\tilde{a}_{m+2}}{\tilde{b}_{m+2}} + \cdots + \frac{\tilde{a}_{m+p}}{\tilde{b}_{m+p} + w} \tag{2.1.5}$$

for all $m \geq 0$, and $y^{(m)}$ be a second fixed point of $\tilde{S}_p^{(m)}$ if it exists, and

$$x^{(m-1)} = \tilde{s}_m(x^{(m)}), \quad y^{(m-1)} = \tilde{s}_m(y^{(m)}) \tag{2.1.6}$$

for all m with obvious notation. We shall also let $x_n^{(m)}$, $y_n^{(m)}$ and $\Re_n^{(m)}$ denote two fixed points (coinciding if necessary) and the ratio for $S_p^{(np+m)}$, where $x_n^{(m)}$ is attracting or indifferent. Here, as always,

$$S_k^{(n)}(w) := S_n^{-1} \circ S_{n+k}(w) = \frac{a_{n+1}}{b_{n+1}} + \cdots + \frac{a_{n+k}}{b_{n+k} + w}\,. \tag{2.1.7}$$

4.2.2 Finite limits, loxodromic case

Let $\mathbf{K}(a_n/b_n)$ with approximants $S_n(w)$ satisfy (2.1.1) with finite limits $\tilde{a}_m$, $\tilde{b}_m$, where $\tilde{S}_p$ is a loxodromic transformation from $\mathcal{M}$. Then

$$\lim_{n\to\infty} x_n^{(m)} = x^{(m)}, \quad \lim_{n\to\infty} y_n^{(m)} = y^{(m)} \text{ and } \lim_{n\to\infty} \Re_n^{(m)} = \Re \tag{2.2.1}$$

where $\Re$ is the ratio for $\tilde{S}_p$ and thus for $\tilde{S}_p^{(m)}$. In particular

$$x^{(m)} \neq y^{(m)} \text{ for all } m \text{ and } |\Re| < 1. \tag{2.2.2}$$

We shall also allow that one or more $\tilde{a}_m = 0$, as long as the limits (2.1.1) still exist in $\widehat{\mathbb{C}}$ and (2.2.2) holds. Then $\tilde{S}_p^{(m)}$ is a singular transformation, but we still say that $\mathbf{K}(a_n/b_n)$ is a limit p-periodic continued fraction of loxodromic type.

Theorem 4.13. *Let* $\mathbf{K}(a_n/b_n)$ *be a limit p-periodic continued fraction of loxodromic type with finite limits (2.1.1).*

A. *Then* $\mathbf{K}(a_n/b_n)$ *converges generally to some value* $f \in \widehat{\mathbb{C}}$ *with the p-periodic exceptional sequence* $\{y^{(m)}\}_{m=0}^{\infty}$.

B. $\mathbf{K}(a_n/b_n)$ *converges in the classical sense if all* $y^{(m)} \neq 0$.

C. $\lim_{n\to\infty} S_{np+m}^{-1}(t_0) = \begin{cases} x^{(m)} & \text{if } t_0 = f \\ y^{(m)} & \text{if } t_0 \neq f \end{cases}$ *for* $m = 1, \dots, p$.

To prove this theorem we shall use a couple of lemmas on transformations $\{\tau_n\}$ from $\mathcal{M}$. In the first one we consider compositions

$$\mathcal{T}_n := \tau_1 \circ \tau_2 \circ \cdots \circ \tau_n \quad \text{and} \quad \widehat{\mathcal{T}}_n := \tau_n \circ \tau_{n-1} \circ \cdots \circ \tau_1$$

and show that $\{\mathcal{T}_n\}$ and $\{\widehat{\mathcal{T}}_n\}$ converge generally:

Lemma 4.14. *Let* $\tau_n \in \mathcal{M}$ *for all* $n \in \mathbb{N}$ *have ratios* $\Re_n \to \Re$ *and fixed points* $x_n \to x$ *and* $y_n \to y$, *where* $|\Re| < 1$, $x \neq y$ *and* y_n *is the repelling fixed point of* τ_n *from some n on. Then* $\mathcal{T}_n \to f$ *for some* $f \in \widehat{\mathbb{C}}$, *and* $\widehat{\mathcal{T}}_n \to x$.

Proof : Without loss of generality we assume that $x = 0$ and $y = \infty$. (Otherwise we choose a $\psi \in \mathcal{M}$ with $\psi(x) = 0$ and $\psi(y) = \infty$, and consider transformations $\psi \circ \tau_n \circ \psi^{-1}$.) Then $x_n \neq \infty$ from some n on, and thus $\tau_n'(x_n) = \Re_n \to \Re$. Since $x_n \to 0$ and $|\Re| < 1$, there exist constants $n_0 \in \mathbb{N}$ and $\mu, \rho > 0$ such that $|\tau_n'(w)| \leq \mu < 1$ for all $n \geq n_0$ and $|w| \leq \rho$. That is, $\tau_n(\rho\overline{\mathbb{D}}) \subseteq \rho\overline{\mathbb{D}}$ for $n \geq n_0$. Without loss of generality we assume that $n_0 = 1$. (Otherwise we look at $\mathcal{T}_n^{(k)} := \tau_{k+1} \circ \cdots \circ \tau_{k+n}$ and $\widehat{\mathcal{T}}_n^{(k)} := \tau_{k+n} \circ \cdots \circ \tau_{k+1}$ for sufficiently large $k \in \mathbb{N}$, and use that $\mathcal{T}_{k+n} = \mathcal{T}_k \circ \mathcal{T}_n^{(k)}$ and $\widehat{\mathcal{T}}_{k+n} = \widehat{\mathcal{T}}_n^{(k)} \circ \widehat{\mathcal{T}}_k$.) The chain rule leads to

$$\mathcal{T}_n'(w_n) = \tau_1'(w_1) \cdot \tau_2'(w_2) \cdots \tau_n'(w_n),$$
$$\widehat{\mathcal{T}}_n'(\widehat{w}_0) = \tau_n'(\widehat{w}_{n-1}) \cdot \tau_{n-1}'(\widehat{w}_{n-2}) \cdots \tau_1'(\widehat{w}_0)$$

where $w_{k-1} := \tau_k(w_k)$ and $\widehat{w}_k := \tau_k(\widehat{w}_{k-1})$ for all k.

Now, $w_n \in \rho\overline{\mathbb{D}}$ implies that $w_k \in \rho\overline{\mathbb{D}}$ for all $k \leq n$, and $\widehat{w}_0 \in \rho\overline{\mathbb{D}}$ implies that all $\widehat{w}_k \in \rho\overline{\mathbb{D}}$. Hence $|\mathcal{T}_n'(w)| \leq \mu^n \to 0$ and $|\widehat{\mathcal{T}}_n'(w)| \leq \mu^n \to 0$ for $|w| \leq \rho$. Therefore $\operatorname{rad} \mathcal{T}_n(\rho\mathbb{D}) \to 0$ and $\operatorname{rad} \widehat{\mathcal{T}}_n(\rho\mathbb{D}) \to 0$.

The nestedness $\mathcal{T}_{n+1}(\rho\overline{\mathbb{D}}) \subseteq \mathcal{T}_n(\rho\overline{\mathbb{D}})$ implies that $\{\mathcal{T}_n\}$ converges generally to some $f \in \rho\overline{\mathbb{D}}$. The fact that for each $\rho > 0$ sufficiently small, $|\widehat{\mathcal{T}}_n(w)| \leq \rho$ from some n on at (more than) two points $w = w_1$ and $w = w_2$, implies that $\widehat{\mathcal{T}}_n \to 0$. □

Remark. From this proof it also follows that there exists a neighborhood V of x such that $\tau_n(V) \subseteq V$ for all n from some n on.

Lemma 4.15. *Let τ_n be as in Lemma 4.14. Then $\{\mathcal{T}_n\}$ converges generally with exceptional sequence $\{y\}$.*

Proof : That $\{\mathcal{T}_n\}$ converges generally follows from Lemma 4.14. Now, also $\tau_k^{-1} \in \mathcal{M}$ with ratio $\Re_n \to \Re$ and fixed points $x_n \to x$ and $y_n \to y$, only this time x_n is the repelling one for large n. Since $\mathcal{T}_n^{-1} = \tau_n^{-1} \circ \tau_{n-1}^{-1} \circ \cdots \circ \tau_1^{-1}$, it follows from Lemma 4.14 that $\{\mathcal{T}_n^{-1}\}$ converges generally to y. Hence $\{y\}$ is an exceptional sequence for $\{\mathcal{T}_n\}$. □

Proof of Theorem 4.13: A. By Lemma 4.15 it follows that for each given $m \geq 0$, $\{S_{np}^{(m)}\}_{n=0}^{\infty}$ converges generally to some $f^{(m)} \in \widehat{\mathbb{C}}$ with constant exceptional sequence $\{y^{(m)}\}$. We need to prove that $S_m(f^{(m)}) = f^{(0)}$ for all $m \in \mathbb{N}$. This clearly holds if $\{f^{(np+m)}\}_{n=0}^{\infty}$ has no limit point at $y^{(m)}$ for $m = 1, 2, \ldots, p$. Since by Lemma 4.14 $\lim_{n\to\infty} f^{(np+m)} = x^{(m)} \neq y^{(m)}$, the conclusion in A follows with $f := f^{(0)}$.

B. This is a straight forward consequence of the result in A.

C. Since $\{S_{np}^{(m)}\}_{n=1}^{\infty}$ converges generally to $f^{(m)}$ with constant exceptional sequence $\{y^{(m)}\}$, the sequence $\{(S_{np}^{(m)})^{-1}\}_{n=1}^{\infty}$ converges generally to $y^{(m)}$ with constant exceptional sequence $\{f^{(m)}\}$. Since $f^{(np+m)} \to x^{(m)}$ as $n \to \infty$, the result follows. □

Corollary 4.16. *A limit 1- or limit 2-periodic continued fraction* $\mathbf{K}(a_n/b_n)$ *of loxodromic type with limits $\tilde{a}_m \neq \infty$ and $\tilde{b}_m \neq 0$, converges in the classical sense.*

Proof : Let first $\mathbf{K}(a_n/b_n)$ be limit 1-periodic with $a_n \to \tilde{a} \neq \infty$ and $b_n \to \tilde{b} \neq 0$. For sufficiently large n, $b_n \neq 0$, and the fixed points of $s_n(w)$ are

$$x_n := \frac{b_n}{2}(-1 + u_n), \quad y_n := \frac{b_n}{2}(-1 - u_n)$$

where $u_n := \sqrt{1 + 4a_n/b_n^2}$ with $\operatorname{Re} u_n \geq 0$. Hence, $y_n \to y \neq 0$, and the result follows from Theorem 4.13B.

Next, let $\mathbf{K}(a_n/b_n)$ be limit 2-periodic with $\tilde{a}_1, \tilde{a}_2 \neq \infty$ and $\tilde{b}_1, \tilde{b}_2 \neq 0$. By (1.5.2) on page 181 and the subsequent arguments, the fixed points of $S_2^{(2n+m)}$ are

$$x_n^{(m)} = \frac{2a_{2n+m+1} - \lambda_n^{(m)} + \lambda_n^{(m)} u_n^{(m)}}{2b_{2n+m+1}},$$

$$y_n^{(m)} = \frac{2a_{2n+m+1} - \lambda_n^{(m)} - \lambda_n^{(m)} u_n^{(m)}}{2b_{2n+m+1}}$$

for sufficiently large n, where $\lambda_n^{(m)} \to \tilde{a}_1 + \tilde{a}_2 + \tilde{b}_1\tilde{b}_2 \neq 0$ and $y_n^{(m)} \to y^{(m)} \neq 0$ for $m = 0, 1$. □

Example 8. The continued fraction $\mathbf{K}(a_n/1)$ where $a_n \to 0$ is limit 1-periodic of loxodromic type, since $s_n(w) := a_n/(1+w)$ has fixed points x_n, y_n and ratio $\mathfrak{R}_n$ given by

$$x_n = \frac{-1+u_n}{2}, \quad y_n = \frac{-1-u_n}{2}, \quad \mathfrak{R}_n = \frac{1-u_n}{1+u_n}$$

where $u_n := \sqrt{1+4a_n} \to 1$. Hence $\mathbf{K}(a_n/1)$ converges to some value $f \in \widehat{\mathbb{C}}$. Its tail values $f^{(n)} \to 0$, and every tail sequence $\{t_n\}$ with $t_0 \neq f$ converges to -1. ◇

Example 9. The continued fraction

$$\underset{n=1}{\overset{\infty}{\mathrm{K}}} \frac{a_n}{1} := \frac{3+1/1^2}{1} + \frac{4+3/2^2}{1} + \frac{3+1/3^2}{1} + \frac{4+3/4^2}{1} + \frac{3+1/5^2}{1} + \cdots$$

in Example 4 on page 13 is limit 2-periodic with corresponding 2-periodic continued fraction

$$\underset{n=1}{\overset{\infty}{\mathrm{K}}} \frac{\tilde{a}_n}{1} := \frac{3}{1} + \frac{4}{1} + \frac{3}{1} + \frac{4}{1} + \frac{3}{1} + \cdots .$$

Since

$$\tilde{S}_2(w) = \cfrac{3}{1 + \cfrac{4}{1+w}} = \frac{3(1+w)}{5+w}$$

has fixed points $x^{(0)} = 1$ and $y^{(0)} = -3$ where $|5 + y^{(0)}| < |5 + x^{(0)}|$, it follows that $\mathbf{K}(a_n/1)$ is of loxodromic type, and that $x^{(0)}$ is the attracting fixed point for $\tilde{S}_2$. Therefore: $\mathbf{K}(a_n/1)$ converges to some value $f \in \widehat{\mathbb{C}}$. Since

$$x^{(1)} := \frac{4}{1+x^{(0)}} = 2 \quad \text{and} \quad y^{(1)} := \frac{4}{1+y^{(0)}} = -2,$$

every tail sequence $\{t_n\}$ for $\mathbf{K}(a_n/1)$ satisfies

$$\lim_{n\to\infty} t_{2n} = \begin{cases} 1 & \text{if } t_0 = f, \\ -3 & \text{if } t_0 \neq f, \end{cases} \quad \lim_{n\to\infty} t_{2n-1} = \begin{cases} 2 & \text{if } t_0 = f, \\ -2 & \text{if } t_0 \neq f. \end{cases}$$

◇

Example 10. Let $\mathbf{K}(a_n/1)$ be a limit 3-periodic continued fraction with

$$\lim a_{3n+1} = 2, \quad \lim a_{3n+1} = 1, \quad \lim a_{3n} = -1.$$

We recognize the corresponding periodic continued fraction

$$\mathop{\mathrm{K}}_{n=1}^{\infty} \frac{\tilde{a}_n}{1} = \frac{2}{1} + \frac{1}{1} - \frac{1}{1} + \frac{2}{1} + \frac{1}{1} - \frac{1}{1} + \cdots$$

from Example 6 on page 179 and Example 2 on page 59. $\tilde{S}_3^{(m)}$ is loxodromic with attracting fixed point $x^{(m)}$ and repelling fixed point $y^{(m)}$ given by

$$x^{(0)} = \tfrac{1}{2}, \ x^{(1)} = 3, \ x^{(2)} = -\frac{2}{3}, \ y^{(0)} = 0, \ y^{(1)} = \infty, \ y^{(2)} = 1.$$

Hence $\mathbf{K}(a_n/1)$ converges generally to some $f \in \widehat{\mathbb{C}}$, and its tail sequences satisfy

$$\lim_{n\to\infty} t_{3n+m} = \begin{cases} x^{(m)} & \text{if } t_0 = f, \\ y^{(m)} & \text{if } t_0 \neq f, \end{cases}$$

just as the tail sequences for $\mathbf{K}(\tilde{a}_n/1)$.

Now, $\mathbf{K}(\tilde{a}_n/1)$ diverges by Thiele oscillation. This does not necessarily imply that $\mathbf{K}(a_n/1)$ diverges. It depends on *how* $\{a_{3n+m}\}_{n=1}^{\infty}$ approaches its limit $\tilde{a}_m$ for $m = 1, 2, 3$. It may actually be very complicated to determine whether $\mathbf{K}(a_n/1)$ converges or diverges in the classical sense. ◇

Example 11. Let $\mathbf{K}(a_n/b_n)$ satisfy $a_n \to 0$, $b_n \to 0$, but $a_n/b_nb_{n-1} \to 2$. Then $b_n \neq 0$ from some n on, so $\mathbf{K}(a_n/b_n)$ is equivalent to $\mathbf{K}(c_n/d_n)$ where $c_n := a_n/b_nb_{n-1}$ and $d_n := 1$ from some n on. That is, $\mathbf{K}(c_n/d_n)$ is limit 1-periodic of loxodromic type, and thus converges in the classical sense to some $f \in \widehat{\mathbb{C}}$. Therefore also $\mathbf{K}(a_n/b_n)$ converges to f. Since $s(w) := 2/(1+w)$ has attracting fixed point 1 and repelling fixed point -2, the tail sequences $\{\tilde{t}_n\}$ for $\mathbf{K}(c_n/d_n)$ satisfy

$$\lim_{n\to\infty} \tilde{t}_n = \begin{cases} 1 & \text{if } \tilde{t}_0 = f \\ -2 & \text{if } \tilde{t}_0 \neq f. \end{cases}$$

The tail sequences $\{t_n\}$ for $\mathbf{K}(a_n/b_n)$ satisfy $t_n = b_n\tilde{t}_n$ from some n on. Hence $t_n \to 0$, or more precisely,

$$t_n \sim \begin{cases} b_n & \text{if } t_0 = f \\ -2b_n & \text{if } t_0 \neq f. \end{cases}$$

◇

4.2.3 Finite limits, parabolic case

Also in this section we consider limit p-periodic continued fractions $\mathbf{K}(a_n/b_n)$ where the limits $\tilde{a}_m$ and $\tilde{b}_m$ in (2.1.1) are finite, but now we assume that $\tilde{S}_p$ given by (2.1.3) is a parabolic transformation from $\mathcal{M}$. We shall not allow $\tilde{a}_m = 0$ in this section.

The situation is significantly different from the loxodromic case. Periodic continued fractions of parabolic type converge (Theorem 4.5 on page 179), but among their closest neighbors are the elliptic ones which diverge generally. This is reflected in the convergence behavior of limit p-periodic continued fractions of parabolic type. We shall see that they converge or diverge generally depending on *how* the elements (a_n, b_n) approach their limit points (2.1.1).

$\mathbf{K(a_n/b_n)}$ limit 1-periodic.

$\mathbf{K}(a_n/b_n)$ is limit 1-periodic of parabolic type if $a_n \to \tilde{a}$ and $b_n \to \tilde{b}$ where $\tilde{a}/\tilde{b}^2 = -\frac{1}{4}$. We can therefore assume that all $b_n \neq 0$, without loss of generality. Then $\mathbf{K}(a_n/b_n) \sim \mathbf{K}(c_n/1)$ where $c_n \to -\frac{1}{4}$. This is of course also so for continued fractions $\mathbf{K}(a_n/b_n)$ with $a_n/b_{n-1}b_n \to -\frac{1}{4}$ more generally.

So let $\mathbf{K}(a_n/1)$ be a continued fraction with $a_n \to -\frac{1}{4}$. What can we say about the convergence of $\mathbf{K}(a_n/1)$? It turns out that the Parabola Sequence Theorem on page 154 is very useful in this situation. It concerns continued fractions $\mathbf{K}(a_n/1)$ from $\{E_{\alpha,n}\}$ given by

$$E_{\alpha,n} :=\{a \in \mathbb{C};\ |a| - \operatorname{Re}(a\, e^{-2i\alpha}) \leq 2g_{n-1}(1-g_n)\cos^2\alpha\} \quad \text{where} \quad -\tfrac{\pi}{2} < \alpha < \tfrac{\pi}{2} \text{ and } 0 < \varepsilon \leq g_n \leq 1-\varepsilon < 1. \tag{2.3.1}$$

Without loss of generality we assume that

$$\widehat{\Sigma}_\infty := \sum_{n=0}^{\infty} \widehat{P}_n = \infty \quad \text{where} \quad \widehat{P}_n := \prod_{k=1}^{n} \frac{1-g_k}{g_k}. \tag{2.3.2}$$

(Otherwise we replace $\{g_n\}$ by a sequence $\{g_n^*\}$ as explained in the remarks on page 136.) Since $a_n \to -\frac{1}{4}$, the limit point case then occurs for $S_n(V_{\alpha,n})$ where $V_{\alpha,n} := -g_n + e^{i\alpha}\overline{\mathbb{H}}$. Hence $\mathbf{K}(a_n/1)$ converges. We can say even more (the proof is postponed to page 209):

Theorem 4.17. *Let $\mathbf{K}(a_n/1)$ with $a_n \to -\frac{1}{4}$ be a continued fraction from $\{E_{\alpha,n}\}_{n=1}^{\infty}$ given by (2.3.1). If $g_n \to \frac{1}{2}$, then $\{S_n(w)\}$ converges to some finite value $f \in -1+\varepsilon+e^{i\alpha}\overline{\mathbb{H}}$ for all $w \in \widehat{\mathbb{C}}$ with $w \neq -\frac{1}{2}$, and every tail sequence for $\mathbf{K}(a_n/1)$ converges to $-\frac{1}{2}$. If $g_{n-1}(1-g_n) \geq \frac{1}{4}$ from some n on, then also $S_n(-\frac{1}{2}) \to f$.*

Example 12. Let $|a_n| - \operatorname{Re} a_n \leq \frac{1}{2}$ for all n for $\mathbf{K}(a_n/1)$ with $a_n \to -\frac{1}{4}$. Then $a_n \in E_{0,n}$ with $g_n := \frac{1}{2}$ for all n. Hence every tail sequence for $\mathbf{K}(a_n/1)$ converges to $-\frac{1}{2}$, so naturally $\{S_n(w)\}$ converges to the value f of $\mathbf{K}(a_n/1)$ for every $w \neq -\frac{1}{2}$. But according to Theorem 4.17B also $S_n(-\frac{1}{2}) \to f$.

Let $\mathbf{K}(c_n/d_n)$ with approximants $T_n(w_n)$ be equivalent to $\mathbf{K}(a_n/1)$. Then every tail sequence $\{t_n\}$ of $\mathbf{K}(c_n/d_n)$ behaves like $t_n \sim -d_n/2$ and $T_n(d_n w) \to f$ for every $w \in \widehat{\mathbb{C}}$. ◇

We have a mixture of two kinds of restrictions on a_n in Theorem 4.17: $\mathbf{K}(a_n/1)$ converges if either

(i) $\{a_n\}$ approaches $-\frac{1}{4}$ from the "right direction"; i.e., from within a sequence $\{E_{\alpha,n}\}$ of parabolic regions for a fixed $|\alpha| < \frac{\pi}{2}$. By Lemma 4.12 on page 185, the tangent of ∂E_α at $a = -\frac{1}{4} = -(-\frac{1}{2})^2$ is the line $-\frac{1}{4} + i\,e^{i\alpha}\mathbb{R}$. Hence $\mathbf{K}(a_n/1)$ converges in particular if $a_n \to -\frac{1}{4}$ through a compact subset of the open half plane $-\frac{1}{4} + e^{i\alpha}\mathbb{H}$,

(ii) or $\{a_n\}$ approaches $-\frac{1}{4}$ fast enough. That is, if

$$|a_n + \tfrac{1}{4}| \leq g_{n-1}(1 - g_n) - \tfrac{1}{4} \quad \text{for all } n, \tag{2.3.3}$$

since then $a_n \in E_{\alpha,n}$ with $\alpha = 0$. (Then $\{a_n\}$ also satisfies condition (2.5.3) in the extended Worpitzky Theorem on page 136.)

Condition (2.3.3) is best in the following sense:

Theorem 4.18. *Let* $E_n := \{a \in \mathbb{C};\ |a + \frac{1}{4}| \leq \tilde{\rho}_n\}$ *with*

$$\tilde{\rho}_n := (g_{n-1}(1 - g_n) - \tfrac{1}{4})(1 + \delta) > 0 \quad \textit{for } n = 1, 2, 3, \ldots \tag{2.3.4}$$

for some fixed numbers $0 < g_n < 1$ *and* $\delta > 0$. *Then there exist divergent continued fractions* $\mathbf{K}(a_n/1)$ *from* $\{E_n\}$.

To prove this, we first observe that $g_n \to \frac{1}{2}$:

Lemma 4.19. *Let*

$$\rho_n := g_{n-1}(1 - g_n) - \tfrac{1}{4} > 0 \quad \textit{for } n = 1, 2, 3, \ldots . \tag{2.3.5}$$

where $0 < g_n < 1$ *for all* n. *Then* $\frac{1}{2} < g_n \to \frac{1}{2}$ *strictly monotonely.*

Proof : With such a choice of $\{g_n\}$, we have $-\frac{1}{4} \in E_{0,n}^{\circ}$ for all n, where $E_{0,n}$ is given by (2.3.1). Since $\mathbf{K}((-\frac{1}{4})/1)$ converges, its tail values $-\frac{1}{2} \in V_{0,n}^{\circ} = -g_n + \mathbb{H}$ for all n. Hence $g_n > \frac{1}{2}$ for $n \geq 0$.

It also follows from (2.3.5) that $(1 - g_n) > (\frac{1}{4})/g_{n-1}$, and thus

$$g_n < 1 - \frac{1/4}{g_{n-1}} < 1 - \frac{1/4}{1} \;{}_{-}\; \frac{1/4}{g_{n-2}} < \cdots < 1 + S_n(g_0)$$

where $S_n(g_0) \to -\frac{1}{2}$ (Theorem 4.5). Hence $\limsup g_n \leq \frac{1}{2}$. This shows that $g_n \to \frac{1}{2}$ The monotonicity follows since

$$g_n < 1 - \frac{1/4}{g_{n-1}} < g_{n-1} \quad \Leftrightarrow \quad g_{n-1} - \frac{1}{4} < g_{n-1}^2 \quad \Leftrightarrow \quad 0 < (g_{n-1} - \tfrac{1}{2})^2.$$

□

Remark. It follows similarly that if $\rho_n \geq 0$ for all n, then $\frac{1}{2} \leq g_n \to \frac{1}{2}$ monotonely.

Proof of Theorem 4.18: Without loss of generality we assume that (2.3.2) holds, so that $\{-g_n\}$ is the sequence of tail values for $\mathbf{K}(\widehat{a}_n/1)$ from $\{E_n\}$ with $\widehat{a}_n := -g_{n-1}(1 - g_n)$ for all n. Since $g_n > \frac{1}{2}$, the sequence $\{\widehat{P}_n\}$ is a monotonely decreasing sequence of positive numbers with $\sum \widehat{P}_n = \infty$. Let $z \neq 1$ be some Nth root of unity; i.e., $z \neq 1$ with $z^N = 1$ for some $N \in \mathbb{N}$. Then $\sum \widehat{P}_n z^n$ diverges by oscillation. Let further

$$t_n := \frac{-g_n}{g_n + z - g_n z}, \quad \text{so that} \quad \frac{1 + t_n}{-t_n} = \frac{1 - g_n}{g_n} z,$$

and thus $\sum_{n=0}^{\infty} \prod_{k=1}^{n} \frac{1+t_k}{-t_k}$ diverges by oscillation. Since $\{t_n\}$ is a tail sequence for the continued fraction $\mathbf{K}(a_n/1)$ given by

$$a_n := t_{n-1}(1 + t_n) = \frac{-g_{n-1}}{g_{n-1} + z - g_{n-1} z} \cdot \frac{(1 - g_n) z}{g_n + z - g_n z},$$

this continued fraction diverges (Corollary 2.7 on page 68). On the other hand, $\lim_{z \to 1} a_n = g_{n-1}(1 - g_n)$, so $\mathbf{K}(a_n/1)$ is a continued fraction from $\{E_n\}$ for z close enough to 1, which can be achieved for N large enough. □

$\mathbf{K}(\mathbf{a_n}/\mathbf{b_n})$ limit 2-periodic.

Also here we assume that all $b_n \neq 0$, so that an equivalence transformation brings the continued fraction to the form $\mathbf{K}(c_n/1)$. We therefore study continued fractions $\mathbf{K}(a_n/1)$. Our analysis thereby covers continued fractions $\mathbf{K}(a_n/b_n)$ where

$$\lim_{n\to\infty} \frac{a_{2n+1}}{b_{2n} b_{2n+1}} = \tilde{a}_1 \quad \text{and} \quad \lim_{n\to\infty} \frac{a_{2n}}{b_{2n-1} b_{2n}} = \tilde{a}_2,$$
$$\text{with } \tilde{S}_2 := \tilde{s}_1 \circ \tilde{s}_2 \text{ parabolic when } \tilde{s}_m(w) := \tilde{a}_m/(1 + w)\,. \tag{2.3.6}$$

$\mathbf{K}(a_n/1)$ is limit 2-periodic of parabolic type with finite limits $\tilde{a}_1$, $\tilde{a}_2$ if

$$a_{2n-1} \to \tilde{a}_1 = -x^2 \quad \text{and} \quad a_{2n} \to \tilde{a}_2 = -(1+x)^2 \tag{2.3.7}$$

for some $x \neq -1,\ 0,\ \infty$; that is, if $(1+\tilde{a}_1+\tilde{a}_2)^2 = 4\tilde{a}_1\tilde{a}_2 \neq 0$. The transformations $\tilde{S}_2$ and $\tilde{S}_2^{(1)}$ are then parabolic with attracting fixed points x and $-(1+x)$ respectively. In this situation, Theorem 3.4 on page 105 is very useful: if the even and odd parts of $\mathbf{K}(a_n/1)$ converge, then $\mathbf{K}(a_n/1)$ itself converges. And the even and odd parts of $\mathbf{K}(a_n/1)$ are limit 1-periodic continued fractions of parabolic type.

But we can also use the Parabola Sequence Theorem directly. As noted in Remark 1 on page 184 the boundary of E_α is made up of parabolic pairs. Indeed, to every parabolic pair $(-x^2, -(1+x)^2)$ with $x \in \mathbb{C} \setminus (-\infty, -\frac{1}{2}]$, there exists a parabolic region E_α from the Parabola Theorem, with boundary passing through both $-x^2$ and $-(1+x)^2$ (Remark 2 and 3 on page 184). Moreover, we can always find $\{E_{\alpha,n}\}$ such that $g_{n-1}(1-g_n) > \frac{1}{4}$ and thus $E_\alpha \subseteq E_{\alpha,n}$ for all n.

Theorem 4.20. *Let $\mathbf{K}(a_n/1)$ be a limit 2-periodic continued fraction of parabolic type (with finite limits) from $\{E_{\alpha,n}\}$ given by (2.3.1). If $g_n \to \frac{1}{2}$, then $\mathbf{K}(a_n/1)$ converges to some value $f \in -1+\varepsilon+e^{i\alpha}\overline{\mathbb{H}}$, and every tail sequence $\{t_n\}$ for $\mathbf{K}(a_n/1)$ satisfies $\lim t_{2n} = x$, $\lim t_{2n+1} = -1-x$ where x is given by (2.3.7). If $g_{n-1}(1-g_n) \geq \frac{1}{4}$ from some n on, then $S_n(w) \to f$ for every $w \in \widehat{\mathbb{C}}$.*

The proof of this theorem is similar to the proof of Theorem 4.17 and will be omitted.

More general cases.

For longer periods than $p = 2$, p-contractions can often give some answers, for example in combination with the following lemma:

Lemma 4.21. *Let $\mathbf{K}(a_n/1)$ be limit p-periodic with finite limits $\tilde{a}_m := \lim_{n\to\infty} a_{np+m} \neq 0$ for $m = 1, 2, \ldots, p$. If $\{S_{np}^{(m)}\}_{n=0}^{\infty}$ converges generally for $m = 0, 1, \ldots, p-1$, then $\mathbf{K}(a_n/1)$ converges generally to $f^{(0)}$.*

Proof : Since $\{a_m\}$ is bounded and bounded away from 0, there always exist sequences $\{u_n\}$ and $\{v_n\}$ from $\widehat{\mathbb{C}}$ such that

$$\begin{aligned} f &= \lim_{n\to\infty} S_{np+m}(w_{np+m}) \\ &= \lim_{n\to\infty} S_{np+m}\Bigl(\frac{a_{np+m+1}}{1+w_{np+m+1}}\Bigr) = \lim_{n\to\infty} S_{np+m+1}(w_{np+m+1}) \end{aligned}$$

for both $\{w_n\} := \{u_n\}$ and $\{w_n\} := \{v_n\}$, where

$$\liminf \mathfrak{m}(u_n, v_n) > 0 \quad \text{and} \quad \liminf \mathfrak{m}\Bigl(\frac{a_{n+1}}{1+u_{n+1}}, \frac{a_{n+1}}{1+v_{n+1}}\Bigr) > 0.$$

□

4.2.4 Finite limits, elliptic case

Since periodic continued fractions of elliptic type diverge, it is easy to believe that also limit periodic ones diverge. However, as pointed out by Gill ([Gill73]) there also exist convergent ones. Here is a rather simple example (compared to other examples in the literature) of this phenomenon:

Example 13. A limit 1-periodic continued fraction $\mathbf{K}(a_n/1)$ is of elliptic type if

$$a_n \to \tilde{a} < -\tfrac{1}{4}\,. \tag{2.4.1}$$

Then $\tilde{S}_1(w) = \tilde{a}/(1+w)$ has the two indifferent fixed points

$$x = -\tfrac{1}{2} + iq, \quad y = -\tfrac{1}{2} - iq \quad \text{where } q := \sqrt{|\tilde{a}| - \tfrac{1}{4}} > 0. \tag{2.4.2}$$

In particular $y = \overline{x}$, $1 + x = -\overline{x}$ and $\tilde{a} = -|x|^2$. Let $t_n := x - 2|x|^2/(n+1)$ for all $n \ge 0$. Then $\{t_n\}$ is a tail sequence for the limit 1-periodic continued fraction $\mathbf{K}(a_n/1)$ of elliptic type given by $a_n := t_{n-1}(1+t_n)$. Now,

$$t_0 - f_n = \frac{t_0}{\Sigma_n} \quad \text{where} \quad \Sigma_n := \sum_{k=0}^{n} \prod_{j=1}^{k} \frac{1+t_j}{-t_j}. \tag{2.4.3}$$

(See formula (1.5.7) on page 66.) We have

$$\begin{aligned} \frac{1+t_{j-1}}{-t_{j-1}} &= \frac{1+x-2|x|^2/j}{-x+2|x|^2/j} \\ &= \frac{-\overline{x}-2|x|^2/j}{-x+2|x|^2/j} = \frac{1+2x/j}{1-2\overline{x}/j}\cdot\frac{\overline{x}}{x} = \frac{j-1+2iq}{j+1+2iq}\cdot\frac{\overline{x}}{x} \end{aligned}$$

and thus

$$\prod_{j=1}^{k} \frac{1+t_j}{-t_j} = \frac{(1+2iq)(2+2iq)}{(k+1+2iq)(k+2+2iq)} \cdot \left(\frac{\overline{x}}{x}\right)^k$$

which means that

$$\sum_{k=0}^{\infty}\prod_{j=1}^{k}\left|\frac{1+t_j}{-t_j}\right| = \sum_{k=0}^{\infty}\prod_{j=2}^{k+1}\left|\frac{1+t_{j-1}}{-t_{j-1}}\right| = 1 + |1+2iq|\cdot|2+2iq|\sum_{k=1}^{\infty}\frac{1}{k^2+\mathcal{O}(k)}.$$

That is, Σ_n converges to a finite number, and $\mathbf{K}(a_n/1)$ converges to some $f \neq t_0$. ◇

Still, $\mathbf{K}(a_n/b_n)$ diverges if $\{S_p^{(np)}\}$ converges fast enough to an elliptic transformation.

Lemma 4.22. *Let $\tau_n \in \mathcal{M}$ have ratio $\Re_n$ and fixed points x_n, y_n for all $n \in \mathbb{N}$, where*

$$\sum |x_n| < \infty, \quad \sum 1/|y_n| < \infty \ \text{ and } \sum |\Re_n - \Re| < \infty \tag{2.4.4}$$

for some $\Re \neq 1$ with $|\Re| = 1$, and let $\mathcal{T}_n := \tau_1 \circ \cdots \circ \tau_n$ for all $n \in \mathbb{N}$. Then $\{\mathcal{T}_n\}$ is totally non restrained, and $\{\mathcal{T}_n(\Re^{-n}w)\}$ converges to $\mathcal{T}(w)$ for all $w \in \widehat{\mathbb{C}}$ for a $\mathcal{T} \in \mathcal{M}$.

To prove this we shall use the following auxiliary result:

Lemma 4.23. *For a given sequence $\{r_n\}$ of positive numbers with $r := \sum_{n=1}^{\infty} r_n < \infty$, let $\{(X_n, Y_n)\}_{n=0}^{\infty}$ satisfy*

$$\begin{aligned} &0 \le X_n \le (1+r_n)X_{n-1} + r_n Y_{n-1}, \\ &0 \le Y_n \le r_n X_{n-1} + (1+r_n)Y_{n-1} \ \text{ for } n = 1,2,3,\dots \end{aligned} \tag{2.4.5}$$

with $X_0 \ge 0$ and $Y_0 \ge 0$. Then $\{(X_n, Y_n)\}$ is bounded.

Proof : The worst case occurs when we have equality for all n in the two inequalities in (2.4.5); i.e., when

$$\begin{pmatrix} X_n \\ Y_n \end{pmatrix} = M_n \begin{pmatrix} X_{n-1} \\ Y_{n-1} \end{pmatrix} \ \text{ where } M_n := \begin{pmatrix} 1+r_n & r_n \\ r_n & 1+r_n \end{pmatrix};$$

i.e., when

$$\begin{pmatrix} X_n \\ Y_n \end{pmatrix} = M_{n,1} \begin{pmatrix} X_0 \\ Y_0 \end{pmatrix} \quad \text{where } M_{n,1} := M_n M_{n-1} \cdots M_1.$$

Now, the matrix product $M_{k+1}M_k$ is

$$M_{k+1}M_k = \begin{pmatrix} 1 + r_{k+1,k} & r_{k+1,k} \\ r_{k+1,k} & 1 + r_{k+1,k} \end{pmatrix}$$
$$\text{where } r_{k+1,k} := r_k + r_{k+1} + 2r_k r_{k+1},$$

so all $M_{n,1}$ have the same form with some $r_{n,1}$ for which $r_{1,1} = r_1$ and

$$r_{k+1,1} = r_{k,1} + r_{k+1} + 2r_{k,1}r_{k+1} = r_{k,1}(1 + 2r_{k+1}) + r_{k+1}.$$

We shall prove by induction that

$$r_{n,1} = \frac{1}{2}\prod_{j=1}^{n}(1 + 2r_j) - \frac{1}{2}. \tag{2.4.6}$$

Evidently $r_{1,1} = r_1 = \frac{1}{2}(1 + 2r_1) - \frac{1}{2}$, and if (2.4.6) holds for $n = k$, then

$$\begin{aligned} r_{k+1,1} &= r_{k,1}(1 + 2r_{k+1}) + r_{k+1} \\ &= \frac{1}{2}\prod_{j=1}^{k+1}(1 + 2r_j) - \frac{1}{2}(1 + 2r_{k+1}) + r_{k+1} = \frac{1}{2}\prod_{j=1}^{k+1}(1 + 2r_j) - \frac{1}{2} \end{aligned}$$

which proves (2.4.6). Since $\sum r_k < \infty$, it follows that $\{r_{n,1}\}$ is bounded. □

Proof of Lemma 4.22: Let $\tau_n \in \mathcal{M}$ be given, with ratio $\Re_n$ and fixed points x_n, y_n (coinciding if necessary), where $x_n \to x$, $y_n \to y$ and $\Re_n \to \Re$ with $|\Re| = 1$, $\Re \neq 1$. Without loss of generality we assume that $x = 0$, $y = \infty$, and that $x_n \neq \infty$, $y_n \neq 0$ and $\Re_n \neq 1$ for all n. (Otherwise we consider $\widehat{\tau}_n := \varphi \circ \tau_n \circ \varphi^{-1}$ where $\varphi(x) = 0$ and $\varphi(y) = \infty$, and look at $\widehat{\mathcal{T}}_n := \widehat{\tau}_{m+1} \circ \cdots \circ \widehat{\tau}_{m+n} = \widehat{\mathcal{T}}_m^{-1} \circ \widehat{\mathcal{T}}_{m+n}$ for sufficiently large m.)

Since all $\Re_n \neq 1$, we have all $x_n \neq y_n$, and by Theorem 4.2 on page 174, τ_n can be written

$$\tau_n(w) = \frac{(x_n - \Re_n y_n)w - (1 - \Re_n)x_n y_n}{(1 - \Re_n)w + \Re_n x_n - y_n} = \frac{(\Re_n - \frac{x_n}{y_n})w + (1 - \Re_n)x_n}{-\frac{1}{y_n}(1 - \Re_n)w - \Re_n \frac{x_n}{y_n} + 1}$$

with the natural limit form if $y_n = \infty$. That is,

$$\tau_n(w) = \frac{(1 + \delta_n^{(1)})\Re w + \delta_n^{(2)}}{\delta_n^{(3)} w + 1 + \delta_n^{(4)}} \quad \text{for } n = 1, 2, 3, \ldots \tag{2.4.7}$$

where $\sum |\delta_n^{(k)}| < \infty$ for $k = 1, 2, 3, 4$ under our conditions. Let $\{r_n\}$ be a sequence of positive numbers such that

$$|\delta_n^{(k)}| \le r_n \quad \text{for all } n \text{ and } k = 1, 2, 3, 4 \quad \text{and } r := \sum r_n < \infty.$$

We want to prove that

$$\mathcal{T}_n(w) = \frac{\mathcal{A}_n w + \mathcal{B}_n}{\mathcal{C}_n w + \mathcal{D}_n} \text{ where } \begin{array}{ll} \lim \mathcal{A}_n \Re^{-n} = \mathfrak{C}^{(1)}, & \lim \mathcal{B}_n = \mathfrak{C}^{(2)} \\ \lim \mathcal{C}_n \Re^{-n} = \mathfrak{C}^{(3)}, & \lim \mathcal{D}_n = \mathfrak{C}^{(4)} \end{array} \tag{2.4.8}$$

with $\mathfrak{C}^{(1)}\mathfrak{C}^{(4)} - \mathfrak{C}^{(2)}\mathfrak{C}^{(3)} \neq 0$ for r sufficiently small. Since $\mathcal{T}_n = \mathcal{T}_{n-1} \circ \tau_n$, we find that

$$\begin{array}{ll} \mathcal{A}_n = (1 + \delta_n^{(1)})\Re\mathcal{A}_{n-1} + \delta_n^{(3)}\Re\mathcal{B}_{n-1}, & \mathcal{B}_n = \delta_n^{(2)}\mathcal{A}_{n-1} + (1 + \delta_n^{(4)})\mathcal{B}_{n-1}, \\ \mathcal{C}_n = (1 + \delta_n^{(1)})\Re\mathcal{C}_{n-1} + \delta_n^{(3)}\Re\mathcal{D}_{n-1}, & \mathcal{D}_n = \delta_n^{(2)}\mathcal{C}_{n-1} + (1 + \delta_n^{(4)})\mathcal{D}_{n-1} \end{array} \tag{2.4.9}$$

for $n \ge 2$. It follows therefore from Lemma 4.23 that $\{\mathcal{A}_n\}$, $\{\mathcal{B}_n\}$, $\{\mathcal{C}_n\}$ and $\{\mathcal{D}_n\}$ are bounded sequences.

From the recurrence relations (2.4.9) we find that

$$\begin{aligned} |\mathcal{A}_n \Re^{-n} - \mathcal{A}_{n-1}\Re^{-(n-1)}| &= |\delta_n^{(1)}\Re^{-(n-1)}\mathcal{A}_{n-1} + \delta_n^{(3)}\Re^{-(n-1)}\mathcal{B}_{n-1}| \\ &\le r_n|\mathcal{A}_{n-1}| + r_n|\mathcal{B}_{n-1}| \end{aligned}$$

since $|\Re| = 1$, where $\{\mathcal{A}_n\}$ and $\{\mathcal{B}_n\}$ are bounded. Hence $\{\mathcal{A}_n\Re^{-n}\}$ converges absolutely to some finite constant $\mathfrak{C}^{(1)}$ as $n \to \infty$. Similarly $\mathcal{C}_n\Re^{-n} \to \mathfrak{C}^{(3)} \neq \infty$. Moreover, $|\mathcal{B}_n - \mathcal{B}_{n-1}| \le r_n(|\mathcal{A}_{n-1}| + |\mathcal{B}_{n-1}|)$ and $|\mathcal{D}_n - \mathcal{D}_{n-1}| \le r_n(|\mathcal{C}_{n-1}| + |\mathcal{D}_{n-1}|)$ which proves that the finite limits $\mathfrak{C}^{(2)} := \lim \mathcal{B}_n$ and $\mathfrak{C}^{(4)} := \lim \mathcal{D}_n$ also exist. Finally, by (2.4.9)

$$\begin{aligned} \Delta_n &:= \mathcal{A}_n\Re^{-n}\mathcal{D}_n - \mathcal{B}_n\mathcal{C}_n\Re^{-n} \\ &= [(1 + \delta_n^{(1)})\mathcal{A}_{n-1} + \delta_n^{(3)}\mathcal{B}_{n-1}]\Re^{-(n-1)} \cdot [\delta_n^{(2)}\mathcal{C}_{n-1} + (1 + \delta_n^{(4)})\mathcal{D}_{n-1}] \\ &\quad - [(1 + \delta_n^{(1)})\mathcal{C}_{n-1} + \delta_n^{(3)}\mathcal{D}_{n-1}]\Re^{-(n-1)} \cdot [\delta_n^{(2)}\mathcal{A}_{n-1} + (1 + \delta_n^{(4)})\mathcal{B}_{n-1}] \\ &= [(1 + \delta_n^{(1)})(1 + \delta_n^{(4)}) - \delta_n^{(2)}\delta_n^{(3)}]\Delta_{n-1} \\ &= \cdots = \Delta_1 \prod_{k=2}^{n} [(1 + \delta_k^{(1)})(1 + \delta_k^{(4)}) - \delta_k^{(2)}\delta_k^{(3)}] \end{aligned}$$

which stays bounded away from 0 when $\sum r_n < \infty$ and all $r_n < 1$ with $(1 - r_n)^2 - r_n^2 = 1 - 2r_n > 0$; i.e., all $r_n < \frac{1}{2}$. □

Our main theorem follows immediately:

Theorem 4.24. *Let $\tau_n \in \mathcal{M}$ have ratio $\Re_n$ and fixed points x_n, y_n, and let $\mathcal{T}_n := \tau_1 \circ \tau_2 \circ \cdots \circ \tau_n$ for all $n \in \mathbb{N}$. If $\tau_n \to \tau \in \mathcal{M}$ where τ is elliptic with ratio $\Re$ and fixed points x, y, and*

$$\sum |\Re_n - \Re| < \infty, \quad \sum \mathfrak{m}(x_n, x) < \infty \quad \text{and} \quad \sum \mathfrak{m}(y_n, y) < \infty,$$

then $\mathcal{T}_n(\Re^{-n}w) \to \mathcal{T}(w)$ for a $\mathcal{T} \in \mathcal{M}$. In particular $\{\mathcal{T}_n\}$ is totally non-restrained.

Therefore, a limit periodic continued fraction of elliptic type diverges if $\{S_p^{(np)}\}$ approaches the elliptic transformation fast enough. For instance:

Corollary 4.25. *Let $\mathbf{K}(a_n/b_n)$ be limit 1-periodic of elliptic type with $a_n \to a \neq 0, \infty$ and $b_n \to b \neq \infty$. If $\sum |a_n - a| < \infty$ and $\sum |b_n - b| < \infty$, then $\mathbf{K}(a_n/b_n)$ diverges generally.*

Proof : The linear fractional transformations

$$s_n(w) := \frac{a_n}{b_n + w} \quad \text{and} \quad s(w) := \frac{a}{b + w}$$

have finite fixed points

$$x_n, y_n = -\tfrac{1}{2}(b_n \pm \sqrt{b_n^2 + 4a_n}), \quad x, y = -\tfrac{1}{2}(b \pm \sqrt{b^2 + 4a}).$$

Therefore $\sum |x_n - x| < \infty$ and $\sum |y_n - y| < \infty$. Moreover, the ratios of s_n and s are given by

$$\Re_n = \frac{b_n + y_n}{b_n + x_n} = \frac{x_n}{y_n}, \quad \Re = \frac{x}{y},$$

where $y_n, y \neq 0$ since $a_n, a \neq 0$, so also $\sum |\Re_n - \Re| < \infty$. □

Example 14. Let $0 \neq a_n \to a < -\frac{1}{4}$ so fast that $\sum |a_n - a| < \infty$. Then $\mathbf{K}(a_n/1)$ is a generally divergent limit periodic continued fraction of elliptic type. ◇

4.2.5 Infinite limits

Let $\mathbf{K}(a_n/b_n)$ be limit p-periodic where one or more of the limits (2.1.1) are equal to ∞. Then it may be a good idea to use an equivalence transformation to bring it over to the form $\mathbf{K}(c_n/1)$ or $\mathbf{K}(1/d_n)$ if possible. The cases $c_n \to 0$ and $d_n \to \infty$ are very simple: the continued fraction converges. The cases $c_n \to \infty$ and $d_n \to 0$ can sometimes be resolved by use of a Parabola Theorem or Van Vleck's Theorem, but far from always. In this section we shall consider the case

$$\mathbf{K}(a_n/1) \quad \text{where} \quad a_n \to \infty\,. \tag{2.5.1}$$

We know that $\mathbf{K}(a_n/1)$ diverges if its Stern-Stolz series

$$\mathcal{S} := \sum_{n=0}^{\infty} \prod_{k=1}^{n} |a_k|^{(-1)^{n+k+1}} \tag{2.5.2}$$

converges to a finite value (the Stern-Stolz Theorem on page 100). Hence we concentrate on the case $\mathcal{S} = \infty$. We follow an idea from [JaWa86], and analyze the even and odd parts of (2.5.1). They are given by

$$\frac{a_1}{1+a_2} - \frac{a_2 a_3}{1+a_3+a_4} - \frac{a_4 a_5}{1+a_5+a_6} - \cdots \qquad \text{(even part)}$$

$$a_1 - \frac{a_1 a_2}{1+a_2+a_3} - \frac{a_3 a_4}{1+a_4+a_5} - \cdots \qquad \text{(odd part).}$$

If they both converge to the same value, then $\mathbf{K}(a_n/1)$ converges. This is in particular the case in the following situation:

Theorem 4.26. $\mathbf{K}(a_n/1)$ *converges if* $a_n \to \infty$ *with*

$$\lim_{n\to\infty} \frac{a_{2n-1}}{a_{2n}} =: \alpha \in \widehat{\mathbb{C}}, \qquad \lim_{n\to\infty} \frac{a_{2n}}{a_{2n+1}} =: \beta \in \widehat{\mathbb{C}} \ \textit{where}\ |\alpha|, |\beta| \neq 1.$$

Proof : Let $a_n \to \infty$ with $(a_{2n-1}/a_{2n}) \to \alpha$ and $(a_{2n}/a_{2n+1}) \to \beta$, and let $S_n(w)$ denote the approximants of $\mathbf{K}(a_n/1)$. Assume first that $|\alpha| < 1$. The even part of $\mathbf{K}(a_n/1)$ is equivalent to

$$\frac{a_1/a_2}{1/a_2+1} - \frac{a_3/a_4}{1/a_4+a_3/a_4+1} - \frac{a_5/a_6}{1/a_6+a_5/a_6+1} - \cdots\,. \tag{2.5.3}$$

The transformation $s(w) := \dfrac{-\alpha}{\alpha+1+w}$ has fixed points $x = -\alpha$, $y = -1$ and ratio $\Re = \dfrac{\alpha+1+y}{\alpha+1+x} = \alpha$. Hence s is loxodromic with repelling fixed point -1. Therefore (2.5.3) converges generally to some $f^{(e)} \in \widehat{\mathbb{C}}$ with constant exceptional sequence $\{-1\}$. Hence it also converges to $f^{(e)}$ in the classical sense.

Next, let $|\alpha| > 1$. The even part of $\mathbf{K}(a_n/1)$ is also equivalent to

$$\frac{1}{1/a_1+a_2/a_1} - \frac{a_2/a_1}{1/a_3+1+a_4/a_3} - \frac{a_4/a_3}{1/a_5+1+a_6/a_5} - \cdots\,. \tag{2.5.4}$$

The transformation $\tilde{s}(w) := \dfrac{-1/\alpha}{1/\alpha + 1 + w}$ is also loxodromic with repelling fixed point $y = -1$. Hence $\{S_{2n}\}$ still converges generally and in the classical sense to some $f^{(e)} \in \widehat{\mathbb{C}}$ with exceptional sequence $\{-1\}$.

Similarly, $\{S_{2n+1}\}$ converges generally to some $f^{(o)} \in \widehat{\mathbb{C}}$ with exceptional sequence $\{-1\}$ if $|\beta| \neq 1$. It remains to prove that $f^{(o)} = f^{(e)}$. By (2.5.4) on page 89

$$S_n^{(e)}(-a_{2n}) = S_{2n-1}(0).$$

Since $(-a_{2n}) \to \infty$, and thus stays asymptotically away from the exceptional sequence $\{-1\}$, it follows that $S_n^{(e)}(-a_{2n}) \to f^{(e)}$. Since $S_{2n-1}(0) \to f^{(o)}$, the result follows. □

If $|\alpha| = 1$, then (2.5.3) and (2.5.4) are limit 1-periodic of parabolic ($\alpha = 1$) or elliptic ($\alpha \neq 1$) type, and the question of convergence is more subtle.

Example 15. The continued fraction

$$\mathcal{R}_\eta(a,b) := \frac{a}{\eta +}\,\frac{1^2b^2}{\eta}\,\underset{+}{}\,\frac{2^2a^2}{\eta}\,\underset{+}{}\,\frac{3^2b^2}{\eta}\,\underset{+}{}\,\frac{4^2a^2}{\eta}\,\underset{+}{}\,\frac{5^2b^2}{\eta}\,\underset{+}{}\cdots$$

where a, b, η are complex numbers $\neq 0$, is called *Ramanujan's AGM-fraction.* The reason for this is the surprising identity

$$\mathcal{R}_\eta(\tfrac{a+b}{2}, \sqrt{ab}) = \tfrac{1}{2}(\mathcal{R}_\eta(a,b) + \mathcal{R}_\eta(b,a))$$

for $a, b, \eta > 0$. That is, AGM is an abbreviation for the arithmetic-geometric mean.

Since $\mathcal{R}_\eta(a,b) \sim \mathcal{R}_1(a/\eta, b/\eta)$, we may without loss of generality assume that $\eta = 1$. It follows from the Parabola Theorem on page 151 that $\mathcal{R}_1(a,b)$ converges if $a^2 = b^2 \in \mathbb{C} \setminus [-\infty, 0]$. Let $a^2 \neq b^2$. Then $\mathcal{R}_1(a,b)$ has the form $\mathbf{K}(a_n/1)$ where $a_n \to \infty$ and

$$\lim_{n\to\infty} \frac{a_{2n}}{a_{2n-1}} = \frac{b^2}{a^2} \quad \text{and} \quad \lim_{n\to\infty} \frac{a_{2n-1}}{a_{2n}} = \frac{a^2}{b^2}.$$

Therefore Theorem 4.26 shows that $\mathcal{R}_1(a,b)$ also converges whenever $|a| \neq |b|$. This was first proved by J. Borwein et al ([BoCF04], [BoCR04]) and later on by Lorentzen ([Lore08b]).

Let $|a| = |b|$, but $a^2 \neq b^2$. Then (2.5.3) and (2.5.4) are limit periodic continued fractions of elliptic type. Since

$$\sum \left|\frac{a_{2n+2}}{a_{2n+1}} - \frac{b^2}{a^2}\right| = \sum \left|\frac{b}{a}\right|^2 \left(\frac{2n+1}{2n} - 1\right)^2 = \sum \frac{1}{4n^2} < \infty,$$

it follows from Corollary 4.25 that (2.5.4) diverges generally. Hence $\mathcal{R}_1(a,b)$ diverges generally in this case. This was also first proved by J. Borwein et al ([BBCM07]) and later on by Lorentzen ([Lore08b]). ◇

4.3 Continued fractions with multiple limits

4.3.1 Periodic continued fractions with multiple limits

A p-periodic continued fraction of elliptic type diverges, of course. Still, the asymptotic behavior of $\{S_n\}$ is quite interesting.

Example 16. The 1-periodic continued fraction $\mathbf{K}(-1/1)$ is of elliptic type since $S_1(w) = s_1(w) = -1/(1+w)$ has the two fixed points $(-1 \pm i\sqrt{3})/2$, and thus the ratio

$$\Re = \frac{1 - i\sqrt{3}}{1 + i\sqrt{3}} = e^{-2i\pi/3}.$$

Since $\Re^3 = 1$, the sequence $\{S_n\}$ of linear fractional transformations is 3-periodic. Indeed,

$$\begin{aligned} S_1(w) &= s_1(w) = \frac{-1}{1+w}, \\ S_2(w) &= s_1 \circ s_1(w) = -\frac{1+w}{w}, \\ S_3(w) &= s_1 \circ s_1 \circ s_1(w) = S_2 \circ s_1(w) = w, \\ S_4(w) &= S_3 \circ s_1(w) = S_1(w) \quad \text{and so on.} \end{aligned}$$

Therefore, even though $\{S_n\}$ is totally non-restrained, $\{S_n(0)\}$ is the 3-periodic sequence

$$\{f_n\}: \quad -1,\ \infty,\ 0,\ -1,\ \infty,\ 0,\ -1, \dots .$$

We follow Bowman and Mc Laughlin ([BoML06]) and say that $\mathbf{K}(-1/1)$ has 3 limits. Also $\{S_n^{-1}\}$ is 3-periodic, so every tail sequence for $\mathbf{K}(-1/1)$ is 3-periodic. For instance, starting with $t_0 = 1$ we get

$$\{t_n\}: \quad 1,\ -2,\ -\tfrac{1}{2},\ 1,\ -2,\ -\tfrac{1}{2},\ 1,\ -2,\ -\tfrac{1}{2}, \dots .$$

Indeed, the continued fraction $\mathbf{K}(a_n/1)$ can be regarded as Np periodic of identity type. ◇

This is part of a general picture. The essential thing is that $\Re^N = 1$; i.e., the ratios $\{\Re^n\}$ for $\{\tau^{[n]}\}$ are N-periodic. Since $\tau^{[n]}$ has the same fixed points as τ itself, this means that $\{\tau^{[n]}\}$ is an N-periodic sequence of linear fractional transformations. (See Property 2 on page 54.)

Lemma 4.27. *Let $\mathbf{K}(a_n/b_n)$ be a p-periodic continued fraction with ratio $\Re \neq 1$. If $\Re^N = 1$ for some $N \in \mathbb{N}$, then $\{S_n\}$ is an (Np)-periodic sequence of linear fractional transformations. If moreover $\mathbf{K}(a_n/b_n)$ has a tail sequence $\{t_n\}$ with all $t_n \neq \infty$ and $\prod_{n=1}^{p}(-t_n)^N = 1$, then also $\{A_n\}$ and $\{B_n\}$ are (Np)-periodic.*

Proof : Since $\{S_{np}^{(m)}\}_n$ is N-periodic for $m = 0, 1, \ldots, p-1$, it follows that $\{S_n\}$ is (Np)-periodic. (In principle it may also have a shorter period length, of course.) Indeed, $S_{Np}(w) = I(w) \equiv w$ is the identity transformation, and thus $S_{nNp} = I$ for all $n \in \mathbb{N}$, so there exists a $\gamma \neq 0, \infty$ such that

$$A_{Np} = 0, \quad A_{Np-1} = \gamma, \quad , B_{Np} = \gamma \quad \text{and} \quad B_{Np-1} = 0.$$

On the other hand it follows from Theorem 2.6 on page 66 that

$$A_{Np} - B_{Np} t_0 = \prod_{n=0}^{Np} (-t_n); \quad \text{i.e.,} \quad -\gamma \, t_0 = -t_0 \prod_{n=1}^{Np} (-t_n).$$

Hence, if $\prod_{n=1}^{Np}(-t_n) = 1$, then $\gamma = 1$, and thus also $\{A_n\}$ and $\{B_n\}$ are (Np)-periodic. □

This was proved by Bowman and Mc Laughlin who studied such continued fractions in a series of papers ([BoML06], [BoML07], [BoML08]).

Example 17. The constant sequence $\{x\}$ with $x := (-1 + i\sqrt{3})/2 = e^{2i\pi/3}$ is a tail sequence for $\mathbf{K}(-1/1)$ from Example 16. Since $\mathfrak{R}^3 = 1$ and $p = 1$, this means that $\{A_n/B_n\}$ is 3-periodic. Moreover, $(-x)^6 = 1$ and $\mathfrak{R}^6 = 1$. Therefore $\{A_n\}$ and $\{B_n\}$ are 6-periodic. Straight forward computation confirms this:

n	-1	0	1	2	3	4	5	6	7	8	9	$\cdots$
a_n			-1	-1	-1	-1	-1	-1	-1	-1	-1	$\cdots$
A_n	1	0	-1	-1	0	1	1	0	-1	-1	0	$\cdots$
B_n	0	1	1	0	-1	-1	0	1	1	0	-1	$\cdots$
A_n/B_n			-1	∞	0	-1	∞	0	-1	∞	0	$\cdots$

◇

4.3.2 Limit periodic continued fractions with multiple limits

A similar situation occurs if $\tau_n \to \tau \in \mathcal{M}$ fast enough, where τ has ratio $\mathfrak{R} \neq 1$ with $\mathfrak{R}^N = 1$. Then $\{\mathcal{T}_n\}$ given by $\mathcal{T}_n := \tau_1 \circ \tau_2 \circ \cdots \circ \tau_n$ is a limit N-periodic (totally non-restrained) sequence. To see this, we let τ have fixed points $x = 0$ and $y = \infty$, without loss of generality. Then we can use Lemma 4.22 on page 197 to prove:

Lemma 4.28. *Let $\{\tau_n\}$ and $\{\mathcal{T}_n\}$ be as in Lemma 4.22. If $\mathfrak{R} \neq 1$, but $\mathfrak{R}^N = 1$ for some $N \in \mathbb{N}$, then $\{\mathcal{T}_{nN+m}\}_n$ converges to some $\tilde{\mathcal{T}}_m \in \mathcal{M}$ for $m = 1, 2, \ldots, N$.*

Proof : From Lemma 4.22 we know that $\mathcal{T}_n(\Re^{-n}w) \to \mathcal{T}(w)$ for a $\mathcal{T} \in \mathcal{M}$. Since $\Re^N = \Re^{-N} = 1$, this means that $\lim_{n\to\infty} \mathcal{T}_{nN+m}(w) = \mathcal{T}(\Re^m w) =: \widehat{\mathcal{T}}_m(w)$ for each fixed $1 \le m \le N$. □

Remark: This means that $\{\mathcal{T}_{nN+m}(w)\}_n$ converges for each $m \in \mathbb{N}$, but to values depending on w.

Example 18. The limit 1-periodic continued fraction $\mathbf{K}(1/b_n)$ where $b_n \to \tilde{b}i$ is of elliptic type if $\tilde{b} \in \mathbb{R}$ with $|\tilde{b}| < 2$. In particular this holds for $\tilde{b} = 0$ when $\tilde{S}_1(w) := 1/(\tilde{b} + w) = 1/w$, and thus has the ratio $\Re = -1$. That is, $\Re^2 = 1$. Therefore $\mathbf{K}(1/b_n)$ with $\sum |b_n| < \infty$ is totally non-restrained and $\{S_{2n-1}\}$ and $\{S_{2n}\}$ converge to functions $T_1 \in \mathcal{M}$ and $T_2 \in \mathcal{M}$ respectively in this case. This was exactly what we proved in Van Vleck's Theorem on page 142. ◇

4.4 Fixed circles

4.4.1 Introduction

In order to get a more geometric idea of the asymptotic behavior of $\{S_n(w)\}$ and $\{S_n^{-1}(w)\}$, we shall look at fixed circles for S_p for p-periodic continued fractions. This will also give us a better idea of what happens in the limit periodic case.

A circle C in $\widehat{\mathbb{C}}$ is called a *fixed circle* for $\tau \in \mathcal{M}$ if it is invariant under the mapping τ; i.e., if $\tau(C) = C$. If $\infty \in C$, it may also be called a *fixed line* for τ. The crucial point is that if w is a point on such a circle C, then iterations of τ will move this point along C in the sense that $\tau^{[n]}(w) \in C$ for all n.

Not every $\tau \in \mathcal{M}$ has fixed circles, but if τ has one, then it has infinitely many. Indeed, we shall see that if τ has fixed circles, and $\tau \neq I$ and $\Re \neq -1$, then for each $w \in \widehat{\mathbb{C}}$ not a fixed point for τ, there is a *unique* fixed circle for τ passing through w. We let C_w denote this fixed circle.

4.4.2 Fixed circles for $\tau \in \mathcal{M}$

If C is a fixed circle for τ, then $\varphi(C)$ is a fixed circle for $\tilde{\tau} := \varphi \circ \tau \circ \varphi^{-1}$. Indeed,

$$\varphi(C_w) = \tilde{C}_{\varphi(w)} \tag{4.2.1}$$

where $\tilde{C}_v$ denotes the fixed circle for $\tilde{\tau}$ through v. We shall use this fact to describe the fixed circles for τ.

If $\tau = I$; i.e., $\tau(w) \equiv w$, then the case is clear: every circle C on $\widehat{\mathbb{C}}$ is a fixed circle for τ. But this fact is of minor interest. Nothing happens to w under iterations of

I. Not much happens if $\Re = -1$ either. Then τ is conjugate to $\tilde{\tau}(w) = -w$ which moves a point from w to $-w$, and back again to w in the sense that $\tilde{\tau}^{[2n-1]}(w) = -w$ and $\tilde{\tau}^{[2n]}(w) = w$ for all n. In the following we assume that $\tau \neq I$ and $\Re \neq -1$; i.e., $\tau^{[2]} \neq I$.

The elliptic case: Let $\tau \in \mathcal{M}$ be elliptic with fixed points x, y and ratio $\Re \neq -1$. Then $|\Re| = 1$ with $\Re \neq \pm 1$; i.e., $\Re = e^{i\theta}$ for some $0 < \theta < 2\pi$ with $\theta \neq \pi$, and τ is conjugate to $\tilde{\tau}(w) := \Re w$ which has fixed points at 0 and ∞. $\tilde{\tau}$ moves a point $w \neq 0, \infty$ the angle θ around on the circle $\tilde{C}_w$ through w centered at the origin. Hence every circle $\tilde{C}$ with center at the origin is a fixed circle for $\tilde{\tau}$, and no other circle has this property. A circle with center at 0 can also be seen as a circle with center at ∞.

Let $\varphi \in \mathcal{M}$ be the transformation which makes $\tau = \varphi^{-1} \circ \tilde{\tau} \circ \varphi$. Then the circles $C := \varphi^{-1}(\tilde{C})$ on the Riemann sphere are fixed circles for τ. They are no longer necessarily concentric, but they are disjoint. For

$$\tau(w) := \frac{aw+b}{cw+d} \quad \text{with } \Delta := ad - bc \neq 0, \ \ c \neq 0,$$

$\varphi(w) := (w-x)/(w-y)$, and thus $\varphi^{-1}(w) = (x - yw)/(1-w)$. The fixed circle for $\tilde{\tau}$ through $\varphi(\infty) = 1$ is the unit circle $\partial\mathbb{D}$. Therefore the fixed line L for τ is given by $L = \varphi^{-1}(\partial\mathbb{D})$. That is, L is given by

$$\varphi^{-1}(e^{it}) = \frac{x - y\,e^{it}}{1-e^{it}} = \frac{x+y}{2} + \frac{x-y}{2}\,\frac{1+e^{it}}{1-e^{it}} \quad \text{for } 0 \le t < 2\pi,$$

where

$$\frac{1+e^{it}}{1-e^{it}} = \frac{e^{-it/2}+e^{it/2}}{e^{-it/2}-e^{it/2}} = \frac{\cos t/2}{-i\sin t/2} = it_1 \quad \text{for } t_1 \in \mathbb{R} \cup \{\infty\} =: \widehat{\mathbb{R}}.$$

That is, L is the perpendicular bisector of the line segment connecting x and y. The other fixed circles $C = \varphi^{-1}(\tilde{C})$ for τ have their centers on the line through x and y, and each one of them winds around x or y, and none of them intersects or touches L or each other. Moreover, since 0 and ∞ are symmetric points with respect to every circle $\tilde{C}$ centered at the origin, it follows that the fixed points x and y are symmetric with respect to each fixed circle (remarks on page 109). We distinguish between the following two cases:

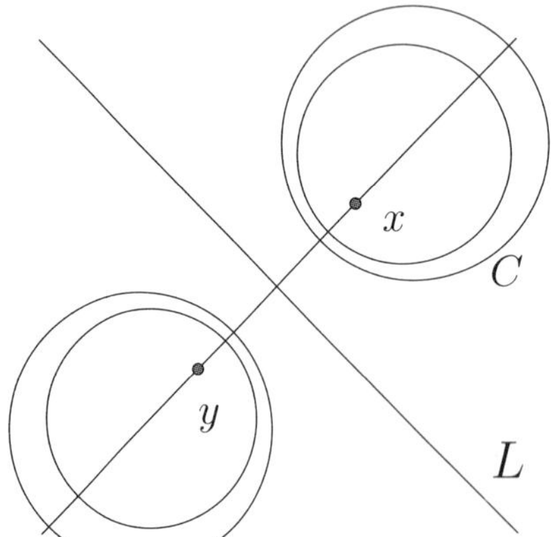

Case 1: $\Re^N \neq 1$ for all $N \in \mathbb{N}$. Then $\{\Re^n\}$ is a sequence of distinct points dense on $\partial\mathbb{D}$, and thus $\{\tilde{\tau}^{[n]}\}$ is a sequence of distinct elliptic transformations $\Re^n w$ from $\mathcal{M}$.

Hence also $\{\tau^{[n]}\}$ is a sequence of distinct elliptic transformations from $\mathcal{M}$. Iterations of τ move a point $w \neq x, y$ around and around on C_w, indefinitely, without ever hitting the same point twice. The limit points for $\{\tau^{[n]}(w)\}$ are dense on C_w.

Case 2: $\Re^N = 1$ where $N > 2$ is the smallest positive integer for which this holds. Then $\{\Re^n\}$ is an N-periodic sequence of equidistant points on $\partial\mathbb{D}$ with $\Re^{nN} = 1$ for all n. Hence $\{\tilde{\tau}^{[n]}\}$ is an N-periodic sequence from $\mathcal{M}$ with $\tilde{\tau}^{nN} = I$ for all n. All other transformations $\tilde{\tau}^{[n]}$ are elliptic. For $w \neq x, y$, $\{\tilde{\tau}^{[n]}(\varphi(w))\}$ is an N-periodic sequence of equidistant points on the circle $\tilde{C}_{\varphi(w)}$ with center at the origin, and so $\{\tau^{[n]}(w)\}$ is an N-periodic sequence of (not necessarily equidistant) points on the fixed circle $\varphi^{-1}(\tilde{C}_{\varphi(w)}) = C_w$ for τ.

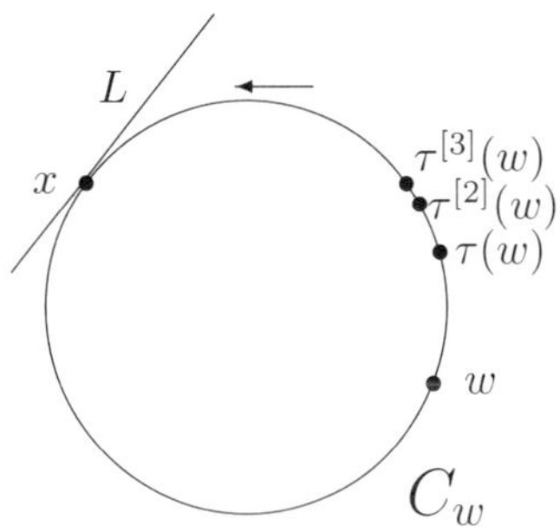

The parabolic case: Let τ be parabolic with fixed point x. If $x = \infty$, then $\tau(w) = w + q$ for some $q \neq 0$. Since $\tau(qt + \lambda) = q(t+1) + \lambda$, it follows that every line $\lambda + q\widehat{\mathbb{R}}$ with $\lambda \in \mathbb{C}$ is invariant under τ. No other circle on $\widehat{\mathbb{C}}$ has this property. If $x \neq \infty$, then τ is conjugate to $\tilde{\tau}(w) := w + q$, and the line L through x and the point $\zeta := \tau^{-1}(\infty) = -d/c$ is a fixed line for τ. This means that the fixed circles for τ are exactly L plus all circles tangent to L at x. Iterations of τ move a point $w \neq x$ monotonely along C_w towards x without ever passing x.

The loxodromic case: Let $\tau \in \mathcal{M}$ be loxodromic. Then τ is conjugate to $\tilde{\tau}(w) = \Re w$ with $0 < |\Re| < 1$ and attracting fixed point $\tilde{x} = 0$ and repelling fixed point $\tilde{y} = \infty$.

Let first $\Re \in \mathbb{R}$. Then $-1 < \Re < 1$, and we say that τ is *hyperbolic*. In this case $\widehat{\mathbb{R}}$ is a fixed line for $\tilde{\tau}$. Indeed, every line $e^{i\alpha}\widehat{\mathbb{R}}$ through the origin is invariant under $\tilde{\tau}$, and these are the only fixed circles for $\tilde{\tau}$. If $\Re \notin \mathbb{R}$, then there are no fixed circles for $\tilde{\tau}$. Roughly speaking, $\tilde{\tau}^{[n]}(w)$ spirals in towards the attracting fixed point at the origin.

This means that a loxodromic transformation τ has fixed circles if and only if it is hyperbolic. In the hyperbolic case, its fixed circles are exactly all the circles passing through both its fixed points x, y. If $\Re > 0$, then iterations of τ move a point w monotonely along the fixed circle C_w towards x. If $\Re < 0$, then iterations of τ also move a point w along C_w towards x, but this time in an alternating manner.

4.4.3 Fixed circles and periodic continued fractions

A p-periodic continued fraction $\mathbf{K}(a_n/b_n)$ has approximants

$$S_{np+m}(w) = S_p^{[n]} \circ S_m(w) = S_m \circ (S_p^{(m)})^{[n]}(w). \tag{4.3.1}$$

If S_p has fixed circles C, then $S_p^{(m)}$ has the fixed circles $S_m^{-1}(C)$. We let $C_w^{(m)}$ denote the fixed circle of $S_p^{(m)}$ through w, and $x^{(m)}$, $y^{(m)}$ be the fixed points of $S_p^{(m)}$ and $x := x^{(0)}$.

The parabolic case.

The fixed circles for $\tilde{\tau}(w) := w + q$ are the lines $\lambda + q\widehat{\mathbb{R}}$ for $\lambda \in \mathbb{C}$. Iterations $\tilde{\tau}^{[n]}(w_0) = w_0 + nq$ of $\tilde{\tau}$ move a point w_0 monotonely along this line towards ∞. Iterations of $\tilde{\tau}^{-1}$ also move w_0 monotonely along this line towards ∞, but in the opposite direction since $(\tilde{\tau}^{-1})^{[n]}(w_0) = w_0 - nq$. Therefore, iterations $S_{np}^{(m)}(w)$ of the parabolic transformation $S_p^{(m)}$ moves a point $w \neq x^{(m)}$ monotonely towards $x^{(m)}$ along the fixed circle $C_w^{(m)}$ of $S_p^{(m)}$. Similarly, ${S_{np}^{(m)}}^{-1}$ moves w monotonely along $C_w^{(m)}$ towards $x^{(m)}$ from the other direction.

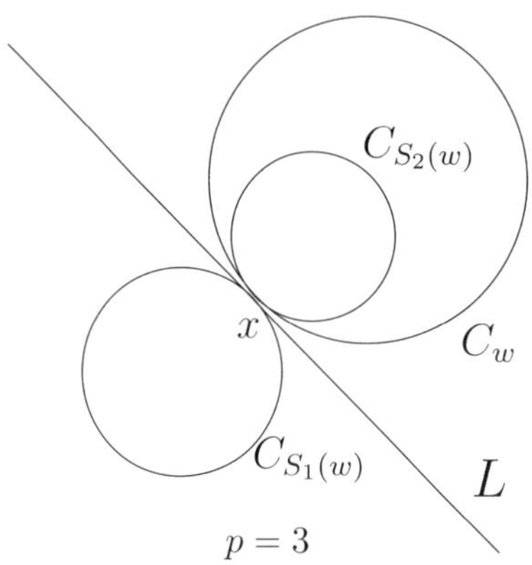

Now, $S_{np+m} = S_m \circ S_{np}^{(m)}$, and $C_{S_m(w)} = S_m(C_w^{(m)})$ is the corresponding fixed circle for S_p for each $m \in \{1, 2, \dots, p\}$. Hence, $S_n(w)$ jumps periodically between the p fixed circles $C_{S_m(w)}$ for S_p, gradually moving towards x, and a tail sequence $\{t_n\}$ with $t_0 \neq x$ jumps periodically between the p circles $C_{t_m}^{(m)}$, since $t_{np+m} := S_{np+m}^{-1}(t_0) = {S_{np}^{(m)}}^{-1} \circ S_m^{-1}(t_0) = {S_{np}^{(m)}}^{-1}(t_m) \in C_{t_m}^{(m)}$. We have proved:

Theorem 4.29. *Let $\mathbf{K}(a_n/b_n)$ be p-periodic of parabolic type. For each $m \in \{1, 2, \dots, p\}$ and $w \in \widehat{\mathbb{C}} \setminus \{x\}$, $\{S_{np+m}(w)\}_n$ is a monotone sequence on $C_{S_m(w)}$ and $\{S_{np+m}^{-1}(w)\}_n$ is a monotone sequence on $C_{S_m^{-1}(w)}^{(m)}$. $\{S_{np}^{(m)}(w)\}_n$ and $\{{S_{np}^{(m)}}^{-1}(w)\}_n$ approach $x^{(m)}$ monotonely along $C_w^{(m)}$ from opposite sides.*

Example 19. The 1-periodic continued fraction $\mathbf{K}((-\frac{1}{4})/1)$ is of parabolic type, and S_1 has the fixed point $x = -\frac{1}{2}$. The fixed circles for S_1 are the line $\widehat{\mathbb{R}}$ and every circle tangent to the real line at $x = -\frac{1}{2}$. Iterations of S_1 move every point $w \neq -\frac{1}{2}$ monotonely along C_w towards $-\frac{1}{2}$, and $\mathbf{K}((-\frac{1}{4})/1)$ converges to $-\frac{1}{2}$. $\{-\frac{1}{2}\}$ is the sequence of tail values for $\mathbf{K}((-\frac{1}{4})/1)$. Hence it follows from Remark 2 on page 182 that $A_n = -n(\frac{1}{2})^{n+1}$ and $B_n = (n+1)(\frac{1}{2})^n$, so that

$$S_n(w) = \frac{-(n-1)(\frac{1}{2})^n w - n(\frac{1}{2})^{n+1}}{n(\frac{1}{2})^{n-1} w + (n+1)(\frac{1}{2})^n} = -\frac{1}{2} + \frac{1}{2n} - \frac{1/2}{2n^2 w + n^2 + n}.$$

Hence $S_n(w)$ approaches $-\frac{1}{2}$ from the right, and a tail sequence $\{t_n\}$ with $t_0 \neq x$ approaches $-\frac{1}{2}$ monotonely along C_{t_0} from the left. ◇

This observation, combined with the following lemma, makes it easy to prove Theorem 4.17 on page 192.

Lemma 4.30. *Let $\mathbf{K}(a_n/b_n)$ be a p-periodic continued fraction of parabolic type with value x, and let $V \subseteq \widehat{\mathbb{C}}$ be a closed set $\neq \widehat{\mathbb{C}}$, containing at least two points. If $S_p(V) \subseteq V$, then $x \in \partial V$.*

Proof : $x \in V$ since $\mathbf{K}(a_n/b_n)$ converges to x. Assume that $x \in V^\circ$. Then every fixed circle C of S_p contains an arc $\mathcal{A} \subseteq V$ with x an inner point in $\mathcal{A}$. For every tail sequence $\{t_n\}$ for $\mathbf{K}(a_n/b_n)$, the points t_{np} lies on such an arc from some n on. Since $S_p(V) \subseteq V$, this means that $t_0 = S_{np}(t_{np}) \in V$ for all $t_0 \in \widehat{\mathbb{C}}$ which is impossible. Hence $x \in \partial V$. □

Proof of Theorem 4.17: Since $a_n \to -\frac{1}{4}$, the convergence of $S_n(w_n)$ with

$$w_n \in V_{\alpha,n} := -g_n + e^{i\alpha}\overline{\mathbb{H}} \quad \text{for } n = 0, 1, 2, \dots \tag{4.3.2}$$

follows from the Parabola Sequence Theorem on page 154. Therefore $\{S_n\}$ and thus also $\{S_n^{-1}\}$ are restrained. Let $g_n \to \frac{1}{2}$. Then $V_{\alpha,n} \to V_\alpha := -\frac{1}{2} + e^{i\alpha}\overline{\mathbb{H}}$. Assume that $\mathbf{K}(a_n/1)$ has a tail sequence $\{t_n\}$ with a limit point $\widehat{x} \neq -\frac{1}{2}$. The fixed circle $C_{\widehat{x}}$ of $\tilde{s}(w) := -\frac{1}{4}/(1+w)$ through $\widehat{x}$ is a circle tangent to the real line at $-\frac{1}{2}$ or the line itself. And every term in the two sequences $\{\tilde{s}^{[n]}(\widehat{x})\}$ and $\{(\tilde{s}^{-1})^{[n]}(\widehat{x})\}$ on $C_{\widehat{x}}$ is also a limit point for $\{t_n\}$ since

$$\begin{aligned} t_{n_k} \to \widehat{x} &\implies t_{n_k-1} = s_{n_k}(t_{n_k}) \to \tilde{s}(\widehat{x}), & t_{n_k-2} &\to \tilde{s}^{[2]}(\widehat{x}), \cdots \\ t_{n_k} \to \widehat{x} &\implies t_{n_k+1} = s_{n_k}^{-1}(t_{n_k}) \to \tilde{s}^{-1}(\widehat{x}), & t_{n_k+2} &\to (\tilde{s}^{-1})^{[2]}(\widehat{x}), \cdots \end{aligned}$$

Since $\tilde{s}^{[n]}(\widehat{x}) \to -\frac{1}{2}$ from one side and $(\tilde{s}^{-1})^{[n]}(\widehat{x}) \to -\frac{1}{2}$ from the other side of $-\frac{1}{2}$ along $C_{\widehat{x}}$, a subarc of $C_{\widehat{x}}$ with $-\frac{1}{2}$ as an inner point must be contained in V_α. This is a contradiction, and $t_n \to -\frac{1}{2}$.

This also means that $S_n(w) \to f$ for every $w \neq -\frac{1}{2}$ since the exceptional sequences for $\{S_n\}$ converge to $-\frac{1}{2}$.

Let $\rho_n := g_{n-1}(1 - g_n) - \frac{1}{4} \geq 0$ from some n on. Without loss of generality we assume that

$$\Sigma_\infty := \sum_{n=0}^{\infty} \prod_{k=1}^{n} \frac{1 - g_k}{g_k} = \infty \quad \text{(remarks on page 136).} \tag{4.3.3}$$

Then the limit point case occurs for $S_n(V_{\alpha,n})$ (the Parabola Sequence Theorem). We know from the remark on page 194 that $\frac{1}{2} \leq g_n \to \frac{1}{2}$. Therefore $-\frac{1}{2} \in V_{\alpha,n}$ from some n on, so also $S_n(-\frac{1}{2}) \to f$. $\square$

The loxodromic case.

If S_p is hyperbolic, then $\{S_{np+m}(0)\}_n$ approaches x along $C_{S_m(0)}$, and $\{t_{np+m}\}_n$ with $t_0 \neq x, y$ approaches $y^{(m)}$ along $C_{t_m}^{(m)}$. If $\Re > 0$, then this convergence is monotone on $C_{S_m(0)}$ and $C_{t_m}^{(m)}$ respectively. If $\Re < 0$, they converge in an alternating manner.

If S_p is loxodromic with $\Re \notin \mathbb{R}$, then $\{S_{np+m}(0)\}_n$ and $\{t_{np+m}\}_n$ with $t_0 \neq x$ spiral in towards their limits x and $y^{(m)}$ respectively.

The elliptic case.

A p-periodic continued fraction of elliptic type diverges, of course, and $\{S_n\}$ is totally non-restrained (Theorem 4.3). But elliptic transformations have fixed circles, so:

- the points $S_{np+m}(w) \in C_{S_m(w)}$ and $S_{np+m}^{-1}(w) \in C_{S_m^{-1}(w)}^{(m)}$ for all n, and they move around and around indefinitely as n increases.
- If the ratio $\Re$ of S_p is an Nth root of unity, then $\{S_{np+m}\}_{n=1}^\infty$ and $\{S_{np+m}^{-1}\}_{n=1}^\infty$ are N-periodic for each $m \in \{1, 2, \dots, p\}$. Otherwise the points $\{S_{np+m}(w)\}_{n=1}^\infty$ and $\{S_{np+m}^{-1}(w)\}_{n=1}^\infty$ are dense on $C_{S_m(w)}$ and $C_{S_m^{-1}(w)}^{(m)}$ respectively.

Example 20. The continued fraction $\mathbf{K}(-2/1)$ is 1-periodic of elliptic type since $S_1(w) = s_1(w) = -2/(1+w)$ has the two fixed points $x := (-1 + i\sqrt{7})/2$ and $y = (-1 - i\sqrt{7})/2$ which gives the ratio $\Re = (1+y)/(1+x) = (1 - i\sqrt{7})/(1 + i\sqrt{7})$. Now, $\Re$ seems to be no Nth root of unity. If this is as it seems, then the sequence $\{S_n(w)\}$ of approximants and the tail sequence $\{S_n^{-1}(w)\}$ with $w \neq x, y$ consist of distinct points dense on the fixed circle C_w for S_1. For instance, $\{f_n\}$ and $\{t_n\}$ with $t_0 \in \widehat{\mathbb{R}}$ are dense on $\widehat{\mathbb{R}}$. Below we have recorded the first few terms of $\{f_n\}$ and $\{t_n\}$ with $t_0 := 1$:

$$\{f_n\}: \quad -2, 2, -\frac{2}{3}, -6, \frac{2}{5}, -\frac{10}{7}, \frac{14}{3}, -\frac{6}{17}, -\frac{34}{11}, \frac{22}{23}, \dots$$

$$\{t_n\}: \quad 1, -3, -\frac{1}{3}, 5, -\frac{7}{5}, \frac{3}{7}, -\frac{17}{3}, -\frac{11}{17}, \frac{23}{11}, -\frac{45}{23}, \dots .$$

$\diamond$

The identity case.

Let $\mathbf{K}(a_n/b_n)$ be a p-periodic continued fraction of identity type. That is, $S_p(w) \equiv w$. Then $S_{np+m} = S_m$ for all $n \in \mathbb{N}$ and $m \in \{1, 2, \dots, p\}$, and both $\{S_n(w)\}$ and $\{S_n^{-1}(w)\}$ are p-periodic. This can not happen with $p = 1$ since $S_1(w) = a_1/(b_1 + w)$. Indeed, since $S_2(w) = a_1(b_2 + w)/(b_1 b_2 + a_2 + b_1 w) \equiv w$ only if $b_1 = b_2 = 0$ and $a_1 = a_2$, and thus $\mathbf{K}(a_n/b_n)$ is 1-periodic of elliptic type, we need $p \geq 3$.

Example 21. For the 3-periodic continued fraction

$$\operatorname*{K}_{n=1}^{\infty} \frac{a_n}{b_n} = -\frac{2}{2} - \frac{2}{1} - \frac{1}{1} - \frac{2}{2} - \frac{2}{1} - \frac{1}{1} - \frac{2}{2} - \cdots,$$

$S_3(w) = w$, so $\mathbf{K}(a_n/b_n)$ is of identity type with

$$S_{3n}(w) = w, \quad S_{3n+1}(w) = \frac{-2}{2+w}, \quad S_{3n+2}(w) = -\frac{1+w}{w}$$

for all n. Hence $\{S_n(0)\}$ is the 3-periodic sequence $-1, \infty, 0, -1, \infty, 0, -1, \dots$, and the tail sequence starting with $t_0 := 1$ is the 3-periodic sequence given by $1, -4, -\frac{1}{2}, 1, -4, \dots$. ◇

4.5 Remarks

1. **Periodic continued fractions.** According to Jones and Thron ([JoTh80], p 1), the first continued fractions in the literature were, as far as we know, terminating periodic continued fractions for approximating square roots. But the first study of infinite periodic continued fractions seems to be due to Daniel Bernoulli ([Berno75]) who studied the 1-periodic continued fraction $\mathbf{K}(1/b)$. Periodic continued fractions was also the topic of E. Galois' first published paper ([Galo28]). Here he studied reversed regular periodic continued fractions. The first comprehensive study of periodic continued fractions in general seems to be due to Stolz ([Stolz86]). Later on, a number of authors have added to the theory, in particular Thiele ([Thie79]), Pringsheim ([Prin00]) and Perron ([Perr05]).

2. **Ratio and trace for $\tau \in \mathcal{M}$.** We have chosen to use the ratio $\Re$ to classify the transformations τ from $\mathcal{M}$. An equivalent criterion is based on the trace $\operatorname{tr}(\tau) := a + d$ of
$$\tau(w) := \frac{aw + b}{cw + d} \quad \text{where } \Delta := ad - bc = 1.$$
The connection between $\Re$ and $\operatorname{tr}(\tau)$ when $\Delta = 1$ is
$$\Re = \frac{1-u}{1+u} \quad \text{where } u := \sqrt{1 - \frac{\Delta}{\operatorname{tr}(\tau)^2}} = \sqrt{1 - \operatorname{tr}(\tau)^{-2}}.$$

3. **Limit periodic continued fractions.** The systematic work on limit periodic continued fractions started with Van Vleck ([VanV04]). He studied S-fractions $\mathbf{K}(a_n z/1)$ where $0 < a_n \to a$. Later on Pringsheim ([Prin10]) and Perron ([Perr13]) made substantial contributions to the theory. This was later simplified in ([Perr29]).

4. **Tail sequences and the linear recurrence relation.** A sequence $\{t_n\}$ is a tail sequence for $\mathbf{K}(a_n/b_n)$ if and only if $t_n = -X_n/X_{n-1}$ for all n where $\{X_n\}$ is a non-trivial solution of the recurrence relation

$$X_n = b_n X_{n-1} + a_n X_{n-2} \quad \text{for } n = 1, 2, 3, \ldots.$$

Theorem 4.13 on page 188 can therefore also be seen as a consequence of Poincaré's work on recurrence relations ([Poin85]) if all $\tilde{a}_m \neq 0$.

5. **Periodic sequences of element sets.** The remark on page 189 shows that if $\mathbf{K}(\tilde{a}_n/\tilde{b}_n)$ is a p-periodic continued fraction of loxodromic type, then there exists a p-periodic sequence $\{\Omega_n\}$ of element sets such that
   - $\mathbf{K}(\tilde{a}_n/\tilde{b}_n)$ is from $\{\Omega_n^\circ\}$, and
   - every $\mathbf{K}(a_n/b_n)$ from $\{\Omega_n\}$ converges generally.

   For the case $p = 2$, several such sequences $\{\Omega_n\}$ have been derived by Thron and coauthors in a number of papers ([Thron43], [Thron59], [LaTh60], [Lange66], [JoTh68], [JoTh70], [Lore08a]. Results for larger p can be found in [Jaco82] and [Jaco87].

6. **Continued fractions and conjugation.** In Remark 3 on page 47 we indicated that our definition of continued fractions, as a sequence $\{S_n\}$ of linear fractional transformations with $S_n(\infty) = S_{n-1}(0)$, could be generalized to give definitions invariant under conjugation.

4.6 Problems

1. ♠ **Diagonalization of matrices.** In Problem 12 on page 49 we saw that if

$$\tau_n := \frac{a_n w + b_n}{c_n w + d_n} \quad \text{was identified by} \quad T_n := \begin{pmatrix} a_n & b_n \\ c_n & d_n \end{pmatrix},$$

then $\tau_1 \circ \tau_2$ was identified by the matrix product $T_1 T_2$.

   (a) Prove that if τ_1 is parabolic, then the composition $\varphi \circ \tau \circ \varphi^{-1}(w) = q + w$ in (1.2.6) on page 173 correspond to diagonalization of the matrix $T_1 := \begin{pmatrix} a & b \\ c & d \end{pmatrix}$.

   (b) Prove that if τ is elliptic or loxodromic, then the composition $\varphi \circ \tau \circ \varphi^{-1}(w) = \Re w$ in (1.2.8) on page 173 also corresponds to diagonalization of T_1.

2. ♠ **Ratio and derivative.** Let x be a finite fixed point for $\tau \in \mathcal{M}$ with ratio $\Re$. Prove that then $\tau'(x) = \Re$ or $\tau'(x) = 1/\Re$.

3. **Convergence of periodic continued fractions.** Prove that $\mathbf{K}(a_n/b_n)$ converges if $a_n = 2$ and $b_n = 4e^{i\theta}$ for all n.

4. **Convergence of periodic continued fractions.** Given the continued fraction

$$b + \operatorname*{K}_{n=1}^{\infty} \frac{a}{2b} = b + \frac{a}{2b} \genfrac{}{}{0pt}{}{}{+} \frac{a}{2b} \genfrac{}{}{0pt}{}{}{+} \frac{a}{2b} \genfrac{}{}{0pt}{}{}{+} \cdots$$

(a) Prove that if $a > 0$ and $b > 0$ then $b + \mathbf{K}(a/2b)$converges to $\sqrt{a+b^2}$. Use this to find a rational approximation to $\sqrt{13}$ with an error less than 10^{-4}.

(b) For which values of $(a, b) \in \mathbb{C} \times \mathbb{C}$ does $b + \mathbf{K}(a/2b)$ converge/diverge?

5. ♠ **Convergence neighborhood of loxodromic transformation.** For given $0 \neq a \in \mathbb{C}$, let $\mathbb{A}$ and $\mathbb{B}$ be the two statements:

$\mathbb{A}$: There exists a region (open, connected set) E with $a \in E$ such that every continued fraction $\mathbf{K}(a_n/1)$ from E converges.

$\mathbb{B}$: $s(w) := a/(1+w)$ is a loxodromic transformation.

Show that $\mathbb{A}$ and $\mathbb{B}$ are equivalent.

6. ♠ **Reversed continued fraction.** The reversed continued fraction of a p-periodic continued fraction

$$\mathop{\mathrm{K}}_{n=1}^{\infty} \frac{a_n}{b_n} = \frac{a_1}{b_1} \mathbin{+} \frac{a_2}{b_2} \mathbin{+} \cdots \mathbin{+} \frac{a_p}{b_p} \mathbin{+} \frac{a_1}{b_1} \mathbin{+} \frac{a_2}{b_2} \mathbin{+} \cdots \mathbin{+} \frac{a_p}{b_p} \mathbin{+} \frac{a_1}{b_1} \mathbin{+} \cdots$$

is the p-periodic continued fraction

$$b_p + \frac{a_p}{b_{p-1}} \mathbin{+} \frac{a_{p-1}}{b_{p-2}} \mathbin{+} \cdots \mathbin{+} \frac{a_1}{b_p} \mathbin{+} \frac{a_p}{b_{p-1}} \mathbin{+} \frac{a_{p-1}}{b_{p-2}} \mathbin{+} \cdots \mathbin{+} \frac{a_1}{b_p} \mathbin{+} \frac{a_p}{b_{p-1}} \mathbin{+} \cdots .$$

Let $\{S_n(w)\}$ and $\{\widehat{S}_n(w)\}$ be the approximants of $\mathbf{K}(a_n/b_n)$ and its reverse, respectively. Prove that $\mathbf{K}(a_n/b_n)$ converges generally if and only if its reversed continued fraction converges generally.

7. ♠ **Approximants of periodic continued fractions.**

(a) Show that the approximants $S_n(w)$ for the 1-periodic continued fraction $\mathbf{K}(-\frac{1}{4}/1)$ can be written

$$S_n(w) = -\frac{1}{2} + \frac{w + 1/2}{2nw + n + 1} \quad \text{for } n = 1, 2, 3, \ldots .$$

Draw a picture of the fixed line for S_n and the fixed circle for S_n through $w_0 := -\frac{1}{2} + 2i$, and mark the approximants $S_n(0)$ and $S_n(w_0)$ for $n = 1, 2, 3$.

(b) Find a closed expression for the nth term of a tail sequence $\{t_n\}$ for $\mathbf{K}(-\frac{1}{4}/1)$, expressed in terms of t_0 and n, and verify that $\{t_n\}$ approaches $-\frac{1}{2}$ from the left when $t_0 := -1$.

8. ♠ **Approximants for 2-periodic continued fractions $\mathbf{K}(a_n/1)$.** For given $x \neq -1, 0, \infty$, find general formulas for the classical approximants of the 2-periodic continued fraction

$$\frac{-x^2}{1} \mathbin{-} \frac{(1+x)^2}{1} \mathbin{-} \frac{x^2}{1} \mathbin{-} \frac{(1+x)^2}{1} \mathbin{-} \frac{x^2}{1} \mathbin{-} \cdots .$$

9. ♠ **Divergence of limit periodic continued fractions of parabolic type.** Let $\mathbf{K}(a_n/1)$ have a tail sequence $\{t_n\}$ given by

$$t_n := -\frac{1}{2} - \frac{1/4 + i\,\mu}{n+1} \quad \text{for } n = 0, 1, 2, \ldots$$

for some $\mu > 0$. Prove that then $\mathbf{K}(a_n/1)$ is a divergent limit 1-periodic continued fraction of parabolic type with $a_n = -\frac{1}{4} - \frac{1/16+\mu^2}{n(n+1)}$.

10. **General and classical convergence.**

    (a) Show that if $\mathbf{K}(a/b)$ is a 1-periodic continued fraction which converges generally to f, then it converges to f also in the classical sense.

    (b) Show that if $\mathbf{K}(a_n/1)$ is a 2-periodic continued fraction which converges generally to f, then it converges to f also in the classical sense.

    (c) Give an example of a periodic continued fraction which converges in the general sense but not in the classical sense.

11. **Thiele oscillation.** For which values of z does the 4-periodic continued fraction

$$\mathop{\mathrm{K}}_{n=1}^{\infty} \frac{a_n}{1} = \frac{1}{1} + \frac{1}{1} - \frac{2}{1} - \frac{2z}{1} + \frac{1}{1} + \frac{1}{1} - \frac{2}{1} - \frac{2z}{1} + \cdots$$

    (a) converge/diverge generally?

    (b) oscillate by Thiele oscillation?

    (c) converge in the classical sense?

    What is the value of $\mathbf{K}(a_n/1)$ when it converges generally? Determine the limiting behavior of its tail sequences in this case.

12. **2-periodic continued fractions and the Parabola Theorem.** Prove that $\mathbf{K}(a_n/1)$ with all $a_{2n-1} := a$ and $a_{2n} := \overline{a}$ converges if and only if $|a| - \mathrm{Re}\, a \le \frac{1}{2}$. (Here $\overline{a}$ is the complex conjugate of a.)

13. **Convergence of (limit) periodic continued fractions.**

    (a) For which values of $z \in \mathbb{C}$ with $|z| = 1$ does the periodic continued fraction $\mathbf{K}(\frac{5z}{4}/1)$ (i) converge? (ii) diverge?

    (b) Find the periodic tail sequences for $\mathbf{K}(\frac{5z}{4}/1)$.

    (c) For which values of $z \in \mathbb{C}$ with $|z| = 1$ does the continued fraction $\mathbf{K}(a_n z/1)$ converge when $a_n := (5n^2+1)/(4(n+1)^2)$ for all n?

14. ♠ **Convergence of limit periodic continued fractions.** For given non-negative integers p, q and r, let

$$a_{2n-1} := \sum_{k=0}^{p} \alpha_k n^k, \qquad a_{2n} := \sum_{k=0}^{p} \gamma_k n^k,$$

$$b_{2n-1} := \sum_{k=0}^{q} \beta_k n^k, \qquad b_{2n} := \sum_{k=0}^{r} \delta_k n^k$$

be polynomials in n for $n = 1, 2, 3, \ldots$, where all α_k, γ_k, β_k and δ_k are complex numbers with $\alpha_p \gamma_p \beta_q \delta_r \neq 0$. Prove that $\mathbf{K}(a_n/b_n)$ converges if either (a), (b), (c) or (d) holds.

(a) $q + r > p$.

(b) $q + r = p$ and $\dfrac{\alpha_p \gamma_p}{(\alpha_p + \gamma_p + \beta_q \delta_r)^2} \notin \left[\frac{1}{4}, \infty\right]$.

(c) $q + r = p - 1$, $\alpha_p = \gamma_p$ and $\beta_q \delta_r / \alpha_p \notin [-\infty, 0]$.

(d) $q + r = p - 2$, $\alpha_p = \gamma_p$ and $\arg \alpha_p = \arg \left(\alpha_{p-1} - \alpha_p \left(\frac{\beta_{q-1}}{\beta_q} + \frac{\delta_{r-1}}{\delta_r} \right) \right)$ and $\beta_q \delta_r / \alpha_p \notin [-\infty, 0]$.

Hint: You may use the Parabola Theorem on page 151 and Theorem 3.4 on page 105.

15. **Convergence of a limit periodic continued fraction.** Prove that $\mathbf{K}(a_n/1)$ with

$$a_{2n} := 1 - 4n^2, \quad a_{2n+1} := -4n^2$$

converges. (Hint: $t_n := (-1)^n n$ is a tail sequence for $\mathbf{K}(a_n/1)$.) What is the value of $\mathbf{K}(a_n/1)$? Does $\mathbf{K}(-a_n/1)$ converge?

16. **Convergence of a limit periodic continued fraction.** Prove that $\mathbf{K}(a_n z/1)$ with

$$a_{2n-1} := a + n, \quad a_{2n} := b + n \quad \text{for } n = 1, 2, 3, \ldots$$

converges for $a, b, z \in \mathbb{C}$ with $|\arg z| < \pi$ (Wall [Wall45]).

17. ♠ **Fixed circles for parabolic transformations.** Let $\tau(w) := (aw + b)/(cw + d)$ be a parabolic transformation from $\mathcal{M}$ with fixed point $x \neq \infty$. Prove that $x + ((a + d)/2c)\widehat{\mathbb{R}}$ is the fixed line for τ.

18. ♠ **Fixed circles for parabolic transformations.** Let $\tau(w) := (aw + b)/(cw + d)$ be a parabolic transformation from $\mathcal{M}$ with finite fixed point x and pole ζ. Prove that the circle with center $\Gamma \in \mathbb{C}$ and radius $\rho > 0$ is a fixed circle for τ if and only if there exists a real number $t \neq 0$ such that

$$\Gamma = x + i(\zeta - x)t, \qquad \rho = |(\zeta - x)t| .$$

19. ♠ **Fixed circles for parabolic transformations.** Let $\{t_n\}$ be a tail sequence for $\mathbf{K}((-\frac{1}{4})/1)$.

(a) Prove that if $t_0 = -\frac{1}{2}$, then $t_n = -\frac{1}{2}$ for all n, and that if $t_0 \neq -\frac{1}{2}$, then all t_n are distinct points on the fixed circle C_{t_0}, approaching $-\frac{1}{2}$ monotonely from the left.

(b) Let $x \in \mathbb{C} \setminus \{0, -1\}$ be arbitrarily chosen. Prove that the 2-periodic continued fraction

$$\mathop{\mathrm{K}} \frac{a_n}{1} = \frac{-x^2}{1} + \frac{-(1+x)^2}{1} + \frac{-x^2}{1} + \frac{-(1+x)^2}{1} + \cdots$$

is of parabolic type with tail values $f^{(2n)} = x$ and $f^{(2n+1)} = -(1 + x)$.

(c) Let $S_n(w)$ be the approximants of $\mathbf{K}(a_n/1)$ in (b), and let C_w and $C_w^{(1)}$ denote the fixed circles through w for S_2 and $S_2^{(1)}$ respectively. Explain why all $S_{2n}(w) \in C_w$ and $S_{2n+1}(w) \in C_{s_1(w)} = C_{-x^2/(1+w)}$.

(d) Let $\{t_n\}$ be a tail sequence for $\mathbf{K}(a_n/1)$. Explain why $t_{2n} \to x$, $t_{2n+1} \to -(1+x)$, and $t_{2n} \in C_{t_0}$, $t_{2n+1} \in C_{t_1}^{(1)}$ for $t_0 \neq x$.

(e) Prove that the fixed line for S_2 in (b) is the line through x and $S_2^{-1}(\infty) = x(2+x)$.

(f) Sketch some fixed circles for S_2 in (b) with $x := i$.

20. **Fixed circle for elliptic transformation.** Prove that the circle with center at γ and radius ρ is a fixed circle for the elliptic transformation $s(w) := -2/(1+w)$ when

(a) $\gamma := -\frac{1}{2} + 2i,\ \ \rho := \frac{3}{2}$. (b) $\gamma := -\frac{1}{2} + \frac{2}{3}\sqrt{7}\,i,\ \ \rho := \frac{7}{6}$.

21. **Fixed circle for elliptic transformation.** Let C be the circle with center at $-\frac{1}{2} + iq$ and radius r where $0 < r < q$. Prove that C is a fixed circle for $s(w) := (-\frac{1}{4} - t^2)/(1+w)$ with $0 \neq t \in \mathbb{R}$ if and only if $q^2 - t^2 = r^2$.

Chapter 5

Numerical computation of continued fractions

If $\mathbf{K}(a_n/b_n)$ converges generally to f, then its approximants $\{S_n(w_n)\}$ converge to f as long as $\{w_n\}$ stays asymptotically away from its exceptional sequence. Therefore we can use $S_n(w_n)$ as an approximation to f. However, the choice of $\{w_n\}$ makes quite a difference. For one thing, if one is looking for, say, a *rational* approximant, then this limits the choice for $\{w_n\}$. But often one just wants fast convergence to f. In this chapter we suggest a number of ideas for how to choose $\{w_n\}$ to this aim.

If we still insist on $S_n(w_n)$ being rational, we chose the elements a_n and b_n of $\mathbf{K}(a_n/b_n)$ to be polynomials or rational functions, and use a rational approximation $\widehat{w}_n$ to w_n to get the rational approximants $S_n(\widehat{w}_n)$.

Even more important from a practical point of view is the control of the truncation error $|f - S_n(w_n)|$. One needs to have explicit bounds for this error in terms of n, in order to take full advantage of this faster convergence. The second part of this chapter deals with truncation error analysis, and we show how one can derive good (i.e., tight) and reliable truncation error bounds.

We also need a method to compute $S_n(w_n)$ in a stable manner. Luckily the backward algorithm is stable in most cases, and good choices for $\{w_n\}$ even stabilize it further. In the last part of this chapter we indicate why and how this works.

5.1 Choice of approximants

5.1.1 Fast convergence

Let $\mathbf{K}(a_n/b_n)$ converge generally to f and let $\{f^{(n)}\}$ be its sequence of tail values. For simplicity we assume that $f \neq \infty$. We can do so without loss of generality, since if $f = \infty$, then $f^{(1)} \neq \infty$, and

$$\left|\frac{1}{S_n(w)}\right| = \left|\frac{1}{f} - \frac{1}{S_n(w)}\right| = \left|\frac{f^{(1)} - S_{n-1}^{(1)}(w)}{a_1}\right| \tag{1.1.1}$$

is a measure for the truncation error. Here, as always, $S_n^{(k)}(w)$ denotes the nth approximant of the kth tail of $\mathbf{K}(a_n/b_n)$. For $f \neq \infty$ we want bounds for the truncation error $|f - S_n(w_n)|$, and the sequence $\{\zeta_n\}$ with $\zeta_n := S_n^{-1}(\infty) = -B_n/B_{n-1}$ is an exceptional sequence for $\mathbf{K}(a_n/b_n)$.

We know that $S_n(w_n) \to f$ whenever $\{w_n\}$ stays asymptotically away from an exceptional sequence $\{w_n^\dagger\}$. But some choices of $\{w_n\}$ give faster convergence than others. Since $f = S_n(f^{(n)})$, it seems reasonable to expect that $S_n(w_n)$ converges faster the better w_n approximates $f^{(n)}$. Since by Lemma 1.1 on page 6

$$\begin{aligned}\frac{f - S_n(w)}{f - S_n(v)} &= \frac{S_n(f^{(n)}) - S_n(w)}{S_n(f^{(n)}) - S_n(v)} \\ &= \frac{B_{n-1}v + B_n}{B_{n-1}w + B_n} \cdot \frac{w - f^{(n)}}{v - f^{(n)}} = \frac{v - \zeta_n}{w - \zeta_n} \cdot \frac{w - f^{(n)}}{v - f^{(n)}}\end{aligned} \tag{1.1.2}$$

(with the natural limit forms if $f^{(n)} = \infty$ or $\zeta_n = \infty$), we have

$$\frac{f - S_n(w_n)}{f - S_n(v_n)} \to 0 \quad \Longleftrightarrow \quad \frac{\kappa_n(w_n)}{\kappa_n(v_n)} \to 0 \quad \text{where } \kappa_n(w) := \frac{w - f^{(n)}}{w - \zeta_n}. \tag{1.1.3}$$

We say that $\{S_n(w_n)\}$ *accelerates* the convergence of $\mathbf{K}(a_n/b_n)$ when

$$\lim_{n\to\infty} \frac{f - S_n(w_n)}{f - S_n(0)} = 0. \tag{1.1.4}$$

The quantity $\kappa_n(w_n)$ is a measure for how well $S_n(w_n)$ approximates f for a given continued fraction. It is essentially the ratio of w_n's distances to the "good tail sequence" $\{f^{(n)}\}$ and the "bad tail sequence" $\{\zeta_n\}$. Hence even a rather lousy approximation to $f^{(n)}$ may work well. In particular we do not need high precision in the computation of w_n to obtain a reasonably good approximation $S_n(w_n)$ to f. But $\{f^{(n)}\}$ is in general unknown. Still, if we have some information on its asymptotic behavior, as we for example often have for limit periodic continued fractions, then this can be used to find favorable w_n. To determine such asymptotic behavior one may need to transform the continued fraction. The good thing about transformations like equivalence transformations or contractions, is that we know exactly what happens to tail sequences. It is also clear that if $\{w_n\}$ accelerates the convergence of the kth tail of $\mathbf{K}(a_n/b_n)$, then it also accelerates the convergence of $\mathbf{K}(a_n/b_n)$.

5.1.2 The fixed point method

This method gives the "obvious" approximants for limit periodic continued fractions.

The loxodromic case.

Let $\mathbf{K}(a_n/b_n)$ be limit p-periodic of loxodromic type with attracting fixed points $(x^{(0)}, \dots, x^{(p-1)})$ and repelling fixed points $(y^{(0)}, \dots, y^{(p-1)})$. Then we know from Theorem 4.13 on page 188 that

- $\mathbf{K}(a_n/b_n)$ converges generally to some $f \in \widehat{\mathbb{C}}$,
- $\lim_{n\to\infty} f^{(np+m)} = x^{(m)}$ for $m = 1, 2, \dots, p$
- $\lim_{n\to\infty} w^\dagger_{np+m} = y^{(m)} \neq x^{(m)}$ for $m = 1, 2, \dots, p$ for the exceptional sequences $\{w^\dagger_n\}$ for $\mathbf{K}(a_n/b_n)$.

Therefore it is a very good idea to use the approximants $S_{np+m}(x^{(m)})$. Indeed, by (1.1.3)

$$\lim_{n\to\infty} \frac{f - S_{np+m}(x^{(m)})}{f - S_{np+m}(w_{np+m})} = 0 \quad \text{for } f \neq \infty \tag{1.2.1}$$

for every other sequence $\{w_n\}$ with $\liminf_{n\to\infty} \mathfrak{m}(w_{np+m}, x^{(m)}) > 0$. In particular,

$$\lim_{n\to\infty} \frac{f - S_{np+m}(x^{(m)})}{f - S_{np+m}(0)} = 0 \text{ for } x^{(m)} \neq 0,\ f \neq \infty. \tag{1.2.2}$$

(If $x^{(m)} = 0$, then $S_{np+m}(0)$ can already be seen as a result of the fixed point method.) The effect of this choice of approximants has been demonstrated in a number of examples in Chapter 1. Of course, it works better the faster $f^{(np+m)}$ approaches $x^{(m)}$, but there is always a positive effect in the long run. Since there is no extra work to speak of to compute $S_{np+m}(x^{(m)})$ instead of $S_{np+m}(0)$, we strongly recommend the use of $S_{np+m}(x^{(m)})$ (or even faster converging approximants to be suggested later in this chapter) whenever convenient.

The parabolic case.

In the parabolic case the question is much more subtle, since $\{f^{(np+m)}\}_n$ and $\{\zeta_{np+m}\}_n$ normally both converge to the attracting fixed point $x^{(m)}$ when $\mathbf{K}(a_n/b_n)$

converges. To get an impression of what to expect, we shall investigate some special cases of the situation

$$\begin{aligned}&\mathbf{K}(a_n/1) \quad \text{with} \quad a_n \to -\tfrac{1}{4} \text{ and } w_n := -\tfrac{1}{2}, \quad \text{and}\\ &a_n \in E_{\alpha,n} := \{a \in \mathbb{C};\ |a| - \operatorname{Re}(a\,e^{-2i\alpha}) \le 2g_{n-1}(1-g_n)\cos^2\alpha\}\end{aligned} \tag{1.2.3}$$

for some fixed $\alpha \in \mathbb{R}$ with $|\alpha| < \pi/2$ and $0 < \varepsilon \le g_n \le 1-\varepsilon < 1$ with $g_{n-1}(1-g_n) \ge \frac{1}{4}$ for all n. It follows then from the Parabola Sequence Theorem on page 154 that $\mathbf{K}(a_n/1)$ converges to a finite value $f \in -g_0 + e^{i\alpha}\overline{\mathbb{H}}$. Moreover $S_n(-\frac{1}{2}) \to f$ (Theorem 4.17 on page 192). The question is now whether

$$\lim_{n\to\infty} \frac{f - S_n(-\frac{1}{2})}{f - S_n(0)} \to 0.$$

By (1.1.2),

$$\frac{f - S_n(-\frac{1}{2})}{f - S_n(0)} = \frac{\kappa_n(-\frac{1}{2})}{\kappa_n(0)} = \frac{f^{(n)} + \frac{1}{2}}{f^{(n)}} \cdot \frac{\zeta_n}{\zeta_n + \frac{1}{2}} \sim \frac{f^{(n)} + \frac{1}{2}}{\zeta_n + \frac{1}{2}}. \tag{1.2.4}$$

To control this ratio, we assume that

$$|a_n + \tfrac{1}{4}| \le \rho_n := g_{n-1}(1-g_n) - \tfrac{1}{4} \quad \text{for } n = 1,2,3,\dots. \tag{1.2.5}$$

Then either $\rho_n = 0$ for all n, or $0 < \rho_n \to 0$ monotonely (Lemma 4.19 on page 193).

Lemma 5.1. *Let $\mathbf{K}(a_n/1)$ satisfy (1.2.5). Then $\{V_n\}_{n=0}^{\infty}$ given by $V_n := \mathfrak{B}(-\frac{1}{2}, g_n - \frac{1}{2})$ is a sequence of value sets for $\mathbf{K}(a_n/1)$, and $f^{(n)} \in V_n$ for all n.*

Proof : We first prove that $\{V_n\}$ is a sequence of value sets for $\mathbf{K}(a_n/1)$: let $w := -\frac{1}{2} + R\,e^{i\psi} \in V_n$; that is, $R \le g_n - \frac{1}{2}$. Since $a_n = -\frac{1}{4} + r\,e^{i\theta}$ for some $0 \le r \le \rho_n$, we get

$$\begin{aligned}\left|\frac{a_n}{1+w} + \frac{1}{2}\right| &= \left|\frac{-\frac{1}{4} + r\,e^{i\theta}}{1 - \frac{1}{2} + R\,e^{i\varphi}} + \frac{1}{2}\right| = \left|\frac{r\,e^{i\theta} + \frac{1}{2}R\,e^{i\varphi}}{\frac{1}{2} + R\,e^{i\varphi}}\right|\\ &\le \frac{\rho_n + \frac{1}{2}(g_n - \frac{1}{2})}{\frac{1}{2} - (g_n - \frac{1}{2})} = \frac{g_{n-1}(1-g_n) - \frac{1}{4} + \frac{1}{2}g_n - \frac{1}{4}}{1 - g_n} = g_{n-1} - \frac{1}{2}.\end{aligned}$$

Hence $a_n/(1+w) \in V_{n-1}$.

Without loss of generality we assume that

$$\sum_{n=0}^{\infty} \widehat{P}_n = \infty \quad \text{for} \quad \widehat{P}_n := \prod_{k=1}^{n} \frac{1 - g_k}{g_k} \tag{1.2.6}$$

(Theorem 3.31 on page 137). Then $\{S_n(V_{0,n})\}$ with $V_{0,n} := -g_n + \overline{\mathbb{H}}$ converges to a limit point f (Theorem 3.45 on page 155). In particular $S_n(w_n) \to f$ for $w_n \in V_n \subseteq V_{0,n}$. Since $S_n(w_n) \in V_0$ for all n, also $f \in V_0$. Similarly $f^{(n)} = \lim_{k\to\infty} S_k^{(n)}(w_{n+k}) \in V_n$ for all n. □

Therefore, if $a_n \in \mathfrak{B}(-\frac{1}{4}, \rho_n)$ for all n, then $|f^{(n)} + \frac{1}{2}| \leq g_n - \frac{1}{2}$ and $|\zeta_n + \frac{1}{2}| > g_n - \frac{1}{2}$ ($\zeta_n \notin V_n$ since $S_n(\zeta_n) = \infty \notin V_0$), and we can guarantee that

$$\limsup_{n\to\infty} \left| \frac{f - S_n(-\frac{1}{2})}{f - S_n(0)} \right| \leq 1. \tag{1.2.7}$$

So, at least we do not loose much in the long run by using $S_n(-\frac{1}{2})$ instead of $S_n(0)$ if $|a_n + \frac{1}{4}| \leq \rho_n$ from some n on. But can we gain? The answer is yes: let $\{\widehat{g}_n\}$ be a second sequence of positive numbers < 1 such that

$$0 < \widehat{\rho}_n := \widehat{g}_{n-1}(1 - \widehat{g}_n) - \tfrac{1}{4} \leq \rho_n \quad \text{and} \quad \frac{\widehat{g}_n - \frac{1}{2}}{g_n - \frac{1}{2}} \to 0. \tag{1.2.8}$$

Then the following holds:

Theorem 5.2. *Let $\{g_n\}$ and $\{\widehat{g}_n\}$ with $\frac{1}{2} < g_n, \widehat{g}_n < 1$ be given such that (1.2.8) holds, where ρ_n is given by (1.2.5). Then $\mathbf{K}(a_n/1)$ with $|a_n + \frac{1}{4}| \leq \widehat{\rho}_n$ for all $n \in \mathbb{N}$, converges to a finite value f and*

$$\lim_{n\to\infty} \left| \frac{f - S_n(-\frac{1}{2})}{f - S_n(0)} \right| = 0. \tag{1.2.9}$$

Proof : This time $|f^{(n)} + \frac{1}{2}| \leq \widehat{g}_n - \frac{1}{2}$ since $\{\mathfrak{B}(-\frac{1}{2}, \widehat{g}_n - \frac{1}{2})\}$ is a sequence of value sets for $\mathbf{K}(a_n/1)$. But also $\{\mathfrak{B}(-\frac{1}{2}, g_n - \frac{1}{2})\}$ is a sequence of value sets for $\mathbf{K}(a_n/1)$ since $|a_n + \frac{1}{4}| \leq \widehat{\rho}_n \leq \rho_n$. Therefore $|\zeta_n + \frac{1}{2}| > g_n - \frac{1}{2}$, and the result follows since

$$\left| \frac{f - S_n(-\frac{1}{2})}{f - S_n(0)} \right| \sim \left| \frac{f^{(n)} + \frac{1}{2}}{\zeta_n + \frac{1}{2}} \right| \leq \frac{\widehat{g}_n - \frac{1}{2}}{g_n - \frac{1}{2}} \to 0.$$

□

The conditions in this theorem are really quite restrictive. Indeed, since $g_n \to \frac{1}{2}$ monotonely, we know from (2.5.10)-(2.5.11) on page 139 that $g_{n-1}(1 - g_n) \sim 1/(16n^2)$ is the best we can do. With the standard little o - notation; i.e., $m_n = o(k_n)$ means that $\lim_{n\to\infty}(m_n/k_n) = 0$, we get:

Corollary 5.3. *Let $\mathbf{K}(a_n/1)$ with $|a_n + \frac{1}{4}| = o(n^{-2})$ have a finite value f. Then (1.2.9) holds.*

Proof : Let $g_n := \frac{1}{2} + 1/(4n+2)$ for $n \geq 0$. Then

$$\begin{aligned} \rho_n &:= g_{n-1}(1-g_n) - \tfrac{1}{4} = \frac{1}{16n^2-4} = (g_{n-1}-\tfrac{1}{2})(g_n-\tfrac{1}{2}) \\ \text{and } g_{n-1} - g_n &= \frac{4}{16n^2-4} = 4(g_{n-1}-\tfrac{1}{2})(g_n-\tfrac{1}{2}) = 4\rho_n. \end{aligned} \tag{1.2.10}$$

Then $\mathfrak{B}(-\frac{1}{4}, \rho_n) \subseteq E_{0,n}$ given by (1.2.3) for all n. Since $\widehat{\rho}_n := |a_n + \frac{1}{4}| = o(\rho_n)$, we have $a_n \in \mathfrak{B}(-\frac{1}{4}, \rho_n)$ from some n on. Without loss of generality we assume that this holds for all $n \geq 1$. Let $\widehat{a}_n := -\frac{1}{4} - |a_n + \frac{1}{4}|$. Then also $\widehat{a}_n \in \mathfrak{B}(-\frac{1}{4}, \rho_n)$ for all n, so $\mathbf{K}(\widehat{a}_n/1)$ converges, and its nth tail value, which we denote by $-\widehat{g}_n$, belongs to $\mathfrak{B}(-\frac{1}{2}, g_n - \frac{1}{2})$ for all n (Lemma 5.1). This means that $1 - g_n \leq \widehat{g}_n \leq g_n$. Indeed, since all $\widehat{a}_n = -\widehat{g}_{n-1}(1-\widehat{g}_n) \leq -\frac{1}{4}$, it follows from Lemma 4.19 on page 193 and the subsequent remark that $\frac{1}{2} \leq \widehat{g}_n \to \frac{1}{2}$ monotonely, so $\frac{1}{2} \leq \widehat{g}_n \leq g_n$.

Now, $a_n \in \mathfrak{B}(-\frac{1}{4}, \widehat{\rho}_n)$ where $(\widehat{\rho}_n/\rho_n) \to 0$. We want to prove that

$$\delta_n := \frac{\widehat{g}_n - \frac{1}{2}}{g_n - \frac{1}{2}} \to 0.$$

Straight forward computation, using (1.2.10), shows that

$$\begin{aligned} \frac{\widehat{\rho}_n}{\rho_n} &= \frac{\widehat{g}_{n-1}(1-\widehat{g}_n) - \frac{1}{4}}{\rho_n} = \frac{\frac{1}{2}(\widehat{g}_{n-1} - \widehat{g}_n) - (\widehat{g}_{n-1} - \frac{1}{2})(\widehat{g}_n - \frac{1}{2})}{(g_{n-1}-\frac{1}{2})(g_n - \frac{1}{2})} \\ &= \frac{\frac{1}{2}(\widehat{g}_{n-1} - \widehat{g}_n)}{(g_{n-1}-\frac{1}{2})(g_n-\frac{1}{2})} - \delta_{n-1}\delta_n = \frac{1}{2}\,\frac{(\widehat{g}_{n-1}-\frac{1}{2}) - (\widehat{g}_n - \frac{1}{2})}{(g_{n-1}-\frac{1}{2})(g_n-\frac{1}{2})} - \delta_{n-1}\delta_n \\ &= \frac{1}{2}\,\frac{\delta_{n-1}\dfrac{g_{n-1}-\frac{1}{2}}{g_n-\frac{1}{2}} - \delta_n}{g_{n-1}-\frac{1}{2}} - \delta_n\delta_{n-1} \end{aligned}$$

where $1 < (g_{n-1} - \frac{1}{2})/(g_n - \frac{1}{2}) \to 1$. Assume that a subsequence $\delta_{n_k-1} \to \delta > 0$. Since $(g_{n_k-1} - \frac{1}{2}) \to 0$, this means that also $\delta_{n_k} \to \delta$. Therefore $\delta_n \to \delta$. However, this implies that

$$\begin{aligned} \widehat{\rho}_n &= \tfrac{1}{2}(\widehat{g}_{n-1} - \widehat{g}_n) - (\widehat{g}_{n-1} - \tfrac{1}{2})(\widehat{g}_n - \tfrac{1}{2}) \\ &\sim \delta \cdot \tfrac{1}{2}(g_{n-1} - g_n) - \delta^2(g_{n-1} - \tfrac{1}{2})(g_n - \tfrac{1}{2}) = \delta(2\rho_n - \delta\rho_n) = \delta(2-\delta)\rho_n \end{aligned}$$

which is impossible since $\widehat{\rho}_n = o(\rho_n)$ and $0 < \delta \leq 1$. Hence $\delta_n \to 0$. □

Remark. If in particular all $|a_n + \frac{1}{4}| \leq C\,n^{-d}$ for some $C > 0$ and $d > 2$, or if $|a_n + \frac{1}{4}| \leq C\,r^n$ for some $C > 0$ and $0 < r < 1$, then we can use

$$g_n := \frac{1}{2} + \frac{1}{4n+2} \quad \text{and} \quad \widehat{g}_n := \frac{1}{2} + \frac{1}{n^{\delta+1}} \quad \text{or} \quad \widehat{g}_n := \frac{1}{2} + q^n$$

from some n on in Theorem 5.2, where $\delta > d - 2$ and $r < q < 1$ are fixed constants. Then $|\zeta_n + \frac{1}{2}| > g_n$ and $|f^{(n)} + \frac{1}{2}| < \widehat{g}_n$ from some n on, and thus it follows from Corollary 5.3 and (1.2.8) that

$$\left|\frac{f - S_n(-\frac{1}{2})}{f - S_n(0)}\right| = \begin{cases} o(n^{-\delta}) & \text{when } |a_n + \frac{1}{4}| \leq C\,n^{-d}, \\ o(\rho^n) & \text{when } |a_n + \frac{1}{4}| \leq C\,r^n \end{cases} \tag{1.2.11}$$

when $f \neq \infty$, as first proved by Thron and Waadeland ([ThWa80a]).

But what if $\mathbf{K}(a_n/b_n)$ has a more general form? If $\mathbf{K}(a_n/b_n)$ is limit 1-periodic with $a_n/b_{n-1}b_n \to -\frac{1}{4}$, it is equivalent to a continued fraction $\mathbf{K}(c_n/d_n)$ where $c_n \to -\frac{1}{4}$ and $d_n = 1$ from some n on.

If $\mathbf{K}(a_n/b_n)$ is limit p-periodic, we can for instance use a p-contraction to approximate the value of $\mathbf{K}(a_n/b_n)$. This contraction is limit 1-periodic, and can possibly be accelerated as described above.

5.1.3 Auxiliary continued fractions

The fixed point method can also be interpreted as follows: let $\mathbf{K}(a_n/b_n)$ be limit p-periodic with finite limits

$$\lim_{n\to\infty} a_{np+m} = \tilde{a}_m \neq 0, \quad \lim_{n\to\infty} b_{np+m} = \tilde{b}_m \quad \text{for } m = 1, 2, \ldots, p.$$

Then its Nth tail looks more and more like the Nth tail of the corresponding periodic continued fraction $\mathbf{K}(\tilde{a}_n/\tilde{b}_n)$ when N increases. We have proved (Theorem 4.13 on page 188) that if $\mathbf{K}(a_n/b_n)$ is of loxodromic type, then the sequence of tail values for $\mathbf{K}(a_n/b_n)$ has the same asymptotic behavior as the sequence $\{\tilde{f}^{(n)}\}$ of tail values for $\mathbf{K}(\tilde{a}_n/\tilde{b}_n)$. Hence the approximants $S_{np+m}(\tilde{f}^{(np+m)}) = S_{np+m}(x^{(m)})$ accelerate the convergence of $\mathbf{K}(a_n/b_n)$ when all $x^{(m)} \neq 0$.

There is no reason why this should only work for pairs of periodic – limit periodic continued fractions. More generally, let $\mathbf{K}(a_n/b_n)$ and $\mathbf{K}(\tilde{a}_n/\tilde{b}_n)$ be two generally convergent continued fractions where

$$\mathfrak{m}(a_n, \tilde{a}_n) \to 0 \quad \text{and} \quad \mathfrak{m}(b_n, \tilde{b}_n) \to 0. \tag{1.3.1}$$

Then their tail values $f^{(n)}$ and $\tilde{f}^{(n)}$ satisfy $\mathfrak{m}(f^{(n)}, \tilde{f}^{(n)}) \to 0$ under proper conditions. We say that $\mathbf{K}(\tilde{a}_n/\tilde{b}_n)$ is an *auxiliary continued fraction* for $\mathbf{K}(a_n/b_n)$, and $\{S_n(w_n)\}$ with $w_n := \tilde{f}^{(n)}$ ought to be a good approximation to the value of $\mathbf{K}(a_n/b_n)$.

Example 1. The continued fraction $\mathbf{K}(n^2/1)$ converges to some $f \neq \infty$ (the Parabola Theorem). Its convergence is slow, so it is a good idea to look for $\{w_n\}$ such that $S_n(w_n) \to f$ faster.

Now, $\mathbf{K}(a_n/1)$ is close to the auxiliary continued fraction $\mathbf{K}((n^2-1)/1)$ which converges with tail values $\tilde{f}^{(n)} := n$. (This follows since $\tilde{f}^{(n-1)}(1+\tilde{f}^{(n)}) = n^2 - 1$ and $\sum_{n=0}^{\infty}\prod_{k=1}^{n}(1+\tilde{f}^{(k)})/(-\tilde{f}^{(k)}) = \infty$.) It seems reasonable to believe that approximants $S_n(w_n)$ with $w_n := n$ is a good choice for $\mathbf{K}(a_n/1)$. The table below indicates that $S_n(n)$ approximates the value $f = 0.4426950409$ (correctly rounded to 10 decimals) reasonably well, and far better than $S_n(0)$. For instance, in order to get an absolute error less than 0.002, we need n to be ≈ 500 for $S_n(0)$, but only ≈ 10 for $S_n(n)$.

n	$S_n(0)$	$S_n(n)$	$\lvert f - S_n(0)\rvert$	$\lvert f - S_n(n)\rvert$
1	1.00000000	0.50000000	$-5.6 \cdot 10^{-1}$	$-5.7 \cdot 10^{-2}$
2	0.20000000	0.42857143	$2.4 \cdot 10^{-1}$	$1.4 \cdot 10^{-2}$
3	0.71428571	0.44827586	$-2.7 \cdot 10^{-1}$	$-5.6 \cdot 10^{-3}$
23	0.48644179	0.44271504	$-4.4 \cdot 10^{-2}$	$-2.0 \cdot 10^{-5}$
24	0.40302141	0.44267739	$4.0 \cdot 10^{-2}$	$1.8 \cdot 10^{-5}$
25	0.48305030	0.44271070	$-4.0 \cdot 10^{-2}$	$-1.6 \cdot 10^{-5}$
501	0.44476903	0.44269504	$-2.1 \cdot 10^{-3}$	$-2.1 \cdot 10^{-9}$
502	0.44063101	0.44269504	$2.1 \cdot 10^{-3}$	$2.0 \cdot 10^{-9}$
503	0.44476080	0.44269504	$-2.1 \cdot 10^{-3}$	$-2.0 \cdot 10^{-9}$

◇

Example 2. The limit periodic continued fraction $\mathbf{K}(\tilde{a}_n/1)$ with $\tilde{a}_n := 110+210/n$ for $n \geq 1$ is of loxodromic type. It converges with tail values $\tilde{f}^{(n)} = 10(n+2)/(n+1)$ for $n \geq 0$. (This can be seen as in the previous example.) To compute the value of $\mathbf{K}((110+215/n)/1)$ we use approximants $S_n(\tilde{f}^{(n)})$. The fixed point method gives approximants $S_n(10)$. The table below shows the values of the approximants $S_n(0)$, $S_n(10)$ and $S_n(\tilde{f}^{(n)})$ for $\mathbf{K}((110+215/n)/1)$ and the corresponding truncation errors. The true value of $\mathbf{K}((110+215/n)/1)$ is $f = 20.18548937$, correctly rounded to 8 decimals.

n	$S_n(0)$	$S_n(10)$	$S_n(10\frac{n+2}{n+1})$
1	325.00000000	29.54545455	20.31250000
2	1.48741419	15.64551422	20.09345794
3	148.35485214	24.22152021	20.25574154
4	3.11185304	17.57667608	20.13065841
97	20.19073608	20.18551791	20.18549003
98	20.18071637	20.18546367	20.18548878
99	20.18983340	20.18551253	20.18548991
100	20.18153746	20.18546851	20.18548889

n	$f - S_n(0)$	$f - S_n(10)$	$f - S_n(10\frac{n+2}{n+1})$
1	-304.8	-9.4	-0.13
2	18.7	4.5	0.092
3	-128.2	-4.0	-0.070
4	17.1	2.6	0.055
97	$-5.2 \cdot 10^{-3}$	$-2.9 \cdot 10^{-5}$	$-6.6 \cdot 10^{-7}$
98	$4.8 \cdot 10^{-3}$	$2.6 \cdot 10^{-5}$	$5.9 \cdot 10^{-7}$
99	$-4.3 \cdot 10^{-3}$	$-2.3 \cdot 10^{-5}$	$-5.4 \cdot 10^{-7}$
100	$4.9 \cdot 10^{-3}$	$2.1 \cdot 10^{-5}$	$4.8 \cdot 10^{-7}$

This continued fraction converges faster than the one in the previous example. Still, the improvement is more than worth the small effort it takes to achieve it. To get an absolute error ≤ 0.005, we need $n \approx 100$ for $S_n(0)$, $n \approx 50$ for $S_n(10)$ and $n \approx 10$ for $S_n(10\frac{n+2}{n+1})$. ◇

But when does $(f^{(n)} - \tilde{f}^{(n)}) \to 0$ hold more generally? Let $\{\Omega_n\}_{n=1}^{\infty}$ be a sequence of element sets for continued fractions $\mathbf{K}(a_n/b_n)$. Let further $\{V_n\}_{n=0}^{\infty}$ be a sequence of value sets for $\{\Omega_n\}$ and let $\operatorname{dist}(x, V)$ denote the euclidean distance between a point $x \in \mathbb{C}$ and a set $V \subseteq \widehat{\mathbb{C}}$. We then follow ([Jaco83], [Jaco87]) and define:

Definition 5.1. *The sequence $\{\Omega_n\}_{n=1}^{\infty}$ of element sets is a tusc with respect to its sequence $\{V_n\}_{n=0}^{\infty}$ of value sets if there exists a sequence $\{\lambda_n\}$; $0 < \lambda_n \to 0$, such that*

$$\operatorname{diam} S_n^{(m)}(V_{m+n}) \leq \lambda_n \quad \text{for all } m, n \in \mathbb{N} \tag{1.3.2}$$

for every continued fraction $\mathbf{K}(a_n/b_n)$ from $\{\Omega_n\}$.

Remarks.

1. The name *tusc* is an abbreviation for "totally uniform sequence of convergence sets".

2. Let $\{\Omega_n\}$ be a tusc with respect to $\{V_n\}$ where $\liminf \operatorname{diam}_{\mathrm{m}}(V_n) > 0$. (As always, $\operatorname{diam}_{\mathrm{m}}(V_n)$ is the chordal diameter of V_n.) Then every continued fraction $\mathbf{K}(a_n/b_n)$ from $\{\Omega_n\}$ converges generally (the limit point case occurs). The convergence of $\{S_n(w_n)\}$ is uniform with respect to $w_n \in V_n$ and $\mathbf{K}(a_n/b_n)$ from $\{\Omega_n\}$.

3. If $\{E_n\}$ is a tusc with respect to $\{V_n\}$ for continued fractions $\mathbf{K}(a_n/1)$, where $\liminf \operatorname{diam}_{\mathrm{m}} V_n > 0$, then $\{E_n\}$ is bounded; i.e., $M := \sup\{|a_n|;\, a_n \in E_n,\, n \in \mathbb{N}\} < \infty$, since the euclidean diameter

$$\operatorname{diam} \frac{a_{m+1}}{1 + V_{m+1}} \leq \lambda_1 \quad \text{for all } m \in \mathbb{N} \cup \{0\}.$$

If moreover $\mathrm{dist}(-1, V_n) \geq \varepsilon > 0$ for all n, then $|S_n^{(m)}(w)| \leq M/\varepsilon$ for all $w \in V_{m+n}$ and $|f^{(m)}| \leq M/\varepsilon$ for every continued fraction from $\{E_n\}$. We can therefore assume without loss of generality that also $\{V_n\}$ is bounded.

4. Let $\{E_{\alpha,n}\}$ and $\{V_{\alpha,n}\}$ be the element sets and value sets in the Parabola Sequence Theorem, where we assume that

$$\widehat{\Sigma}_n^{(m)} := \sum_{k=0}^{n} \widehat{P}_k^{(m)} \to \infty \text{ as } n \to \infty, \text{ uniformly with respect to } m$$

$$\text{for} \quad \widehat{P}_k^{(m)} := \prod_{j=m+1}^{m+n} \frac{1-g_j}{g_j}.$$

Then $\{E_{\alpha,n} \cap \mathfrak{B}(0, M)\}$ is a tusc with respect to $\{V_{\alpha,n}\}$ for every fixed $0 < M < \infty$. This follows since by Remark 2 on page 158

$$\mathrm{diam}\, S_n^{(m)}(V_{\alpha,m+n}) \leq \frac{M/(\varepsilon \cos \alpha}{\prod_{j=1}^{n-1} \left(1 + \frac{\widehat{P}_j^{(m)}}{\widehat{\Sigma}_{j-1}^{(m)}} \cdot \frac{\varepsilon^2 \cos^2 \alpha}{M}\right)}$$

where

$$\prod_{j=1}^{n-1} \left(1 + \frac{\widehat{P}_j^{(m)}}{\widehat{\Sigma}_{j-1}^{(m)}}\right) = \prod_{j=1}^{n-1} \frac{\widehat{\Sigma}_j^{(m)}}{\widehat{\Sigma}_{j-1}^{(m)}} = \widehat{\Sigma}_j^{(m)} \to \infty$$

uniformly with respect to m.

This holds in particular if all $g_n = g$.

Theorem 5.4. *Let $\mathbf{K}(a_n/b_n)$ and $\mathbf{K}(\tilde{a}_n/\tilde{b}_n)$ be two continued fractions from $\{\Omega_n\}$, where $\{\Omega_n\}$ is a tusc with respect to $\{V_n\}$, $\mathrm{dist}(d_n, V_n) \geq \varepsilon > 0$ for all n and all $\mathbf{K}(c_n/d_n)$ from $\{\Omega_n\}$, and $\liminf \mathrm{diam}_{\mathfrak{m}}(V_n) > 0$. If $\{a_n\}$ is bounded, $(a_n - \tilde{a}_n) \to 0$ and $(a_n\tilde{b}_n - \tilde{a}_n b_n) \to 0$, then $(f^{(n)} - \tilde{f}^{(n)}) \to 0$.*

Proof : Let $\{a_n\}$ be bounded, $(a_n - \tilde{a}_n) \to 0$ and $(a_n\tilde{b}_n - \tilde{a}_n b_n) \to 0$. We shall first prove that for each given $n \in \mathbb{N}$,

$$D_n^{(m)} := |S_n^{(m)}(w_{m+n}) - \tilde{S}_n^{(m)}(w_{m+n})| \to 0 \text{ as } m \to \infty \tag{1.3.3}$$

for $w_{m+n} \in V_{m+n}$.

Without loss of generality, $\{V_n\}$ is bounded since

$$|S_n^{(m)}(w)| = \left|\frac{a_{m+1}}{b_{m+1} + S_{n-1}^{(m+1)}(w)}\right| \leq M/\varepsilon \quad \text{for } w \in V_{m+n} \quad \text{where } M := \sup|a_m|.$$

For $n := 1$ and $w_{m+1}, \tilde{w}_{m+1} \in V_{m+1}$,

$$
\begin{aligned}
D_1^{(m)} &= |S_1^{(m)}(w_{m+1}) - \tilde{S}_1^{(m)}(\tilde{w}_{m+1})| = \left| \frac{a_{m+1}}{b_{m+1} + w_{m+1}} - \frac{\tilde{a}_{m+1}}{\tilde{b}_{m+1} + \tilde{w}_{m+1}} \right| \\
&= \left| \frac{a_{m+1}\tilde{b}_{m+1} - \tilde{a}_{m+1}b_{m+1} + a_{m+1}\tilde{w}_{m+1} - \tilde{a}_{m+1}w_{m+1}}{(b_{m+1} + w_{m+1})(\tilde{b}_{m+1} + \tilde{w}_{m+1})} \right|,
\end{aligned}
\tag{1.3.4}
$$

where the denominator is $\geq \varepsilon^2 > 0$ and

$$
a_{m+1}\tilde{w}_{m+1} - \tilde{a}_{m+1}w_{m+1} = a_{m+1}(\tilde{w}_{m+1} - w_{m+1}) + w_{m+1}(a_{m+1} - \tilde{a}_{m+1}).
$$

Hence $D_1^{(m)} \to 0$ if also $(w_{m+1} - \tilde{w}_{m+1}) \to 0$. In particular, $D_1^{(m)} \to 0$ as $m \to \infty$ for $\tilde{w}_{m+1} := w_{m+1}$. We shall prove (1.3.3) by induction on n, so assume that it holds for all $1 \leq n \leq \nu - 1$. For $n := \nu$ we then get

$$
D_n^{(m)} = \left| \frac{a_{m+1}}{b_{m+1} + S_{n-1}^{(m+1)}(w_{n+m})} - \frac{\tilde{a}_{m+1}}{\tilde{b}_{m+1} + \tilde{S}_{n-1}^{(m+1)}(w_{n+m})} \right|
$$

where $v_{m+1} := S_{n-1}^{(m+1)}(w_{m+n}) \in V_m$ and $\tilde{v}_{m+1} := \tilde{S}_{n-1}^{(m+1)}(w_{m+n}) \in V_m$ with $(v_{m+1} - \tilde{v}_{m+1}) \to 0$ as $m \to \infty$ by our induction hypothesis. Hence, also $D_n^{(m)} \to 0$ as $m \to \infty$ by (1.3.4). This proves (1.3.3).

Now, $|f^{(m)} - \tilde{f}^{(m)}| \leq |f^{(m)} - S_n^{(m)}(w_{m+n})| + |\tilde{f}^{(m)} - \tilde{S}_n^{(m)}(w_{m+n})| + D_n^{(m)} \leq 2\lambda_n + D_n^{(m)}$ for every $n \in \mathbb{N}$. Hence, by (1.3.3), $|f^{(m)} - \tilde{f}^{(m)}| \to 0$ as $m \to \infty$. □

5.1.4 The improvement machine for the loxodromic case

Let $\mathbf{K}(a_n/b_n)$ be a limit p-periodic continued fraction of loxodromic type with finite limits, and assume that we have accelerated its convergence by using $S_n(w_n)$ instead of $S_n(0)$, where $0 \neq (w_n - f^{(n)}) \to 0$, for instance by using the fixed point method. If the finite limits

$$
\lim_{n \to \infty} \frac{\lambda_{np+m+1}}{\lambda_{np+m}} =: r_m \quad \text{where } \lambda_n := a_n - w_{n-1}(b_n + w_n) \tag{1.4.1}
$$

exist for $m = 1, 2, \ldots, p$, then we can improve the speed of convergence further. The idea was published by the authors and generalized by Levrie ([Levr89]). We demonstrate how it works for the case $p = 1$ where

$$
a_n \to \tilde{a} \neq \infty, \quad b_n \to \tilde{b} \neq 0, \infty \quad \text{and } |x| < |\tilde{b} + x|,
$$

where x is the attracting fixed point of the loxodromic transformation $\tilde{S}_1(w) := \tilde{a}/(\tilde{b} + w)$ if $\tilde{a} \neq 0$ and $x := 0$ if $\tilde{a} = 0$. (The case $p > 1$ can be found in ([JaWa88], [JaWa90]).)

Theorem 5.5. *Let $\mathbf{K}(a_n/b_n)$ be a limit 1-periodic continued fraction of loxodromic type, with finite limits $\tilde{b} \neq 0$ and $\tilde{a}$, attracting fixed point x, and finite value f. Let further $\{w_n\}$ be a sequence from $\mathbb{C}$ with $w_n \neq 0$ and $0 \neq (w_n - f^{(n)}) \to 0$. If $\lim_{n\to\infty} \lambda_{n+1}/\lambda_n = r$ for $\{\lambda_n\}$ given by (1.4.1), then*

$$\lim_{n\to\infty} \frac{f - S_n(w_n^{(1)})}{f - S_n(w_n)} = 0 \quad \textit{for} \quad w_n^{(1)} := w_n + \frac{\lambda_{n+1}}{(\tilde{b} + x + rx)}.$$

Proof : Let $\lambda_{n+1}/\lambda_n \to r$. Since $f^{(n)} \to x = (-\tilde{b} + \sqrt{\tilde{b}^2 + 4\tilde{a}})/2$ (Theorem 4.13 on page 188), we may without loss of generality assume that all $f^{(n)} \neq \infty$, and $b_n + f^{(n)} \neq 0$. Since $0 \neq \varepsilon_n := (w_n - f^{(n)}) \to 0$, it follows that $\lambda_n \to 0$, and thus that $|r| \leq 1$. Without loss of generality we may therefore assume that $b_n + w_n \neq 0$ and $\lambda_n \neq 0$ for all n (since $\lambda_{n+1}/\lambda_n \to r$). That is,

$$\varepsilon_n := w_n - f^{(n)} \neq 0, \quad f^{(n)} \neq \infty, \quad \lambda_n \neq 0, \quad b_n + f^{(n)} \neq 0, \quad b_n + w_n \neq 0 \quad \text{for all } n.$$

We shall first prove that $\varepsilon_{n+1}/\varepsilon_n \to r$. Since

$$\begin{aligned} \lambda_n &= a_n - w_{n-1}(b_n + w_n) \\ &= f^{(n-1)}(b_n + f^{(n)}) - (f^{(n-1)} + \varepsilon_{n-1})(b_n + f^{(n)} + \varepsilon_n) \\ &= -\varepsilon_{n-1}(b_n + f^{(n)}) - f^{(n-1)}\varepsilon_n - \varepsilon_n\varepsilon_{n-1}, \end{aligned} \tag{1.4.2}$$

it follows that

$$\frac{\lambda_{n+1}}{\lambda_n} = \frac{\varepsilon_n}{\varepsilon_{n-1}} \cdot \frac{b_{n+1} + f^{(n+1)} + w_n\varepsilon_{n+1}/\varepsilon_n}{b_n + f^{(n)} + w_{n-1}\varepsilon_n/\varepsilon_{n-1}}, \tag{1.4.3}$$

so

$$\frac{\varepsilon_n}{\varepsilon_{n-1}} = \frac{\frac{\lambda_{n+1}}{\lambda_n}(b_n + f^{(n)}) + \frac{\lambda_{n+1}}{\lambda_n} w_{n-1}\frac{\varepsilon_n}{\varepsilon_{n-1}}}{b_{n+1} + f^{(n+1)} + w_n\varepsilon_{n+1}/\varepsilon_n}.$$

We multiply this identity by w_{n-1} and solve for $(w_{n-1}\varepsilon_n/\varepsilon_{n-1})$:

$$w_{n-1}\frac{\varepsilon_n}{\varepsilon_{n-1}} = \frac{\frac{\lambda_{n+1}}{\lambda_n}(b_n + f^{(n)})w_{n-1}}{b_{n+1} + f^{(n+1)} - w_{n-1}\frac{\lambda_{n+1}}{\lambda_n} + w_n\frac{\varepsilon_{n+1}}{\varepsilon_n}}.$$

This shows that $\{w_n\varepsilon_{n+1}/\varepsilon_n\}$ is a tail sequence for the continued fraction $\mathbf{K}(c_n/d_n)$ given by

$$\begin{aligned} c_n &:= \frac{\lambda_{n+1}}{\lambda_n}(b_n + f^{(n)})w_{n-1} && \to r(\tilde{b} + x)x = r\tilde{a} =: \tilde{c}, \\ d_n &:= b_{n+1} + f^{(n+1)} - w_{n-1}\frac{\lambda_{n+1}}{\lambda_n} && \to \tilde{b} + x - xr =: \tilde{d}. \end{aligned}$$

Since $\mathbf{K}(c_n/d_n)$ is limit 1-periodic of loxodromic type, where $\tilde{S}_1(w) := \tilde{c}/(\tilde{d}+w)$ has attracting fixed point rx and repelling fixed point $y := -(\tilde{b}+x)$ with $|\tilde{b}+x| > |rx|$, it follows from Theorem 4.13 on page 188 that either $w_{n-1}\varepsilon_n/\varepsilon_{n-1} \to rx$ or $w_{n-1}\varepsilon_n/\varepsilon_{n-1} \to -(\tilde{b}+x)$. However, if $w_{n-1}\varepsilon_n/\varepsilon_{n-1} \to -(\tilde{b}+x)$, then $\varepsilon_n/\varepsilon_{n-1} \to -(\tilde{b}+x)/x$ which is impossible since $0 < \varepsilon_n \to 0$ and $|\tilde{b}+x|/|x| > 1$. Hence $w_{n-1}\varepsilon_n/\varepsilon_{n-1} \to rx$. This proves that $\varepsilon_n/\varepsilon_{n-1} \to r$ when $x \neq 0$.

If $x = 0$, then it follows from (1.4.3) that

$$\frac{\lambda_{n+1}}{\lambda_n} \sim \frac{\varepsilon_n}{\varepsilon_{n-1}} \frac{\tilde{b}}{\tilde{\tilde{b}}} = \frac{\varepsilon_n}{\varepsilon_{n-1}} \quad \text{as } n \to \infty.$$

Hence, also now $\varepsilon_n/\varepsilon_{n-1} \to r$.

Next, we divide (1.4.2) by ε_{n-1} and let $n \to \infty$. Then we find that $\lim_{n\to\infty} \lambda_n/\epsilon_{n-1} = -(\tilde{b}+x+rx) \neq 0$. Since by (1.1.2) on page 218

$$\frac{f - S_n(w_n^{(1)})}{f - S_n(w_n)} = \frac{f^{(n)} - w_n^{(1)}}{f^{(n)} - w_n} \cdot \frac{\zeta_n - w_n}{\zeta_n - w_n^{(1)}},$$

where $\lim(\zeta_n - w_n) = \lim(\zeta_n - w_n^{(1)}) = (y - x) \neq 0, \infty$ ($y := -b - x$ is the repelling fixed point for $\mathbf{K}(a_n/b_n)$) and

$$\frac{f^{(n)} - w_n^{(1)}}{f^{(n)} - w_n} = \frac{-\varepsilon_n - \lambda_{n+1}/(\tilde{b}+x+rx)}{-\varepsilon_n} \to 0,$$

the result follows. □

Remarks.

1. Indeed,
$$\lim_{n\to\infty} \lambda_{n+1}/\lambda_n = r \quad \Longleftrightarrow \quad \lim_{n\to\infty} \epsilon_{n+1}/\epsilon_n = r \tag{1.4.4}$$
under the conditions of Theorem 5.5 since $\Longleftarrow$ follows from (1.4.3).

2. If $w_n = f^{(n)}$ for infinitely many indices, then $\{w_n^{(1)}\}$ does not lead to convergence acceleration, since $f - S_n(w_n) = 0$ for these indices. This is a drawback since $\{f^{(n)}\}$ is not known exactly. On the other hand, this will not disturb the computations too much, since $\{S_n(w_n^{(1)})\}$ still converges reasonably fast to the correct value.

3. The expression for λ_n is also found in the description of the Bauer-Muir transformation. In fact, our improvement machine is closely related to the following idea:
   - Find $\{w_n\}$ such that $(f^{(n)} - w_n) \to 0$.
   - Construct the Bauer-Muir transform $b_0^{(1)} + \mathbf{K}(a_n^{(1)}/b_n^{(1)})$ of $\mathbf{K}(a_n/b_n)$ with respect to $\{w_n\}$.

- Find numbers $\{w_n^{(1)}\}$ which accelerate the convergence of this new continued fraction $b_0^{(1)} + \mathbf{K}(a_n^{(1)}/b_n^{(1)})$, and repeat the process, etc. The main difference is that by using the improvement machine, we do not have to compute the Bauer-Muir transform $\mathbf{K}(a_n^{(1)}/b_n^{(1)})$.

4. A simple, but important special case is the case where all $b_n = 1$ and $a_n \to a$ with $|\arg(a + \frac{1}{4})| < \pi$, and all $w_n = x$, the attractive fixed point for $\mathbf{K}(a_n/1)$. Then $\lambda_n = a_n - a$, and

$$\lim_{n\to\infty} \frac{a_{n+1} - a}{a_n - a} = r \implies \lim_{n\to\infty} \frac{f - S_n(w_n^{(1)})}{f - S_n(x)} = 0$$
$$\text{for} \qquad w_n^{(1)} := x + \frac{a_{n+1} - a}{1 + x + rx}$$

when the value f of $\mathbf{K}(a_n/1)$ is finite.

Example 3. The Stieltjes fraction

$$\mathop{\mathrm{K}}_{n=1}^{\infty} \frac{a_n z}{1} := \cfrac{\frac{1^2 z}{1\cdot 3}}{1} \;+\; \cfrac{\frac{2^2 z}{3\cdot 5}}{1} \;+\; \cfrac{\frac{3^2 z}{5\cdot 7}}{1} \;+\; \cfrac{\frac{4^2 z}{7\cdot 9}}{1} \;+\; \cfrac{\frac{5^2 z}{9\cdot 11}}{1} \;+ \cdots$$

converges to a holomorphic function $f(z)$ in $D := \{z \in \mathbb{C};\, |\arg(z+1)| < \pi\}$ since $a_n \to 1/4 =: \tilde{a}$. Clearly $f(0) = 0$, so let $0 \neq z \in D$. Then $\mathbf{K}(a_n z/1)$ is limit 1-periodic of loxodromic type. Indeed, $\tilde{s}(w) := \tilde{a}z/(1+w)$ has attracting fixed point $x := (u-1)/2$ and repelling fixed point $y := (-u-1)/2$ where $u := \sqrt{1+z}$. Hence the approximants

$$S_n(w_n^{(0)}) \quad \text{where} \quad w_n^{(0)} := x = (u-1)/2; \;\; u := \sqrt{1+z} \tag{1.4.5}$$

can be a starting point for the improvement machine. Since

$$\lambda_n^{(0)} = a_n z - \tilde{a} z = \frac{z/4}{4n^2 - 1} = \frac{z/4}{(2n-1)(2n+1)}$$

in this case, we find that $\lambda_{n+1}^{(0)}/\lambda_n^{(0)} \to 1$, and thus the approximants $S_n(w_n^{(1)})$ with

$$w_n^{(1)} := w_n^{(0)} + \frac{\lambda_{n+1}^{(0)}}{1 + 2x} = x + \frac{a_{n+1}z - \tilde{a}z}{u} = x + \frac{z/(4u)}{(2n+1)(2n+3)}$$

converge faster to $f(z)$ than $S_n(x)$. Now, also $(w_n^{(1)} - f^{(n)}) \to 0$, and

$$\begin{aligned}
\lambda_n^{(1)} &:= a_n z - w_{n-1}^{(1)}(1 + w_n^{(1)}) \\
&= a_n z - \left(x + \frac{z/(4u)}{4n^2-1}\right)\left(1 + x + \frac{z/(4u)}{(2n+1)(2n+3)}\right) \\
&= a_n z - x(1+x) - \frac{xz/(4u)}{(2n+1)(2n+3)} - \frac{(1+x)z/(4u)}{4n^2-1} \\
&\qquad - \frac{z^2/(16u^2)}{(4n^2-1)(2n+1)(2n+3)}
\end{aligned}$$

where $a_n z - x(1+x) = \lambda_n^{(0)}$, so

$$a_n z - x(1+x) - \frac{z/(4u)}{4n^2-1} = \left(1-\frac{1}{u}\right)\frac{z/4}{4n^2-1} = \frac{xz/(2u)}{4n^2-1}\,.$$

Hence

$$\lambda_n^{(1)} = \frac{xz/u}{(4n^2-1)(2n+3)} - \frac{z^2/(16u^2)}{(4n^2-1)(2n+1)(2n+3)}\,,$$

so also $\lambda_{n+1}^{(1)}/\lambda_n^{(1)} \to 1$. Hence $\{S_n(w_n^{(2)})\}$ where

$$w_n^{(2)} := w_n^{(1)} + \frac{\lambda_{n+1}^{(1)}}{1+2x} = w_n^{(1)} + \frac{\lambda_{n+1}^{(1)}}{u}$$

converges even faster. We can continue the process, but for this example we prefer to stop here.

From (2.6.1) in the appendix we know that the principal value of $\arctan z$ has the continued fraction expansion

$$\operatorname{Arctan} z = z/(1+\mathbf{K}(a_n z^2/1)) \quad \text{for } z^2 \in D.$$

That is, $\mathbf{K}(a_n z/1)$ has the value $f(z) = -1+\sqrt{z}/\operatorname{Arctan}\sqrt{z}$. For $z := 1$, this is $f(1) = -1+4/\pi = 0.273239544735$, correctly rounded to 12 decimals. For comparison, the table below shows the first approximants $S_n(0)$, $S_n(x)$, $S_n(w_n^{(1)})$ and $S_n(w_n^{(2)})$ for $\mathbf{K}(a_n z/1)$ for $z := 1$; that is,

$$x := \frac{\sqrt{2}-1}{2}, \quad w_n^{(1)} := x + \frac{1/\sqrt{2}}{4(2n+1)(2n+3)},$$
$$w_n^{(2)} := w_n^{(1)} + \frac{2-\sqrt{2}}{4(2n+1)(2n+3)(2n+5)} - \frac{1/32}{(2n+1)(2n+3)^2(2n+5)}.$$

n	$S_n(0)$	$S_n(x)$	$S_n(w_n^{(1)})$	$S_n(w_n^{(2)})$
5	.2732919254	.2732398207	.2732395555	.2732395411
6	.2732305259	.2732395100	.2732395435	.2732395451
7	.2732410964	.2732395493	.2732395449	.2732395447
8	.2732392779	.2732395441	.2732395447	
9	.2732395906	.2732395448		
10	.2732395368	.2732395447		
11	.2732395461			
12	.2732395448			
13	.2732395448			
14	.2732395447			

Each column stops when we have reached the correctly rounded value with 10 decimals, and this value is repeated for all larger indices. We already see the improvement. Now, the continued fraction converges quite fast for $z := 1$. For values closer to the boundary of D the convergence is slower, and thus the improvement more valuable. For instance, for $z := -2 + i/100$ we get:

n	$\lvert f - S_n(0)\rvert$	$\lvert f - S_n(x)\rvert$	$\lvert f - S_n(w_n^{(1)})\rvert$	$\lvert f - S_n(w_n^{(2)})\rvert$
100	ca 1	$1.0 \cdot 10^{-5}$	$1.5 \cdot 10^{-6}$	$2.9 \cdot 10^{-8}$
1000	$9.2 \cdot 10^{-3}$	$1.2 \cdot 10^{-8}$	$1.6 \cdot 10^{-11}$	$3.5 \cdot 10^{-14}$
5000	$1.9 \cdot 10^{-10}$	$9.5 \cdot 10^{-19}$	$2.7 \cdot 10^{-22}$	$1.1 \cdot 10^{-26}$

This indicates that to obtain a bound for $|f - S_n(w_n)|$ of the order 10^{-8}, we need maybe $n \approx 4000$ for $w_n := 0$, $n \approx 1000$ for $w_n := x$, $n \approx 500$ for $w_n := w_n^{(1)}$ and $n \approx 100$ for $w_n := w_n^{(2)}$. ◇

5.1.5 Asymptotic expansion of tail values

If we can use the improvement machine repeatedly, as we did in the previous example, we actually develop the first terms in some asymptotic expansion of $f^{(n)}$, where the method chooses the terms in the expansion. But we can also take more control over the expansion, and choose its basis functions $u_j(n)$, such that (hopefully)

$$f^{(n)} \sim \sum_{j=\sigma}^{\infty} k_j u_j(n) \quad \text{as } n \to \infty \quad \text{where} \lim_{n\to\infty} \frac{u_{j+1}(n)}{u_j(n)} = 0 \text{ for all } j. \tag{1.5.1}$$

We demonstrate the idea in the following situation: let $\mathbf{K}(a_n/b_n)$ be a convergent continued fraction where $a_n = a(n)$ and $b_n = b(n)$ are polynomials in n. That is, $\mathbf{K}(a_n/b_n)$ is limit periodic. Assume that its tail values $f^{(n)}$ have asymptotic expansions

$$f^{(n)} = f(n) \sim \sum_{j=\sigma}^{\infty} k_j\, n^{-j} \quad \text{as } n \to \infty \tag{1.5.2}$$

for some $\sigma \in \mathbb{N}$ and $k_j \in \mathbb{C}$. That is, we choose an expansion in terms of $u_j(n) := n^{-j}$. The functional equation

$$f(n-1)(b(n) + f(n)) = a(n) \tag{1.5.3}$$

allows us to determine σ and the first coefficients k_j formally. The idea is then to use the approximants $S_n(w_n^{(N)})$ for $\mathbf{K}(a_n/b_n)$ where

$$w_n^{(N)} := \sum_{j=\sigma}^{\sigma+N} k_j\, n^{-j}. \tag{1.5.4}$$

This idea was suggested by Wynn ([Wynn59]).

That $f^{(n)}$ really has an asymptotic expansion of the form (1.5.1) is not always the case. But by the Birkhoff-Trjzinzki theory it follows that if the process of finding σ and k_j works, then $\mathbf{K}(a_n/b_n)$ has a tail sequence $\{t_n\}$ with $t_n \sim \sum_\sigma^\infty k_j u_j(n)$ as $n \to \infty$ where $u_j(n)$ is as specified for $\sigma \le j \le \sigma + N$. We just have to make sure that it is the sequence of tail values. (More information on this can be found in the remark section on page 262.)

Example 4. The continued fraction

$$\mathop{\mathrm{K}}_{n=1}^{\infty} \frac{a_n}{b_n} = \frac{2z}{1} - \frac{1^2z^2}{3} - \frac{2^2z^2}{5} - \frac{3^2z^2}{7} - \cdots$$

converges to $f(z) := \mathrm{Ln}\,\frac{1+z}{1-z}$ in the cut plane D where $0 < \arg(1-z^2) < 2\pi$ (formula (2.4.9) in the appendix on page 271). We assume that (1.5.2) holds. Then (1.5.3) can be written

$$\Big(\sum_{j=\sigma}^{\infty} k_j (n-1)^{-j}\Big)\Big(2n - 1 + \sum_{j=\sigma}^{\infty} k_j\, n^{-j}\Big) = -(n^2 - 2n + 1)z^2 \tag{1.5.5}$$

for $n \geq 2$ where

$$(n-1)^{-j} = n^{-j}\,(1 - 1/n)^{-j} = n^{-j} \sum_{m=0}^{\infty} \binom{-j}{m} \left(-\frac{1}{n}\right)^m . \tag{1.5.6}$$

Hence $\sigma := -1$ is necessary. Comparing the coefficients for n^{-j} on each side of (1.5.5) gives (after some computation) that

$$k_{-1} = u - 1, \quad k_0 = \frac{1-u}{2}, \quad k_1 = \frac{-z^2}{8u}, \ldots \quad \text{where } u^2 = 1 - z^2.$$

That is, we get two possibilities for $w_n^{(2)} := k_{-1}n + k_0 + k_1/n$, depending on our choice for u. To make the right choice we observe that $\mathbf{K}(a_n/1)$ is equivalent to $\mathbf{K}(c_n/1)$ where

$$c_{n+1} = \frac{a_{n+1}}{b_n b_{n+1}} = \frac{-n^2z^2}{4n^2-1} \to -\frac{z^2}{4},$$

so the sequence $\{\widehat{f}^{(n)}\}$ of tail values for $\mathbf{K}(c_n/1)$ converges to the attracting fixed point x of the loxodromic transformation $\widehat{s}(w) := -\frac{z^2}{4}/(1+w)$; i.e., to $x = (u-1)/2$ with $\mathrm{Re}(u) > 0$. Since $f^{(n)} = b_n \widehat{f}^{(n)}$, we let this be our choice for u. Since $\mathbf{K}(a_n/b_n)$ is limit periodic of loxodromic type for $z \in D$, it follows that $S_n(w_n^{(N)})$ converges faster to $\mathrm{Ln}\,\frac{1+z}{1-z}$ than $S_n(0)$.

As an illustration, let $z := 3/4$. Then the continued fraction converges to $\mathrm{Ln}(7) =$ 1.945910149055 correctly rounded to 12 decimals. We compare some low order approximants $S_n(0)$ and $S_n(w_n^{(N)})$ for $N = 0, 1, 2$, where $u := \sqrt{1-z^2}$ and

$$w_n^{(0)} := (u-1)n, \quad w_n^{(1)} := w_n^{(0)} + \frac{1-u}{2}, \quad w_n^{(2)} := w_n^{(1)} - \frac{z^2}{8un}.$$

Each column stops when the value, correctly rounded to 8 decimals, is equal to the rounded value of $\mathrm{Ln}\,\frac{1+z}{1-z}$ for all indices $\geq n$.

n	$S_n(0)$	$S_n(w_n^{(0)})$	$S_n(w_n^{(1)})$	$S_n(w_n^{(2)})$
5	1.94498141	1.94602817	1.94589653	1.94591264
6	1.94571873	1.94593035	1.94590818	1.94591045
7	1.94587082	1.94591370	1.94590985	1.94591019
8	1.94590209	1.94591078	1.94591010	1.94591015
9	1.94590850	1.94591026	1.94591014	
10	1.94590981	1.94591017	1.94591015	
11	1.94591008	1.94591015		
12	1.94591013			
13	1.94591015			

The truncation errors are then for instance

n	$f - S_n(0)$	$f - S_n(w_n^{(0)})$	$f - S_n(w_n^{(1)})$	$f - S_n(w_n^{(2)})$
8	$8.1 \cdot 10^{-6}$	$-6.4 \cdot 10^{-7}$	$4.7 \cdot 10^{-8}$	$-5.6 \cdot 10^{-9}$
9	$1.7 \cdot 10^{-6}$	$-1.2 \cdot 10^{-7}$	$7.7 \cdot 10^{-9}$	$-8.1 \cdot 10^{-10}$
10	$3.4 \cdot 10^{-7}$	$-2.1 \cdot 10^{-8}$	$1.3 \cdot 10^{-9}$	$-1.2 \cdot 10^{-10}$
11	$6.9 \cdot 10^{-8}$	$-4.0 \cdot 10^{-9}$	$2.2 \cdot 10^{-10}$	$-1.9 \cdot 10^{-11}$
12	$1.4 \cdot 10^{-8}$	$-7.4 \cdot 10^{-10}$	$3.7 \cdot 10^{-11}$	$-3.0 \cdot 10^{-12}$

It is more interesting to see what happens for z-values where $\mathbf{K}(a_n/b_n)$ converges more slowly. For $z := -2 + i/100$ the continued fraction has the value

$$f = -1.098567846693 + 3.134926307891\,i,$$

correctly rounded to 12 decimals. The truncation errors are in this case:

n	$\|f - S_n(0)\|$	$\|f - S_n(w_n^{(0)})\|$	$\|f - S_n(w_n^{(1)})\|$	$\|f - S_n(w_n^{(2)})\|$
10	4.3	$1.7 \cdot 10^{-1}$	$5.4 \cdot 10^{-3}$	$3.8 \cdot 10^{-4}$
20	11.3	$8.0 \cdot 10^{-2}$	$1.2 \cdot 10^{-3}$	$4.5 \cdot 10^{-5}$
50	9.1	$2.7 \cdot 10^{-2}$	$1.6 \cdot 10^{-4}$	$2.4 \cdot 10^{-6}$
100	3.1	$1.0 \cdot 10^{-2}$	$3.0 \cdot 10^{-5}$	$2.3 \cdot 10^{-7}$
200	2.6	$2.9 \cdot 10^{-3}$	$4.2 \cdot 10^{-6}$	$1.6 \cdot 10^{-8}$
300	$9.6 \cdot 10^{-1}$	$1.1 \cdot 10^{-3}$	$1.0 \cdot 10^{-6}$	$2.6 \cdot 10^{-9}$
400	$6.2 \cdot 10^{-1}$	$4.5 \cdot 10^{-4}$	$3.3 \cdot 10^{-7}$	$6.2 \cdot 10^{-10}$
500	$3.7 \cdot 10^{-1}$	$2.0 \cdot 10^{-4}$	$1.2 \cdot 10^{-7}$	$1.8 \cdot 10^{-10}$
1000	$1.9 \cdot 10^{-2}$	$5.6 \cdot 10^{-6}$	$1.6 \cdot 10^{-8}$	$1.2 \cdot 10^{-11}$

To get an error less than ca $5 \cdot 10^{-5}$, one actually needs n to be more than 2000 for $S_n(0)$, ca 750 for $S_n(w_n^{(0)})$, ca 100 for $S_n(w_n^{(1)})$ and ca $n = 20$ for $S_n(w_n^{(2)})$. $\diamond$

The method also works in situations where

$$a_{np+m} = a_m(n), \quad b_{np+m} = b_m(n) \quad \text{for } m = 1, 2, \ldots, p.$$

We then try to find the first few terms of expansions $f^{(np+m)} \sim \sum k_j^{(m)} n^{-j}$ for each $m \in \{1, 2, \ldots, p\}$. For further examples we refer to ([Wynn59]).

5.1.6 The square root modification

Let $\mathbf{K}(a_n/b_n)$ be a generally convergent continued fraction with tail values $\{f^{(n)}\}$, where $\{a_n\}$ and $\{b_n\}$ have a monotonic character which makes it natural to replace its nth tail by a periodic continued fraction:

$$f^{(n)} = \frac{a_{n+1}}{b_{n+1}} + \frac{a_{n+2}}{b_{n+2}} + \frac{a_{n+3}}{b_{n+3}} + \cdots \approx \frac{a_{n+1}}{b_{n+1}} + \frac{a_{n+1}}{b_{n+1}} + \frac{a_{n+1}}{b_{n+1}} + \cdots \tag{1.6.1}$$

(or a p-periodic continued fraction). If this last continued fraction converges to some value w_n, then hopefully $f^{(n)}$ is close to this value. Hence we use the approximants $S_n(w_n)$.

This idea was suggested by Gill ([Gill80]). He used it to accelerate continued fractions $\mathbf{K}(a_n/1)$ where $a_n \to -\frac{1}{4}$. In [JaJW87] it was used to accelerate the convergence of continued fractions $\mathbf{K}(a_n/1)$ with $a_n \to \infty$. In both these cases the classical convergence may be rather slow, so a method for convergence acceleration is useful.

Example 5. The continued fraction $\mathbf{K}(a_n/1)$ with

$$a_n = -\frac{1}{4} + \frac{z}{8n} \quad \text{for } n = 1, 2, 3, \ldots$$

is limit 1-periodic of parabolic type. For z in the cut plane $D := \mathbb{C} \setminus (-\infty, 0]$ where $|\arg(z)| < \pi$, it converges to a finite value f by virtue of the Parabola Theorem on page 151, since all $a_n \in E_\alpha$ for $\alpha := \frac{1}{2}\arg z$. Therefore also $\{f^{(n)}\}$ and $\{\zeta_n\}$ converge to $-\frac{1}{2}$, so we do not have high expectations to the speed of convergence of $S_n(-\frac{1}{2})$ as compared to $S_n(0)$. However, $a_n \to -\frac{1}{4}$ monotonely, and its nth tail looks similar to the periodic continued fraction

$$\frac{a_{n+1}}{1} + \frac{a_{n+1}}{1} + \frac{a_{n+1}}{1} + \cdots$$

which converges to

$$w_n := \frac{\sqrt{1 + 4a_{n+1}} - 1}{2} = -\frac{1}{2} + \frac{1}{2}\sqrt{\frac{z}{2(n+1)}}. \tag{1.6.2}$$

We therefore expect $S_n(w_n)$ to converge faster, something which is confirmed by the table below. For $z := 1$, the correct value of $\mathbf{K}(a_n/1)$ is -0.172160228791, rounded to 12 decimals.

n	$S_n(0)$	$S_n(-\frac{1}{2})$	$S_n(w_n)$
1	−0.12500000	−0.25000000	−0.16666667
2	−0.15384615	−0.20000000	−0.17036664
3	−0.16379310	−0.18421053	−0.17144140
4	−0.16793893	−0.17801047	−0.17183361
5	−0.16987898	−0.17523168	−0.17199842
6	−0.17086247	−0.17386837	−0.17207476
7	−0.17139164	−0.17315351	−0.17211279
8	−0.17168993	−0.17275888	−0.17213282
9	−0.17186450	−0.17253187	−0.17214386
10	−0.17196993	−0.17239677	−0.17215017
11	−0.17203530	−0.17231404	−0.17215390
12	−0.17207677	−0.17226212	−0.17215617

For $z := -2 + i/10$ the continued fraction has the value

$$f = -0.634775601464989 + 0.578831734368452\, i,$$

correctly rounded to 15 decimals. The truncation errors for this slower converging continued fractions are for instance:

n	$\lvert f - S_n(0)\rvert$	$\lvert f - S_n(-\frac{1}{2})\rvert$	$\lvert f - S_n(w_n)\rvert$
100	$4.03 \cdot 10^{-1}$	$5.90 \cdot 10^{-1}$	$6.15 \cdot 10^{-3}$
500	$1.28 \cdot 10^{-1}$	$1.64 \cdot 10^{-1}$	$8.03 \cdot 10^{-4}$
1000	$5.47 \cdot 10^{-2}$	$5.95 \cdot 10^{-2}$	$2.25 \cdot 10^{-4}$
5000	$1.15 \cdot 10^{-3}$	$1.15 \cdot 10^{-3}$	$2.03 \cdot 10^{-6}$
10000	$6.15 \cdot 10^{-5}$	$6.16 \cdot 10^{-5}$	$7.69 \cdot 10^{-8}$
100000	$2.52 \cdot 10^{-14}$	$2.52 \cdot 10^{-14}$	$9.98 \cdot 10^{-18}$

In this case there is almost no difference between $|f - S_n(0)|$ and $|f - S_n(-\frac{1}{2})|$, but $|f - S_n(w_n)|$ is considerably smaller. For instance is $|f - S_n(w_n)| \le 2 \cdot 10^{-6}$ for $n = 5000$, whereas we need considerably more than $n = 10\,000$ for $|f - S_n(0)|$ to be so small. ◇

Example 6. Let $a_n := n$ for all $n \in \mathbb{N}$. Then $\mathbf{K}(a_n/1)$ converges to a finite value f (the Seidel-Stern Theorem on page 117). But the convergence of $\{S_n(0)\}$ is relatively slow. The square root modification gives

$$S_n(w_n) \quad \text{with} \quad w_n := \left(\sqrt{1 + 4(n+1)} - 1\right)/2. \tag{1.6.3}$$

The table below shows the approximants $S_n(0)$ and $S_n(w_n)$ and the corresponding truncation errors. Here $f = 0.525135276161$ is the value of $\mathbf{K}(a_n/1)$, correctly rounded to 12 decimals.

n	$S_n(0)$	$S_n(w_n)$	$f - S_n(0)$	$f - S_n(w_n)$
1	1.00000000	0.50000000	$-4.75 \cdot 10^{-1}$	$2.51 \cdot 10^{-2}$
2	0.33333333	0.53518376	$1.92 \cdot 10^{-1}$	$-1.00 \cdot 10^{-2}$
3	0.66666667	0.52051760	$-1.42 \cdot 10^{-1}$	$4.61 \cdot 10^{-3}$
4	0.44444444	0.52752523	$8.07 \cdot 10^{-2}$	$-2.39 \cdot 10^{-3}$
5	0.58333333	0.52380952	$-5.82 \cdot 10^{-2}$	$1.33 \cdot 10^{-3}$
11	0.53312330	0.52504318	$-7.99 \cdot 10^{-3}$	$9.21 \cdot 10^{-5}$
12	0.51912183	0.52519962	$6.01 \cdot 10^{-3}$	$-6.43 \cdot 10^{-5}$
51	0.52514011	0.52513526	$-4.83 \cdot 10^{-6}$	$1.24 \cdot 10^{-8}$
52	0.52513106	0.52513529	$4.21 \cdot 10^{-6}$	$-1.06 \cdot 10^{-8}$
101	0.52513529	0.52513528	$-1.53 \cdot 10^{-8}$	$1.97 \cdot 10^{-11}$
102	0.52513526	0.52513528	$1.39 \cdot 10^{-8}$	$-1.76 \cdot 10^{-11}$

An alternative method is to rather use the even (or odd) part of $\mathbf{K}(a_n/1)$, ([JaWa86]). The even part of $\mathbf{K}(n/1)$ is

$$\frac{1}{1+2} \underset{-}{} \frac{2\cdot 3}{1+3+4} \underset{-}{} \frac{4\cdot 5}{1+5+6} \underset{-}{} \cdots$$

which is equivalent to $\mathbf{K}(c_n/1)$ with $c_1 := 1/3$, $c_2 := -1/4$ and

$$\begin{aligned} c_n &:= \frac{-(2n-2)(2n-1)}{(1+2n-3+2n-2)(1+2n-1+2n)} \\ &= -\frac{n-1/2}{4n} = -\frac{1}{4} + \frac{1}{8n} \to -\frac{1}{4} \quad \text{as } n \longrightarrow \infty\,. \end{aligned}$$

We recognize this continued fraction from the previous example where we also used the square root modification. ◇

Of course, the square root modification also works for limit periodic continued fractions of loxodromic type. In our next example we show a method to compute π:

Example 7. The beautiful continued fraction

$$\mathop{\mathrm{K}}_{n=1}^{\infty} \frac{a_n}{1} \quad \text{where } a_1 := 4,\ \ a_{n+1} := \frac{n^2}{4n^2-1} \ \text{ for } n \geq 1 \tag{1.6.4}$$

converges to π (appendix, page 267). $\mathbf{K}(a_n/1)$ converges rather fast, but it is easy to accelerate its convergence. The fixed point method gives for instance the approximants

$$S_n(x) \quad \text{where } x := \tfrac{1}{2}(\sqrt{1+4\cdot\tfrac{1}{4}}-1) = \tfrac{1}{2}(\sqrt{2}-1),$$

and the square root modification

$$S_n(w_n) \quad \text{where } w_n := \tfrac{1}{2}(\sqrt{1+4a_{n+1}}-1) = \frac{1}{2}\Big(\sqrt{\frac{8n^2-1}{4n^2-1}}-1\Big)$$

ought to approximate π even better. The table below shows that this is indeed the case:

n	$S_n(x)$	$S_n(w_n)$	$\pi - S_n(x)$	$\pi - S_n(w_n)$
1	3.3137084990	3.1651513899	$-1.7 \cdot 10^{-1}$	$-2.4 \cdot 10^{-2}$
2	3.1344464996	3.1409646632	$7.1 \cdot 10^{-3}$	$6.3 \cdot 10^{-4}$
3	3.1421356273	3.1416290537	$-5.4 \cdot 10^{-4}$	$-3.6 \cdot 10^{-5}$
4	3.1415404008	3.1415898074	$5.2 \cdot 10^{-5}$	$2.8 \cdot 10^{-6}$
5	3.1415983790	3.1415929165	$-5.7 \cdot 10^{-6}$	$-2.6 \cdot 10^{-7}$
6	3.1415919726	3.1415926266	$6.8 \cdot 10^{-7}$	$2.7 \cdot 10^{-8}$
7	3.1415927393	3.1415926566	$-8.6 \cdot 10^{-8}$	$-3.0 \cdot 10^{-9}$
8	3.1415926423	3.1415926532	$1.1 \cdot 10^{-8}$	$3.5 \cdot 10^{-10}$
9	3.1415926551	3.1415926536	$-1.5 \cdot 10^{-9}$	$-4.3 \cdot 10^{-11}$
10	3.1415926534	3.1415926536	$2.1 \cdot 10^{-10}$	$5.4 \cdot 10^{-12}$

Both $\{S_n(x)\}$ and $\{S_n(w_n)\}$ are alternating sequences – a fact that gives simple a posteriori truncation error bounds. $\diamond$

5.2 Truncation error bounds

5.2.1 The ideas

Let $\mathbf{K}(a_n/b_n)$ converge generally to a finite value f. We want to find bounds for the truncation error $|f - S_n(w_n)|$, where $\{w_n\}$ is a given sequence of complex numbers, chosen to make $\{S_n(w_n)\}$ converge fast to f. In particular we want the bounds to reflect the fast convergence due to the choice of $\{w_n\}$. We shall therefore look for a sequence $\{V_n\}_{n=0}^{\infty}$ of closed value sets for $\mathbf{K}(a_n/b_n)$ with $f^{(n)} \in V_n$ and $w_n \in V_n$ for all n. Then $f = S_n(f^{(n)}) \in S_n(V_n)$ and $S_n(w_n) \in S_n(V_n)$.

Idea 1: We can use $\{V_n\}$ as in Chapter 3 to derive a priori bounds

$$|f - S_n(w_n)| \leq \operatorname{diam} S_n(V_n). \tag{2.1.1}$$

These bounds are valid for every $w_n \in V_n$.

Idea 2: We can use $\{V_n\}$ to derive a posteriori bounds based on the following lemma:

Lemma 5.6. *Let $\infty \neq f_n^* := S_n(w_n) \to f \neq \infty$ for the generally convergent continued fraction $\mathbf{K}(a_n/b_n)$ with value f, tail values $\{f^{(n)}\}$ and critical tail sequence $\{\zeta_n\}$ with $\zeta_n \neq \infty$. Then*

$$f - f_n^* = \frac{a_n(f^{(n)} - w_n)}{-\lambda_n} \cdot \frac{1 - w_{n-1}/\zeta_{n-1}}{f^{(n)} - \zeta_n}(f_n^* - f_{n-1}^*)$$

when $\lambda_n := a_n - w_{n-1}(b_n + w_n) \neq 0$ and $w_n \neq 0, \infty$ for all n.

Proof : Let $w_n^* := s_n^{-1}(w_{n-1}) = -b_n + a_n/w_{n-1}$. Then $w_n^* \neq \infty$ and

$$\begin{aligned} f_n^* - f_{n-1}^* &= S_n(w_n) - S_{n-1}(w_{n-1}) = S_n(w_n) - S_n(w_n^*) \\ &= \frac{A_{n-1}w_n + A_n}{B_{n-1}w_n + B_n} - \frac{A_{n-1}w_n^* + A_n}{B_{n-1}w_n^* + B_n} = \frac{(A_{n-1}B_n - A_nB_{n-1})(w_n - w_n^*)}{(B_{n-1}w_n + B_n)(B_{n-1}w_n^* + B_n)}, \end{aligned}$$

and thus, since $f = S_n(f^{(n)})$,

$$\frac{f - f_n^*}{f_n^* - f_{n-1}^*} = \frac{S_n(f^{(n)}) - S_n(w_n)}{S_n(w_n) - S_n(w_n^*)} = \frac{f^{(n)} - w_n}{w_n - w_n^*} \cdot \frac{B_{n-1}w_n^* + B_n}{B_{n-1}f^{(n)} + B_n}.$$

Now, $\zeta_n = -B_n/B_{n-1} = -b_n + a_n/\zeta_{n-1}$, and both $w_n \neq \zeta_n$ and $f^{(n)} \neq \zeta_n$ since $f_n^* \neq \infty$ and $f \neq \infty$. Since also $B_{n-1} \neq 0$ (because $\zeta_n \neq \infty$), we get

$$\frac{B_{n-1}w_n^* + B_n}{B_{n-1}f^{(n)} + B_n} = \frac{-b_n + a_n/w_{n-1} - \zeta_n}{f^{(n)} - \zeta_n} = a_n \frac{w_{n-1}^{-1} - \zeta_{n-1}^{-1}}{f^{(n)} - \zeta_n}$$

and the result follows. □

If $f^{(n)} \in V_n$, $w_n \in V_n$ and $\zeta_n \notin V_n$ for all n, then we can hopefully find a bound for $|(w_n - f^{(n)})(1 - w_{n-1}/\zeta_{n-1})|/|f^{(n)} - \zeta_n|$, and thus we have an a posteriori bound for $|f - S_n(w_n)|$.

Idea 3: We can combine an already known truncation error bound $|f - S_n(\widehat{w}_n)| \leq \lambda_n$ for $\mathbf{K}(a_n/b_n)$ with the identity

$$|f - S_n(w_n)| = \left|\frac{\widehat{w}_n - \zeta_n}{w_n - \zeta_n} \cdot \frac{w_n - f^{(n)}}{\widehat{w}_n - f^{(n)}}\right| |f - S_n(\widehat{w}_n)| \tag{2.1.2}$$

((1.1.2) on page 218). If $w_n \in V_n$ and $f^{(n)} \in V_n$, then $|w_n - f^{(n)}| \leq \operatorname{diam} V_n$. If we can establish a bound

$$\left|\frac{\widehat{w}_n - \zeta_n}{(w_n - \zeta_n)(\widehat{w}_n - f^{(n)})}\right| \leq k_n,$$

then $|f - S_n(w_n)| \leq k_n \lambda_n \operatorname{diam} V_n$.

In Chapter 3 we started with the value sets, and the purpose was to prove convergence for a large family of continued fractions. The truncation error bounds we obtained, were then valid for every continued fraction from this family and for every $w_n \in V_n$. What we now want, is to find "small" value sets $\{V_n\}$ for a given continued fraction $\mathbf{K}(a_n/b_n)$ to get "small" error bounds. It takes considerably more effort to establish such bounds, so the truncation error bounds in Chapter 3 are still useful because of their simplicity.

Now, $\{w_n\}$ is a known sequence, so it is easy to make sure that $w_n \in V_n$ for all n. The tail values $f^{(n)}$ on the other hand, are unknown. (That is in fact the whole point! If $f^{(n)}$ was known, then $f = S_n(f^{(n)})$ makes the need for $\{w_n\}$ and truncation error bounds quite obsolete.) So, how can we make sure that also $f^{(n)} \in V_n$ for all n? Well, for one thing, if $f^{(n)} \in V_n$ from some n on, then $f^{(n)} \in V_n$ for all n. But more importantly, the following holds:

Lemma 5.7. *Let $\{V_n\}$ be a sequence of closed value sets for $\mathbf{K}(a_n/b_n)$. If there exists a sequence $\{\tilde{w}_n\}$ with $\tilde{w}_n \in V_n$ for all n such that $S_n(\tilde{w}_n) \to f$, then $f^{(n)} := S_n^{-1}(f) \in V_n$ for all n.*

Proof : Since $S_n(\tilde{w}_n) \in V_0$ for all n, we have $f \in V_0$, and similarly

$$S_k^{(n)}(\tilde{w}_{n+k}) := \frac{a_{n+1}}{b_{n+1}} + \cdots + \frac{a_{n+k}}{b_{n+k} + \tilde{w}_{n+k}} \in V_n \text{ for all } n \text{ and } k \tag{2.1.3}$$

implies that

$$f^{(n)} = S_n^{-1}(f) = \lim_{k\to\infty} S_n^{-1} \circ S_{n+k}(\tilde{w}_{n+k}) = \lim_{k\to\infty} S_k^{(n)}(\tilde{w}_{n+k}) \in V_n.$$

□

This means that if $\mathbf{K}(a_n/b_n)$ actually converges generally to this value f, then its tail values $f^{(n)} \in V_n$. This gives us the tool we need to develop a priori truncation error bounds.

5.2.2 Truncation error bounds

Let $\mathbf{K}(a_n/b_n)$ be a generally convergent continued fraction, and let $\{w_n\}$ be a sequence of complex numbers which approximate its tail values. We want to find a sequence $\{V_n\}_{n=0}^{\infty}$ of "small" value sets for $\mathbf{K}(a_n/b_n)$ where $w_n \in V_n$, at least from some n on.

Since circular disks behave so nicely under linear fractional transformations, we shall let V_n be closed circular disks. As earlier, we use the notation

$$\mathfrak{B}(\Gamma, \rho) := \{w \in \mathbb{C};\ |w - \Gamma| \le \rho\} \quad \text{for } \Gamma \in \mathbb{C} \text{ and } \rho > 0. \tag{2.2.1}$$

We let $\Gamma_n := w_n$ be the center of V_n (or $\Gamma_n \approx w_n$). It then remains to determine $\rho_n > 0$ such that $a_n/(b_n + \mathfrak{B}(\Gamma_n, \rho_n)) \subseteq \mathfrak{B}(\Gamma_{n-1}, \rho_{n-1})$ for all n. We first note the following result due to Lane ([Lane45]):

Lemma 5.8. *Let $V_n := \mathfrak{B}(\Gamma_n, \rho_n)$ for $\Gamma_n \in \mathbb{C}$ and $\rho_n > 0$ for $n \geq 0$, and let Ω_n be the set of all pairs $(a,b) \in \mathbb{C}^2$ for which $|b + \Gamma_n| > \rho_n$ and*

$$\left|\frac{a(\bar{b}+\bar{\Gamma}_n)}{|b+\Gamma_n|^2-\rho_n^2} - \Gamma_{n-1}\right| + \frac{|a|\rho_n}{|b+\Gamma_n|^2-\rho_n^2} \leq \rho_{n-1} \tag{2.2.2}$$

for $n \geq 1$. Then $\{V_n\}$ is a sequence of value sets for every $\mathbf{K}(a_n/b_n)$ from $\{\Omega_n\}$.

Proof : From Lemma 3.6 on page 110 it follows that

$$\begin{aligned} &\frac{a}{b+\mathfrak{B}(\Gamma_n,\rho_n)} = \mathfrak{B}(\Gamma_n^*, \rho_n^*) \text{ where} \\ &\Gamma_n^* := \frac{a(\bar{b}+\bar{\Gamma}_n)}{|b+\Gamma_n|^2-\rho_n^2}, \quad \rho_n^* := \frac{\rho_n|a|}{|b+\Gamma_n|^2-\rho_n^2}. \end{aligned} \tag{2.2.3}$$

Since $\mathfrak{B}(\Gamma_n^*, \rho_n^*) \subseteq \mathfrak{B}(\Gamma_{n-1}, \rho_{n-1})$ if and only if $|b+\Gamma_n| > \rho_n$ and $|\Gamma_n^* - \Gamma_{n-1}| + \rho_n^* \leq \rho_{n-1}$, the result follows. □

Theorem 5.9. *Let $\mathbf{K}(a_n/b_n)$ be a continued fraction from $\{\Omega_n\}$ as given in Lemma 5.8. Then*

$$|S_{n+m}(w) - S_n(\Gamma_n)| \leq \rho_n \frac{|\Gamma_0| + \rho_0}{|b_n + \Gamma_n| - \rho_n} \prod_{k-1}^{n-1} M_k \quad \textit{for } w \in V_{n+m}, \tag{2.2.4}$$

where $M_k := \max\left\{\left|\frac{w}{b_k+w}\right|;\ w \in V_k\right\}$ and $V_n := \mathfrak{B}(\Gamma_n, \rho_n)$ for all n.

Proof : Since $S_n = s_1 \circ s_2 \circ \cdots \circ s_n$ where $s_k(w) := a_k/(1+w)$, the chain rule shows that its derivative at $w \in V_n$ is given by

$$S_n'(w) = s_1'(w_{n,1}) \cdot s_2'(w_{n,2}) \cdots s_n'(w_{n,n}) \tag{2.2.5}$$

where $w_{n,n} := w \in V_n$, $w_{n,k} := s_{k+1}(w_{n,k+1}) \in V_k$ for $k := n-1, n-2, \ldots, 1$, and

$$s_k'(w_{n,k}) = \frac{-a_k}{(b_k + w_{n,k})^2} = \frac{-w_{n,k-1}}{b_k + w_{n,k}}. \tag{2.2.6}$$

Therefore

$$|S_n'(w)| = \prod_{k=1}^{n} \frac{|w_{n,k-1}|}{|b_k + w_{n,k}|} = \frac{|w_{n,0}|}{|b_n + w_{n,n}|} \prod_{k=1}^{n-1} \left| \frac{w_{n,k}}{b_k + w_{n,k}} \right|$$
$$\leq \frac{|\Gamma_0| + \rho_0}{|b_n + \Gamma_n| - \rho_n} \prod_{k=1}^{n-1} M_k,$$

and thus, for $w^* \in V_n$,

$$|S_n(w^*) - S_n(\Gamma_n)| \leq |w^* - \Gamma_n| \cdot \sup_{w \in V_n} |S_n'(w)| \leq \rho_n \cdot \frac{|\Gamma_0| + \rho_0}{|b_n + \Gamma_n| - \rho_n} \prod_{k=1}^{n-1} M_k\,.$$

Since $S_{n+m}(w) = S_n(w^*)$ for a $w^* \in V_n$ when $w \in V_{n+m}$, this proves (2.2.4). □

Remarks:

1. If $S_n(\Gamma_n) \to f \neq \infty$, then (2.2.4) implies that

$$|f - S_n(\Gamma_n)| \leq \rho_n \frac{|\Gamma_0| + \rho_0}{|b_n + \Gamma_n| - \rho_n} \prod_{k=1}^{n-1} M_k.$$

2. It follows from the proof of (2.2.4) that

$$\operatorname{diam} S_n(\mathfrak{B}(\Gamma_n, \rho_n)) \leq 2\rho_n \frac{|\Gamma_0| + \rho_0}{|b_n + \Gamma_n| - \rho_n} \prod_{k=1}^{n-1} M_k.$$

3. For given $\{\rho_n^*\}$ with $\rho_n \leq \rho_n^* < |b_n + \Gamma_n|$ for all n,

$$\rho_n \frac{|\Gamma_0| + \rho_0}{|b_n + \Gamma_n| - \rho_n} \prod_{k=1}^{n-1} M_k \leq \rho_n^* \frac{|\Gamma_0| + \rho_0^*}{|b_n + \Gamma_n| - \rho_n^*} \prod_{k=1}^{n-1} M_k^*$$

where

$$M_k^* := \max \left\{ \left| \frac{w}{b_k + w} \right| ;\ |w - \Gamma_k| \leq \rho_k^* \right\}.$$

4. Just as in Remark 1 on page 163, we find that

$$M_k = \frac{|b_k + \Gamma_k|^2 - (\overline{b}_k + \overline{\Gamma}_k) - \rho_k^2| + \rho_k}{|b_k + \Gamma_k|^2 - \rho_k^2}\,.$$

5.2.3 The Oval Sequence Theorem

The idea is now to let $\Gamma_n := w_n$ and choose ρ_n such that $(a_n, b_n) \in \Omega_n$ given in Lemma 5.8. Now, Lemma 5.8 offers no guarantee for $\Omega_n \neq \emptyset$. Indeed, it would be helpful to have an idea of what Ω_n looks like. Its shape was investigated in [JaTh86] for the special case where all $b_n = 1$. To keep things reasonably simple, we stay with this case. Then we need $\rho_n < |1+\Gamma_n|$, and we replace Ω_n by $E_n \times \{1\}$ where

$$E_n := \left\{a \in \mathbb{C};\ \left|\frac{a(1+\overline{\Gamma}_n)}{|1+\Gamma_n|^2 - \rho_n^2} - \Gamma_{n-1}\right| + \frac{|a|\rho_n}{|1+\Gamma_n|^2 - \rho_n^2} \le \rho_{n-1}\right\}. \qquad (2.3.1)$$

If $\Gamma_{n-1} = 0$, then $a \in E_n$ if

$$0 \le |a|\,\frac{|1+\Gamma_n| + \rho_n}{|1+\Gamma_n|^2 - \rho_n^2} = \frac{|a|}{|1+\Gamma_n| - \rho_n} \le \rho_{n-1}.$$

That is, E_n is the circular disk

$$E_n = \mathfrak{B}(0, r_n) \quad \text{where } r_n := \rho_{n-1}(|1+\Gamma_n| - \rho_n) > 0. \qquad (2.3.2)$$

Let $\Gamma_{n-1} \neq 0$. Then straight forward checking shows that E_n is bounded by the curve $\partial E_n = a_n^* \mathcal{O}_n$ where

$$a_n^* := \frac{\Gamma_{n-1}}{1+\overline{\Gamma}_n} d_n \quad \text{with } d_n := |1+\Gamma_n|^2 - \rho_n^2 \qquad (2.3.3)$$

and

$$\mathcal{O}_n := \{\xi \in \mathbb{C};\ |\xi - 1| + k_n|\xi| = \ell_n\} \quad \text{where } k_n := \frac{\rho_n}{|1+\Gamma_n|},\quad \ell_n := \frac{\rho_{n-1}}{|\Gamma_{n-1}|}. \qquad (2.3.4)$$

Let O_n be the closed set bounded by $\mathcal{O}_n$. Since $k_n < 1$, we find that $O_n = \emptyset$ if $1 \notin O_n$: i.e., if $k_n > \ell_n$. If $k_n = \ell_n$, then O_n consists of the point $\xi = 1$ only. Hence we assume that

$$k_n < 1 \quad \text{and} \quad k_n < \ell_n. \qquad (2.3.5)$$

Then $1 \in O^\circ$, and we recognize $\mathcal{O}_n = \partial O_n$ as a cartesian oval. As a corollary to Theorem 5.9 we now get:

Corollary 5.10. (The Oval Sequence Theorem.) *Let $\mathbf{K}(a_n/1)$ be a continued fraction from $\{E_n\}$ given by (2.3.1). Then*

$$|S_{n+m}(w) - S_n(\Gamma_n)| \le \rho_n \frac{|\Gamma_0| + \rho_0}{|1+\Gamma_n| - \rho_n} \prod_{k=1}^{n-1} M_k \ \text{ for } w \subset V_{n+m}$$

where $M_k := \max\left\{\left|\frac{w}{1+w}\right|;\ w \in V_k\right\}$ *and* $V : n := \mathfrak{B}(\Gamma_n, \rho_n)$ *for all* n.

Remark. From Remark 4 on page 242, we find that

$$M_k = \frac{|\Gamma_k(1+\overline{\Gamma}_k) - \rho_k^2| + \rho_k}{|1+\Gamma_k|^2 - \rho_k^2}. \tag{2.3.6}$$

If $\Gamma_k > 0$ and $\rho_k^2 \leq \Gamma_k(1+\Gamma_k)$, this reduces to

$$M_k = \frac{\Gamma_k + \rho_k}{1+\Gamma_k+\rho_k}. \tag{2.3.7}$$

5.2.4 An algorithm to find value sets for a given continued fraction of form $\mathbf{K}(a_n/1)$

Let $\mathbf{K}(a_n/1)$ be a given generally convergent continued fraction with (unknown) sequence $\{f^{(n)}\}$ of tail values. Let $\{w_n\}$ be a given sequence from $\mathbb{C}\setminus\{-1\}$ approximating $\{f^{(n)}\}$. We are looking for circular disks in $\mathbb{C}$ centered at $\Gamma_n := w_n$ (or at a point Γ_n close to w_n) such that $\{V_n\}_{n=0}^{\infty}$ with

$$w_n \in V_n := \mathfrak{B}(\Gamma_n, \rho_n) \quad \text{for } n \geq n_0 \tag{2.4.1}$$

for some $n_0 \in \mathbb{N}$, is a sequence of value sets for $\mathbf{K}(a_n/1)$. That is, we are looking for radii $\rho_n > 0$ such that $a_n \in E_n$ given by (2.3.1).

The shape of the ovals.

If $\Gamma_{n-1} \neq 0$, then $E_n = a_n^* O$ given by (2.3.3)–(2.3.4) has the following properties:

1. E_n is convex and symmetric about the line $e^{2i\alpha_n}\mathbb{R}$ where

   $$\alpha_n := \tfrac{1}{2}\arg a_n^* = \tfrac{1}{2}\arg(\Gamma_{n-1}(1+\Gamma_n)). \tag{2.4.2}$$

   (Lemma 3.50 on page 161.)

2. The intersection $E_n \cap (e^{2i\alpha_n}\mathbb{R})$ is called the *axis* of E_n. Lemma 3.50 shows that $E_n \cap (e^{2i\alpha_n}\mathbb{R}) = e^{2i\alpha_n} I_n$ where $I_n := [u_n, v_n]$ with $u_n := a_n^*\mu$ and $v_n := a_n^*\nu$; that is,

   $$u_n := \begin{cases} (|\Gamma_{n-1}| - \rho_{n-1})(|1+\Gamma_n| - \rho_n) & \text{if } |\Gamma_{n-1}| \leq \rho_{n-1}, \\ (|\Gamma_{n-1}| - \rho_{n-1})(|1+\Gamma_n| + \rho_n) & \text{if } |\Gamma_{n-1}| \geq \rho_{n-1}, \end{cases} \tag{2.4.3}$$
   $$v_n := (|\Gamma_{n-1}| + \rho_{n-1})(|1+\Gamma_n| - \rho_n),$$

   as essentially pointed out in [JaTh86, p100].

3. Straight forward checking shows that

   $$\mathfrak{B}(a_n^*, r_n^*) \subseteq E_n \quad \text{when } \Gamma_{n-1} \neq 0, \tag{2.4.4}$$

   where

   $$r_n^* := (\rho_{n-1}|1+\Gamma_n| - \rho_n|\Gamma_{n-1}|)(1 - \rho_n/|1+\Gamma_n|). \tag{2.4.5}$$

   The boundary of this disk is tangent to ∂E_n at $v_n e^{2i\alpha}$.

4. It is often easier to work with a disk $\subseteq E_n$ centered at $\widehat{a}_n := \Gamma_{n-1}(1+\Gamma_n)$. This disk must also be tangent to ∂E_n at $v_n e^{2i\alpha}$ if it exists. This happens if and only if
$$\begin{aligned} r_n &:= v_n - |\Gamma_{n-1}(1+\Gamma_n)| \\ &= \rho_{n-1}|1+\Gamma_n| - \rho_n|\Gamma_{n-1}| - \rho_n\rho_{n-1} > 0. \end{aligned} \tag{2.4.6}$$
Then
$$\mathfrak{B}(\Gamma_{n-1}(1+\Gamma_n), r_n) \subseteq E_n. \tag{2.4.7}$$

Strategies.

We can derive several strategies for our algorithm, strategies based on the oval sets $\{E_n\}$, strategies based on the interval $I_n = [u_n, v_n]$ given by (2.4.3), strategies based on the disk $\mathfrak{B}(a_n^*, r_n^*)$ in (2.4.4), and strategies based on the disk
$$\begin{aligned} \mathfrak{B}(\widehat{a}_n, r_n) \quad &\text{where } \widehat{a}_n := \Gamma_{n-1}(1+\Gamma_n) \\ &\text{and } r_n := \rho_{n-1}|1+\Gamma_n| - \rho_n|\Gamma_{n-1}| - \rho_n\rho_{n-1} > 0 \end{aligned} \tag{2.4.8}$$
from (2.4.7). We choose the latter alternative, since it gives an easier algorithm (but not always the best result).

The algorithm.

The algorithm is due to Jacobsen ([Jaco82], [Jaco87], ([Lore03b]).

1. Let $\{\sigma_n\}$ be a sequence of positive numbers such that $\{\Psi_n\}$ given by
$$\Psi_n := \sigma_n|1+\Gamma_n| - \sigma_{n-1}|\Gamma_{n-1}| > 0; \quad n \in \mathbb{N} \tag{2.4.9}$$
is non-decreasing. Such a sequence always exists when $\Gamma_n \neq -1$ for $n \geq 1$. For instance, for arbitrary $\Psi > 0$, the choice
$$\begin{aligned} \sigma_n := \frac{\Psi}{|1+\Gamma_n|}\Big(1 + \Big|\frac{\Gamma_{n-1}}{1+\Gamma_{n-1}}\Big| \\ + \Big|\frac{\Gamma_{n-1}}{1+\Gamma_{n-1}}\Big|\Big|\frac{\Gamma_{n-2}}{1+\Gamma_{n-2}}\Big| + \cdots + \prod_{j=0}^{n-1}\Big|\frac{\Gamma_j}{1+\Gamma_j}\Big|\Big) \end{aligned} \tag{2.4.10}$$
for $n = 0, 1, 2, \ldots$ gives $\Psi_n = \Psi$ for all n.

2. Let
$$\begin{aligned} \rho_n &:= \frac{1}{2\sigma_n}\left(\Psi_{n+1} - \sqrt{\Psi_{n+1}^2 - 4\delta_{n+1}}\right) \quad \text{where} \\ \delta_{n+1} &:= \sup_{m \geq n+1} \{\sigma_{m-1}\sigma_m|a_m - \widehat{a}_m|;\ \widehat{a}_m := \Gamma_{m-1}(1+\Gamma_m)\} \end{aligned} \tag{2.4.11}$$

for $n \geq n_0$, where $n_0 \in \mathbb{N}$ is chosen such that $4\delta_{n_0+1} \leq \Psi_{n_0+1}^2$, and thus $\rho_n \geq 0$ for $n \geq n_0$. Then $\{\rho_n\sigma_n\}_{n=n_0}^{\infty}$ is non-increasing since

$$2\rho_n\sigma_n = \Psi_{n+1} - \sqrt{\Psi_{n+1}^2 - 4\delta_{n+1}} = \frac{4\delta_{n+1}}{\Psi_{n+1} + \sqrt{\Psi_{n+1}^2 - 4\delta_{n+1}}}$$

where $\{\Psi_n\}$ is non-decreasing and $\{\delta_n\}$ is non-increasing..

We want $|a_n - \widehat{a}_n| \leq r_n$ where $\widehat{a}_n$ and r_n are given by (2.4.8). We have

$$\begin{aligned} &r_n\sigma_n\sigma_{n-1} \\ &= (\rho_{n-1}\sigma_{n-1})\sigma_n|1+\Gamma_n| - (\rho_n\sigma_n)\sigma_{n-1}|\Gamma_{n-1}| - (\rho_n\sigma_n)(\rho_{n-1}\sigma_{n-1}) \\ &\geq (\rho_{n-1}\sigma_{n-1})(\Psi_n - \rho_{n-1}\sigma_{n-1}) = \delta_n. \end{aligned}$$

That is, $|a_n - \widehat{a}_n|\sigma_n\sigma_{n-1} \leq \delta_n \leq r_n\sigma_n\sigma_{n-1}$; i.e., $|a_n - \widehat{a}_n| \leq r_n$ for $n \geq n_0 + 1$.

Theorem 5.11. *For a given continued fraction* $\mathbf{K}(a_n/1)$ *and a given sequence* $\{\Gamma_n\}$ *of complex numbers* $\neq -1$*, let* $\{\sigma_n\}$ *be a sequence of positive numbers such that* $\{\Psi_n\}$ *given by (2.4.9) is non-decreasing. Let further* ρ_n *and* δ_n *be given by (2.4.11). If* $4\delta_{n_0+1} \leq \Psi_{n_0+1}^2$ *for some* $n_0 \geq 0$*, then there exists a sequence* $\{V_n\}$ *of value sets for* $\mathbf{K}(a_n/1)$ *with* $V_n := \mathfrak{B}(\Gamma_n, \rho_n)$ *for* $n \geq n_0$.

Remarks.

1. The expression (2.4.11) for ρ_{n-1} satisfies

$$\rho_{n-1} = \frac{2\delta_n/\sigma_{n-1}}{\Psi_n + \sqrt{\Psi_n^2 - 4\delta_n}} \leq \frac{2\delta_n}{\Psi_n\sigma_{n-1}} =: \rho_{n-1}^*. \tag{2.4.12}$$

2. If $n_0 = 0$; i.e., $4\delta_1 \leq \Psi_1^2$, then $\{\mathfrak{B}(\Gamma_n, \rho_n)\}_{n=0}^{\infty}$ is a sequence of value sets for $\mathbf{K}(a_n/1)$.

3. If $n_0 > 0$, then $\{V_n\}_{n=0}^{\infty}$ given by $V_n := \mathfrak{B}(\Gamma_n, \rho_n)$ for $n \geq n_0$ and $V_{n-1} := a_n/(1 + V_n)$ for $n = n_0, n_0 - 1, \ldots, 1$, is a sequence of value sets for $\mathbf{K}(a_n/1)$.

Error bounds.

If the sequence $\{V_n\}$ of value sets derived for $\mathbf{K}(a_n/1)$ in this way consists of bounded sets, then the Oval Sequence Theorem still applies for $n \geq n_0 > 0$ if we replace $|\Gamma_0| + \rho_0$ by $\sup\{|w|; w \in V_0\}$ in the error bound.

Another method to obtain error bounds when $n_0 > 0$, is to use

$$S_{n_0+k}(w) = S_{n_0} \circ S_k^{(n_0)}(w) = \frac{A_{n_0-1}S_k^{(n_0)}(w) + A_{n_0}}{B_{n_0-1}S_k^{(n_0)}(w) + B_{n_0}} = \frac{f_{n_0-1}S_k^{(n_0)}(w) - f_{n_0}\zeta_{n_0}}{S_k^{(n_0)}(w) - \zeta_{n_0}}$$

(Lemma 1.1 on page 6). Then

$$S_{n+m}(w) - S_n(v) = \frac{\zeta_{n_0}(f_{n_0} - f_{n_0-1})(S_{n+m-n_0}^{(n_0)}(w) - S_{n-n_0}^{(n_0)}(v))}{(S_{n+m-n_0}^{(n_0)}(w) - \zeta_{n_0})(S_{n-n_0}^{(n_0)}(v) - \zeta_{n_0})},$$

where $|S_{n+m-n_0}^{(n_0)}(w) - \zeta_{n_0}| > |\Gamma_{n_0} - \zeta_{n_0}| - \rho_{n_0}$ if $\zeta_{n_0} \notin V_{n_0}$, and

$$f_m = \frac{a_1}{1} \, {}_{+} \, \frac{a_2}{1} \, {}_{+} \cdots {}_{+} \, \frac{a_m}{1} \quad \text{and} \quad \zeta_m = -1 - \frac{a_m}{1} \, {}_{+} \, \frac{a_{m-1}}{1} \, {}_{+} \cdots {}_{+} \, \frac{a_2}{1}$$

are easy to compute for low values of m. Hence, for $w \in V_{n+m}$, it follows from the Oval Sequence Theorem that

$$|S_{n+m}(w) - S_n(\Gamma_n)| < \frac{|\zeta_{n_0}(f_{n_0} - f_{n_0-1})| \cdot \rho_n(|\Gamma_{n_0}| + \rho_{n_0})}{(|\Gamma_{n_0} - \zeta_{n_0}| - \rho_{n_0})^2(|1 + \Gamma_n| - \rho_n)} \prod_{j=n_0+1}^{n-1} M_j \qquad (2.4.13)$$

for $w \in V_{n+m}$ and $n > n_0$.

Remarks.

1. For $n_0 = 1$ we have $\zeta_1 = -1$, $f_1 = a_1$ and $f_0 = 0$, and thus

$$|S_{n+m}(w) - S_n(\Gamma_n)| < \frac{|a_1|\rho_n(|\Gamma_1| + \rho_1)}{(|1 + \Gamma_1| - \rho_1)^2(|1 + \Gamma_n| - \rho_n)} \prod_{j=2}^{n-1} M_j \qquad (2.4.14)$$

   for $w \in V_{n+m}$ and $n \geq 2$.

2. For $n_0 = 1$ we can also replace a_1 by some $\tilde{a}_1 \neq 0$ which allows $n_0 = 0$. Then

$$|S_{n+m}(w) - S_n(\Gamma_n)| \leq \left|\frac{a_1}{\tilde{a}_1}\right| \rho_n \frac{|\Gamma_0| + \rho_0}{|1 + \Gamma_n| - \rho_n} \prod_{k=1}^{n-1} M_k \qquad (2.4.15)$$

   where $V_0 = \mathfrak{B}(\Gamma_0, \rho_0) := \tilde{a}_1/(1 + \mathfrak{B}(\Gamma_1 + \rho_1))$.

3. For $n_0 = 2$ we note that $\zeta_2 = -1 - a_2$, $f_2 = a_1/(1 + a_2)$ and $f_1 = a_1$, so

$$|S_{n+m}(w) - S_n(\Gamma_n)| < \frac{|a_1 a_2|\rho_n(|\Gamma_2| + \rho_2)}{|1 + a_2 + \Gamma_2| - \rho_2)^2(|1 + \Gamma_n| - \rho_n)} \prod_{j=3}^{n-1} M_j \qquad (2.4.16)$$

   for $w \in V_{n+m}$ for $n \geq 3$.

4. These formulas easily generalize to continued fractions $\mathbf{K}(a_n/b_n)$.

5.2.5 Value sets and the fixed point method

Let $\mathbf{K}(a_n/1)$ be a limit 1-periodic continued fraction of loxodromic type, such that $a_n \to a$ with $|\arg(a+\frac{1}{4})| < \pi$. To find bounds for the truncation error $|f - S_n(x)|$ where x is the attracting fixed point

$$x := \frac{u-1}{2} \quad \text{where } u := \sqrt{1+4a}, \tag{2.5.1}$$

we shall find a sequence $\{\mathfrak{B}(x, \rho_n)\}$ of value sets for $\mathbf{K}(a_n/1)$ by means of our algorithm.

With $\sigma_n := 1$ for all n, we have $\Psi_n = \Psi = |1+x| - |x|$ for all n, so by (2.4.11)

$$\rho_n := \tfrac{1}{2}(\Psi - \sqrt{\Psi^2 - 4d_{n+1}}) \quad \text{where } d_{n+1} := \sup_{m \geq n+1} |a_m - a|. \tag{2.5.2}$$

That is, by Theorem 5.11, we get the following value sets $\{V_n\}$ for $\mathbf{K}(a_n/1)$:

Corollary 5.12. *Let $\mathbf{K}(a_n/1)$ be limit 1-periodic of loxodromic type with tail values $f^{(n)}$ and attracting fixed point $x \in \mathbb{C}$. Then there exists a sequence $\{V_n\}_{n=0}^{\infty}$ of value sets for $\mathbf{K}(a_n/1)$ with $V_n := \mathfrak{B}(x, \rho_n)$ for $n \geq n_0$, where ρ_n is given by (2.5.2) and $n_0 \in \mathbb{N}$ is chosen such that $4d_{n_0+1} \leq \Psi^2$. In particular, $|f^{(n)} - x| \leq \rho_n$ for $n \geq n_0$.*

Proof : It remains to prove that the tail values $f^{(n)} \in V_n$, but this follows since the constant sequence $\{y\}$ with $y := -1 - x \neq x$ is an exceptional sequence for $\mathbf{K}(a_n/1)$, and thus $f^{(n)} = \lim_{m\to\infty} S_m^{(n)}(x) \in V_n$. □

If $a \geq 0$, this result takes a simpler form, since then $x \geq 0$ and $\Psi = 1$.

Corollary 5.13. *Let $\mathbf{K}(a_n/1)$ be limit 1-periodic with a fixed point $x \geq 0$. Then there exists a sequence $\{V_n\}_{n=0}^{\infty}$ of value sets for $\mathbf{K}(a_n/1)$ with $V_n := \mathfrak{B}(x, \rho_n)$ for $n \geq n_0$, where $\rho_n := \frac{1}{2}(1 - \sqrt{1-4d_{n+1}})$, $d_n := \sup_{m\geq n} |a_m - a|$ and $n_0 \in \mathbb{N}$ is determined by $d_{n_0+1} \leq \frac{1}{4}$. In particular $|f^{(n)} - x| \leq \rho_n$ for $n \geq n_0$.*

Remarks.

1. Since $y := -1 - x$ is the second fixed point for $S_1(w)$, it follows immediately that $\mathbf{K}(a_n/1)$ in Corollary 5.13 is of loxodromic type, and that x is its *attracting* fixed point.

2. If $a_n \to 0$, and thus $V_n = \mathfrak{B}(0, \rho_n)$ in our situation, then the oval region E_n reduces to the disk $E_n = \mathfrak{B}(0, \rho_{n-1}(1-\rho_n))$ (see (2.3.2)), a situation we recognize from the extension of Worpitzky's Theorem on page 136.

3. The radius ρ_n in (2.5.2) can be written

$$\rho_n = \frac{2d_{n+1}}{\Psi + \sqrt{\Psi^2 - 4d_{n+1}}} < \frac{2d_{n+1}}{\Psi} =: \rho_n^*.$$

ρ_n^* is therefore a simpler bound for $|f^{(n)} - x|$ than ρ_n, but $\rho_n^* \sim 2\rho_n$.

4. If $a_n \to -\frac{1}{4}$, then $\mathbf{K}(a_n/1)$ is limit 1-periodic of parabolic type with fixed point $-\frac{1}{2}$. Then Lemma 5.1 on page 220 gives value sets for $\mathbf{K}(a_n/1)$ when $a_n \to -\frac{1}{4}$ fast enough.

Example 8. We shall find a sequence $\{V_n\}$ of value sets for the continued fraction

$$\mathop{\mathrm{K}}_{n=1}^{\infty} \frac{a_n}{1} \quad \text{where } a_1 := 4, \quad a_{n+1} := \frac{n^2}{4n^2-1} \quad \text{for } n \geq 1$$

which converges to π (appendix, page 267). $\mathbf{K}(a_n/1)$ is limit 1-periodic of loxodromic type with $a := \lim a_n = \frac{1}{4}$ and attracting fixed point $x := \frac{1}{2}(\sqrt{1+4a}-1) = \frac{1}{2}(\sqrt{2}-1)$. Now, $d_1 = a_1 - \frac{1}{4} = \frac{15}{4}$ and

$$d_{n+1} := \sup_{m \geq n+1} |a_m - a| = a_{n+1} - a = \frac{n^2}{4n^2-1} - \frac{1}{4} = \frac{1/4}{4n^2-1} \quad \text{for } n \geq 1,$$

so $d_{n+1} \leq \frac{1}{4}$ for $n \geq 1$. Hence it follows from Corollary 5.13 that $\mathbf{K}(a_n/1)$ has a sequence $\{V_n\}_{n=0}^{\infty}$ of value sets given by

$$V_n := \mathfrak{B}(x, \rho_n) \quad \text{where } x := \tfrac{1}{2}(\sqrt{2}-1) \quad \text{and} \quad \rho_n := \tfrac{1}{2}\left(1 - \sqrt{1 - 1/(4n^2-1)}\,\right)$$

for $n \geq 1$ and

$$V_0 := 4/(1+V_1) = \mathfrak{B}(\Gamma_0, \rho_0) \text{ with } \Gamma_0 := \frac{4(1+\overline{x})}{|1+x|^2 - \rho_1^2}, \quad \rho_0 := \frac{4\rho_1}{|1+x|^2 - \rho_1^2}.$$

In particular

$$\left| f^{(n)} - \frac{\sqrt{2}-1}{2} \right| \leq \rho_n = \frac{1}{2} - \sqrt{\frac{n^2 - \frac{1}{2}}{4n^2-1}} < 2\,d_{n+1} = \frac{1/2}{4n^2-1} \quad \text{for } n \geq 1.$$

From the Oval Sequence Theorem it follows for instance that

$$|\pi - S_n(x)| \leq \frac{1/2}{4n^2-1} \cdot \frac{|\Gamma_0| + \rho_0}{|1+x| - \rho_n} \prod_{k=1}^{n-1} \frac{|x(1+\overline{x}) - \rho_k^2| + \rho_k}{|1+x|^2 - \rho_k^2}.$$

That is, $\rho_1 < \frac{1}{6}$, $\Gamma_0 \approx 3.333$, $\rho_0 < 0.47$ and

$$M_k = \frac{|x(1+\overline{x}) - \rho_k^2| + \rho_k}{|1+x|^2 - \rho_k^2} < 0.2 \quad \text{for } k \geq 2,$$

so

$$|\pi - S_n(x)| \leq 0.44 \cdot \frac{(0.2)^{n-2}}{4n^2-1} \quad \text{for } n \geq 2.$$

◇

The positive case.

Of course, Theorem 5.12 holds with $\Psi = 1$ when all $a_n > 0$. But it is a restriction to require that $a_n \in \mathfrak{B}(a, r_n)$, as we have done in Corollary 5.12. (See Property 4 on page 245.) We can do better if we use the oval sets E_n directly:

Theorem 5.14. *Let $\mathbf{K}(a_n/1)$ with $0 < a_n \to a \neq \infty$ have attracting fixed point x. Then $\{\mathfrak{B}(x, \rho_n)\}$ is a sequence of value sets for $\mathbf{K}(a_n/1)$ if for each $n \in \mathbb{N}$ either*

$$\begin{aligned} &\rho_{n-1} \leq x \textit{ and } |a_n - a + \rho_{n-1}\rho_n| \leq \rho_{n-1}(1+x) - \rho_n x \\ &\textit{or} \\ &\rho_{n-1} \geq x \textit{ and } a_n - a + \rho_{n-1}\rho_n \leq \rho_{n-1}(1+x) - \rho_n x. \end{aligned} \tag{2.5.3}$$

Proof : Since $a_n > 0$ and $x = \frac{1}{2}(\sqrt{1+4a} - 1) \geq 0$, it follows from Property 2 on page 244 that $\{\mathfrak{B}(x, \rho_n)\}$ is a sequence of value sets for $\mathbf{K}(a_n/1)$ if $u_n \leq a_n \leq v_n$ for all n, where u_n and v_n are given by (2.4.3) with all $\Gamma_n := x$. That is, if either

$$\begin{aligned} &\rho_{n-1} \leq x, \quad (x - \rho_{n-1})(1 + x + \rho_n) \leq a_n \leq (x + \rho_{n-1})(1 + x - \rho_n) \\ &\text{or if } \rho_{n-1} \geq x, \quad a_n \leq (x + \rho_{n-1})(1 + x - \rho_n), \end{aligned} \tag{2.5.4}$$

which is equivalent to (2.5.3) since $x(1+x) = a$. □

If we are willing to accept a little stronger conditions than (2.5.3), we have the following simpler result:

Corollary 5.15. *Let $\mathbf{K}(a_n/1)$ with $0 < a_n \to a \neq \infty$ have attracting fixed point x. Then $\{\mathfrak{B}(x, \rho_n)\}$ is a sequence of value sets for $\mathbf{K}(a_n/1)$ if*

$$|a_n - a + \rho_{n-1}\rho_n| \leq \rho_{n-1}(1+x) - \rho_n x \quad \textit{for } n \geq 1. \tag{2.5.5}$$

Both in Theorem 5.14 and Corollary 5.15 it is up to us to choose $\{\rho_n\}$. In view of Remark 3 on page 249, on could for instance try $\rho_n := 2d_{n+1}$.

Example 9. Let $\mathbf{K}(a_n/1)$ be given by

$$a_1 := 1, \quad a_{2k} := \frac{k}{4k-2}, \quad a_{2k+1} := \frac{k}{4k+2} \quad \text{for } k \geq 1.$$

Then $a_n \to \frac{1}{4} =: a$, $x = \frac{1}{2}(\sqrt{2}-1)$ and

$$a_1 - a = \frac{3}{4}, \quad a_{2k} - a = \frac{1/4}{2k-1} \quad \text{and } a_{2k+1} - a = \frac{-1/4}{2k+1}$$

for $k \geq 1$. The choice $\rho_n := 2\,d_{n+1}$ gives

$$\rho_0 := \frac{3}{2}, \quad \rho_1 := \frac{1}{2}, \quad \rho_{2k} := \rho_{2k+1} := \frac{1/2}{2k+1} \quad \text{for } k \geq 1.$$

Condition (2.5.5) therefore looks like

$$\begin{aligned}
|\tfrac{3}{4} + \tfrac{3}{2}\cdot\tfrac{1}{2}| &< \tfrac{3}{2}\cdot\tfrac{1}{2}(\sqrt{2}+1) - \tfrac{1}{2}\cdot\tfrac{1}{2}(\sqrt{2}-1) && \text{for } n := 1,\\
\left|\frac{1/4}{2k-1} + \frac{1/4}{4k^2-1}\right| &\leq \frac{(\sqrt{2}+1)/4}{2k-1} - \frac{(\sqrt{2}-1)/4}{2k+1} && \text{for } n := 2k,\\
\left|\frac{-1/4}{2k+1} + \frac{1/4}{(2k+1)^2}\right| &\leq \frac{(\sqrt{2}+1)/4}{2k+1} - \frac{(\sqrt{2}-1)/4}{2k+1} && \text{for } n := 2k+1
\end{aligned}$$

for $k \geq 1$, which actually is satisfied. Hence $\mathbf{K}(a_n/1)$ has a sequence $\{V_n\}$ of value sets given by

$$V_0 := \mathfrak{B}(x, \tfrac{3}{2}), \quad V_1 := \mathfrak{B}(x, \tfrac{1}{2}), \quad V_{2k} := V_{2k+1} := \mathfrak{B}\Big(x, \frac{1/2}{2k+1}\Big); \quad k \geq 1.$$

Hence it follows for instance from the Oval Sequence Theorem that

$$|f - S_n(x)| \leq \rho_n \frac{\frac{1}{2}(\sqrt{2}-1) + \frac{3}{2}}{\frac{1}{2}(\sqrt{2}+1) - \rho_n} \prod_{k=1}^{n-1} M_k \tag{2.5.6}$$

where by (2.3.7)

$$M_{2m} = M_{2m+1} = \frac{\frac{1}{2}(\sqrt{2}-1) + 1/(4m+2)}{\frac{1}{2}(\sqrt{2}+1) + 1/(4m+2)} = \frac{(\sqrt{2}-1)(2m+1)+1}{(\sqrt{2}+1)(2m+1)+1}.$$

This bound is compared to the actual truncation error in the table below. Of course, since $\{M_k\}$ is non-increasing, we can replace M_k by some M_{k_0} for all $k > k_0$ to simplify the computation. This is done in the last column with $k_0 := 4$. This bound is easier to handle if we want to determine the order n of $S_n(x)$ needed to reach a given accuracy. A slight improvement is obtained if we do not insist on having the center Γ_0 of V_0 at x; i.e., if we take $V_0 := 1/(1+V_1) = \mathfrak{B}(1, \sqrt{2}-1)$. Then the Oval Sequence Theorem gives the bound

$$|f - S_n(x)| \le \rho_n \frac{\sqrt{2}}{\frac{1}{2}(\sqrt{2}+1) - \rho_n} \prod_{k=1}^{n-1} M_k. \tag{2.5.7}$$

n	$S_n(x)$	$\lvert f - S_n(x)\rvert$	(2.5.6)	(2.5.6)modified	(2.5.7)
5	.6931772536	$3.0 \cdot 10^{-5}$	$1.1 \cdot 10^{-3}$	$1.1 \cdot 10^{-3}$	$9.2 \cdot 10^{-4}$
6	.6931515697	$4.4 \cdot 10^{-6}$	$1.8 \cdot 10^{-4}$	$1.8 \cdot 10^{-4}$	$1.5 \cdot 10^{-4}$
7	.6931478287	$6.5 \cdot 10^{-7}$	$4.0 \cdot 10^{-5}$	$4.3 \cdot 10^{-5}$	$3.3 \cdot 10^{-5}$
8	.6931472790	$9.8 \cdot 10^{-7}$	$6.6 \cdot 10^{-6}$	$7.1 \cdot 10^{-6}$	$5.5 \cdot 10^{-6}$
9	.6931471956	$1.5 \cdot 10^{-8}$	$1.4 \cdot 10^{-6}$	$1.6 \cdot 10^{-6}$	$1.1 \cdot 10^{-6}$
10	.6931471829	$2.3 \cdot 10^{-9}$	$2.3 \cdot 10^{-7}$	$2.6 \cdot 10^{-7}$	$1.9 \cdot 10^{-7}$
11	.6931471809	$3.7 \cdot 10^{-10}$	$4.7 \cdot 10^{-8}$	$5.5 \cdot 10^{-8}$	$3.9 \cdot 10^{-8}$
12	.6931471806	$5.8 \cdot 10^{-11}$	$7.9 \cdot 10^{-9}$	$9.6 \cdot 10^{-9}$	$6.6 \cdot 10^{-9}$
13	.6931471806	$9.2 \cdot 10^{-12}$	$1.6 \cdot 10^{-9}$	$1.9 \cdot 10^{-9}$	$1.3 \cdot 10^{-9}$
14	.6931471806	$1.5 \cdot 10^{-12}$	$2.7 \cdot 10^{-10}$	$3.2 \cdot 10^{-10}$	$2.2 \cdot 10^{-10}$
15	.6931471806	$2.4 \cdot 10^{-13}$	$5.2 \cdot 10^{-11}$	$6.3 \cdot 10^{-11}$	$4.3 \cdot 10^{-11}$

◇

Limit 1-periodic S-fractions.

Let $z \neq 0$ with $|\arg z| < \pi$ be kept fixed in the S-fraction $\mathbf{K}(a_n z/1)$ where $0 < a_n \to a \ge 0$. Then $\mathbf{K}(a_n z/1)$ is limit 1-periodic of loxodromic type with attracting fixed point

$$x = x(z) = \tfrac{1}{2}(u(z) - 1) \quad \text{where } u(z) := \sqrt{1+4az}.$$

The element sets $\{E_n\}$ for the value sets $\{\mathfrak{B}(x, \rho_n)\}_{n=0}^{\infty}$ are bounded by cartesian ovals with axis along the line $z + z\mathbb{R}$ since $\arg a_n^* = \arg(x(1+x) = \arg z$ in (2.3.3). By (2.4.3) with all $\Gamma_n := x$, the end points of the axis of E_n are $u_n z/|z|$ and $v_n z/|z|$ where $u_n < 0$ if $|x| < \rho_{n-1}$ and

$$\begin{aligned} u_n &= a|z| + |x|\rho_n - |1+x|\rho_{n-1} - \rho_n\rho_{n-1} \quad \text{if } |x| \ge \rho_{n-1}, \\ v_n &= a|z| - |x|\rho_n + |1+x|\rho_{n-1} - \rho_n\rho_{n-1}. \end{aligned}$$

Now, $a_n z \in E_n$ if $u_n \le a_n|z| \le v_n$. Hence, just as for the positive case we get:

Theorem 5.16. *Let $\mathbf{K}(a_n z/1)$ with $0 < a_n \to a \neq \infty$ and fixed z with $|\arg z| < \pi$ have attracting fixed point $x(z)$. Then $\{\mathfrak{B}(x(z), \rho_n(z))\}$ is a sequence of value sets for $\mathbf{K}(a_n z/1)$ if for each $n \in \mathbb{N}$ either*

$$\begin{aligned} &\rho_{n-1} \le |x| \ \textit{and}\ |(a_n - a)|z| + \rho_{n-1}\rho_n| \le \rho_{n-1}|1+x| - \rho_n|x| \\ &\textit{or} \\ &\rho_{n-1} > |x| \ \textit{and}\ (a_n - a)|z| + \rho_{n-1}\rho_n \le \rho_{n-1}|1+x| - \rho_n|x|. \end{aligned} \tag{2.5.8}$$

Example 10. The S fraction $\mathbf{K}(a_n z/1)$ given by

$$a_1 := 1, \quad a_{2k} := \frac{k}{4k-2}, \quad a_{2k+1} := \frac{k}{4k+2} \quad \text{for } k \geq 1,$$

converges to $\mathrm{Ln}(1+z)$ for $|\arg(z+1)| < \pi$ (see page 270). For $z = 0$ its value is clearly 0. In the following we assume that $z \neq 0$ and $|\arg z| < \pi$. Since $a_n \to \frac{1}{4} =: a$ and $d_n := \sup_{m \geq n} |a_m z - az| = |a_n z - az|$ in this case, we find that $x := x(z) = \frac{1}{2}(\sqrt{1+z} - 1)$,

$$d_1 = \frac{3|z|}{4}, \quad d_{2k} = \frac{|z|/4}{2k-1} \text{ and } d_{2k+1} = \frac{|z|/4}{2k+1} \quad \text{for } k \geq 1.$$

As in the previous example we choose $\rho_n := 2\, d_{n+1}$; i.e.,

$$\rho_0 := \frac{3|z|}{2}, \quad \rho_1 := \frac{|z|}{2}, \quad \rho_{2k} := \rho_{2k+1} := \frac{|z|/2}{2k+1} \quad \text{for } k \geq 1.$$

Then condition (2.5.8) looks like

$$\frac{3|z|}{4} + \frac{3|z|^2}{4} \leq \frac{3|z|}{2}|1+x| - \frac{|z|}{2}|x| \quad \text{for } n := 1,$$
$$\frac{|z|/4}{2k-1} + \frac{|z|/2}{2k-1} \cdot \frac{|z|/2}{2k+1} \leq \frac{|z|/2}{2k-1}|1+x| - \frac{|z|/2}{2k+1}|x| \quad \text{for } n := 2k,$$

and for $n := 2k+1 \geq 3$ it only requires that

$$\left| \frac{-|z|/4}{2k+1} + \left(\frac{|z|/2}{2k+1}\right)^2 \right| \leq \frac{|z|/2}{2k+1}(|1+x| - |x|) \quad \text{if } \frac{|z|/2}{2k+1} < |x|.$$

Of course,

$$\Psi := |1+x| - |x| = \tfrac{1}{2}(|\sqrt{1+z}+1| - |\sqrt{1+z}-1|) > 0,$$

so these conditions hold

$$\text{for } n = 1 \text{ if } 1 + |z| \leq 2\Psi,$$
$$\text{for } n = 2k \geq 2 \text{ if } 1 + \frac{|z|}{2k+1} \leq 2\Psi,$$
$$\text{for } n = 2k+1 \geq 3 \text{ if } 2k+1 \leq \frac{|z|}{2|x|} \text{ or } \left| -1 + \frac{|z|}{2k+1} \right| \leq 2\Psi.$$

Hence there exists a sequence $\{V_n\}$ of value sets for $\mathbf{K}(a_n z/1)$ with $V_n := \mathfrak{B}(x, \rho_n)$ for $n \geq n_0$ where

$$n_0 = \begin{cases} 0 \text{ if } 1 + |z| \leq 2|1+x| - \frac{2}{3}|x|, \\ 2k-1 \text{ if } 1 + |z|/(2k+1) \leq 2\Psi \text{ otherwise.} \end{cases}$$

If $z > 0$, then $|1+x| - |x| = 1$. In particular $n_0 = 0$ for $z = 1$, which is exactly what we obtained in Example 9. ◇

Limit 2-periodic continued fractions of loxodromic type.

If $\mathbf{K}(a_n/1)$ is limit 2-periodic of loxodromic type with attracting fixed points $(x^{(0)}, x^{(1)})$, then $\Gamma_{2n} \to x^{(0)}$ and $\Gamma_{2n+1} \to x^{(1)}$ for any natural choice of $\{\Gamma_n\}$. In the particular case where we choose $\Gamma_{2n} := x^{(0)}$ and $\Gamma_{2n+1} := x^{(1)}$ for all n, (2.4.9) holds with equality for all n with the choice

$$\sigma_{2n} := |1 + x^{(1)}| + |x^{(1)}|, \quad \sigma_{2n+1} := |1 + x^{(0)}| + |x^{(0)}| \tag{2.5.9}$$

and $\Psi_n := \Psi = |(1 + x^{(0)})(1 + x^{(1)})| - |x^{(0)}x^{(1)}| > 0$ (Theorem 4.9 on page 182). That is,

$$\rho_{n-1} := \frac{1}{2\sigma_{n-1}}\left(\Psi - \sqrt{\Psi^2 - 4\sigma_1\sigma_2 d_n}\right) \text{ where } d_n := \sup_{m\ge n} |a_m - \widehat{a}_m| \tag{2.5.10}$$

is a sensible choice, where

$$\widehat{a}_{2n-1} := \widehat{a}_1 := \lim a_{2m-1} \quad \text{and} \quad \widehat{a}_{2n} := \widehat{a}_2 := \lim a_{2m}. \tag{2.5.11}$$

With this notation we get:

Corollary 5.17. *Let $\mathbf{K}(a_n/1)$ be limit 2-periodic of loxodromic type with tail values $f^{(n)}$ and finite attracting fixed points $(x^{(0)}, x^{(1)})$. Let further $n_0 \in \mathbb{N}$ be such that $4\sigma_1\sigma_2 d_{n_0+1} \le \Psi^2$. Then there exists a sequence $\{V_n\}$ of value sets for $\mathbf{K}(a_n/1)$ with $V_{2k} := \mathfrak{B}(x^{(0)}, \rho_{2k})$ and $V_{2k+1} := \mathfrak{B}(x^{(1)}, \rho_{2k+1})$ for $n \ge n_0$, and $f^{(n)} \in V_n$ for all n.*

Example 11. We want to find "small" value sets for

$$\mathop{\mathrm{K}}_{n=1}^{\infty} \frac{a_n}{1} := \frac{3+1/1^2}{1} + \frac{4+3/2^2}{1} + \frac{3+1/3^2}{1} + \frac{4+3/4^2}{1} + \frac{3+1/5^2}{1} + \cdots$$

from Example 4 on page 13. $\mathbf{K}(a_n/1)$ is limit 2-periodic of loxodromic type with attracting fixed points $(1, 2)$. Therefore σ_n given by (2.5.9) is $\sigma_{2n} := 3 + 2 = 5$ and $\sigma_{2n+1} := 2 + 1 = 3$, and $\Psi_n = \Psi = 2 \cdot 3 - 1 \cdot 2 = 4$. Hence $\sigma_1\sigma_2 = 15$ and $d_n := \sup_{m\ge n} |a_m - \widehat{a}_m|$ gives $d_1 = 1$, $d_2 = \frac{3}{4}$ and $d_{2n-1} = d_{2n} = 3/(4n^2)$ for $n \ge 2$. Therefore $4\sigma_1\sigma_2 d_n < \Psi^2$ for $n \ge 3$, and thus $\mathbf{K}(a_n/1)$ has a sequence $\{V_n\}$ of value sets with

$$V_{2n} := \mathfrak{B}(2, \rho_{2n}), \quad V_{2n+1} := \mathfrak{B}(1, \rho_{2n+1}) \text{ for } n \ge 1$$

where

$$\rho_{2n} := \frac{1}{2\sigma_2}\left(\Psi - \sqrt{\Psi^2 - 4\sigma_1\sigma_2 d_{2n+1}}\right) = \frac{2}{5} - \frac{2}{5}\sqrt{1 - \frac{45/16}{(n+1)^2}}$$

$$\rho_{2n+1} := \frac{1}{2\sigma_1}\left(\Psi - \sqrt{\Psi^2 - 4\sigma_1\sigma_2 d_{2n+2}}\right) = \frac{2}{3} - \frac{2}{3}\sqrt{1 - \frac{45/16}{(n+1)^2}}.$$

In particular $|f^{(2n)} - 2| \le \rho_{2n}$ and $|f^{(2n+1)} - 1| \le \rho_{2n+1}$ for $n \ge 1$. ◇

Limit p-periodic continued fractions of loxodromic type.

If $\mathbf{K}(a_n/1)$ is limit p-periodic of loxodromic type with attracting fixed points $(x^{(0)}, x^{(1)}, \dots, x^{(p-1)})$, we can choose $\Gamma_{np+m} := \Gamma_m := x^{(m)}$ for all m and n. Then we can choose

$$\sigma_{np+p-1} := \sigma_{p-1} := \sum_{n=-1}^{p-2} \prod_{j=0}^{n} |1+\Gamma_j| \prod_{j=n+1}^{p-2} |\Gamma_j| \tag{2.5.12}$$

and the other σ_ns are determined from σ_{p-1} by cyclic shifts. For instance, for $k = 3$ the three σ_n given by

$$\begin{aligned}\sigma_0 &= |\Gamma_1\Gamma_2| + |1+\Gamma_1||\Gamma_2| + |1+\Gamma_1||1+\Gamma_2|\,,\\ \sigma_1 &= |\Gamma_2\Gamma_0| + |1+\Gamma_2||\Gamma_0| + |1+\Gamma_2||1+\Gamma_0|\,,\\ \sigma_2 &= |\Gamma_0\Gamma_1| + |1+\Gamma_0||\Gamma_1| + |1+\Gamma_0||1+\Gamma_1|\,.\end{aligned}$$

With this choice we get $\Psi_n = \Psi := \prod_{j=0}^{k-1} |1+\Gamma_j| - \prod_{j=0}^{k-1} |\Gamma_j| > 0$. Also now Theorem 5.11 implies that $f^{(np+m)} \in V_{np+m} := \mathfrak{B}(x^{(m)}, \rho_{np+m})$ for $n \ge n_0$ sufficiently large. We refer to [Jaco87] for value sets for more general continued fractions.

5.2.6 Value sets $\mathfrak{B}(w_n \rho_n)$ for limit 1-periodic continued fractions of loxodromic or parabolic type

Let $\mathbf{K}(a_n/1)$ be limit 1-periodic of loxodromic type, and let $w_n \approx f^{(n)}$ be chosen to make $S_n(w_n) \to f \ne \infty$ fast, where f is the value of $\mathbf{K}(a_n/1)$. To take full advantage of this choice, we want $V_n = \mathfrak{B}(w_n, \rho_n)$. If $w_n \to x$, as it normally does, then also $f^{(n)} \in V_n$ in this case (Lemma 5.7 on page 240), so $|f^{(n)} - w_n| \le \rho_n$.

Example 12. We return to $\mathbf{K}(a_n/1)$ with $a_1 := 4$, $a_{n+1} := n^2/(4n^2-1)$ for $n \ge 1$ from Example 7 on page 237. The square root modification gave approximants $S_n(w_n)$ with

$$\begin{aligned} w_n &:= \frac{1}{2}\Bigl(\sqrt{\frac{8n^2-1}{4n^2-1}} - 1\Bigr) = \frac{1}{2}\Bigl(\sqrt{2}\cdot\sqrt{1+\frac{1/2}{4n^2-1}} - 1\Bigr)\\ &\approx \frac{1}{2}\Bigl(\sqrt{2}\Bigl(1+\frac{1/4}{4n^2-1}\Bigr) - 1\Bigr) = \frac{\sqrt{2}-1}{2} + \frac{\sqrt{2}/8}{4n^2-1}.\end{aligned}$$

We therefore choose

$$\Gamma_n := w_n := \frac{\sqrt{2}-1}{2} + \frac{\sqrt{2}/8}{4n^2-1} \quad \text{for } n \ge 1.$$

We first want to find positive σ_ns such that $\Psi_n := \sigma_n(1+w_n) - \sigma_{n-1}w_{n-1} > 0$ is non-decreasing. Since

$$\begin{aligned}1 + w_{n+1} - w_n &= 1 + \frac{\sqrt{2}}{8}\Big(\frac{1}{(2n+1)(2n+3)} - \frac{1}{(2n-1)(2n+1)}\Big)\\ &= 1 - \frac{\sqrt{2}}{8}\,\frac{(2n+3)-(2n-1)}{(2n-1)(2n+1)(2n+3)} = 1 - \frac{\sqrt{2}/2}{(2n-1)(2n+1)(2n+3)}\end{aligned}$$

is monotonely increasing, we can take all $\sigma_n := 1$ to get

$$\Psi_{n+1} = 1 + w_{n+1} - w_n = 1 - \frac{\sqrt{2}/2}{(4n^2-1)(2n+3)} \quad \text{for } n \geq 1.$$

Now,

$$\begin{aligned}\widehat{a}_{n+1} &:= \Gamma_n(1+\Gamma_{n+1})\\ &= \Big(\frac{\sqrt{2}-1}{2} + \frac{\sqrt{2}/8}{4n^2-1}\Big)\Big(\frac{\sqrt{2}+1}{2} + \frac{\sqrt{2}/8}{(2n+1)(2n+3)}\Big)\\ &= \frac{1}{4} + \frac{1/4}{(2n-1)(2n+3)} + \frac{\sqrt{2}/4}{(2n-1)(2n+1)(2n+3)}\\ &\qquad + \frac{1/32}{(2n-1)(2n+1)^2(2n+3)},\end{aligned}$$

so

$$\begin{aligned}a_{n+1} - \widehat{a}_{n+1} &= \frac{(2-\sqrt{2})/4}{(2n-1)(2n+1)(2n+3)} - \frac{1/32}{(2n-1)(2n+1)^2(2n+3)}\\ &< \frac{(2-\sqrt{2})/4}{(2n-1)(2n+1)(2n+3)},\end{aligned}$$

and we choose

$$\rho_n := \tfrac{1}{2}\big(\Psi_{n+1} - \sqrt{\Psi_{n+1}^2 - 4\delta_{n+1}}\,\big) \quad \text{where } \delta_{n+1} := \frac{(2-\sqrt{2})/4}{(2n-1)(2n+1)(2n+3)}$$

for $n \geq 1$. Then $4\delta_{n+1} \leq \Psi_{n+1}^2$ if and only if

$$\frac{2-\sqrt{2}}{(2n-1)(2n+1)(2n+3)} \leq \Big(1 - \frac{\sqrt{2}/2}{(2n-1)(2n+1)(2n+3)}\Big)^2$$

which holds for $n \geq 1$. Hence $\mathbf{K}(a_n/1)$ has a sequence $\{V_n\}$ of value sets with

$$V_n := \mathfrak{B}\Big(\frac{\sqrt{2}-1}{2} + \frac{\sqrt{2}/8}{4n^2-1},\, \rho_n\Big) \quad \text{for } n \geq 1,$$

and

$$|f^{(n)} - w_n| \leq \rho_n < \rho_n^* := 2\frac{\delta_{n+1}}{\Psi_{n+1}} = \frac{(2-\sqrt{2})/2}{(4n^2-1)(2n+3) - \sqrt{2}/2} \quad \text{for } n \geq 1.$$

To obtain a priori truncation error bounds, we set $\mathfrak{B}(\Gamma_0, \rho_0^*) := 1/(1+\mathfrak{B}(\Gamma_1, \rho_1^*))$. Then (Lemma 3.6 on page 110)

$$\Gamma_0 := \frac{4(1+\overline{\Gamma}_1)}{|1+\Gamma_1|^2 - \rho_1^{*2}} \le \frac{4(\frac{\sqrt{2}+1}{2} + \frac{\sqrt{2}}{8\cdot 3})}{(\frac{\sqrt{2}+1}{2} + \frac{\sqrt{2}}{8\cdot 3})^2 - (\frac{(2-\sqrt{2})/2}{3\cdot 5 - \sqrt{2}/2})^2} \approx 3.160305,$$

$$\rho_0^* := \frac{4\rho_1^*}{|1+\Gamma_1|^2 - \rho_1^{*2}} \approx 0.051153.$$

The Oval Sequence Theorem and Remark 3 on page 242 thereby imply that

$$|f - S_n(\Gamma_n)| \le \rho_n^* \frac{|\Gamma_0| + \rho_0^*}{|1+\Gamma_n| - \rho_n^*} \prod_{k=1}^{n-1} M_k^* \le \tag{2.6.1}$$

$$\frac{(2-\sqrt{2})\cdot(3.160305 + 0.051153)\prod_{k=1}^{n-1} M_k^*}{(\sqrt{2}+1)[(4n^2-1)(2n+3) - \frac{\sqrt{2}}{2}] + \frac{\sqrt{2}}{4}(2n+3) - \frac{1/4}{4n^2-1} - 2 + \sqrt{2}}$$

where

$$M_k^* := \frac{\Gamma_k + \rho_k^*}{1+\Gamma_k + \rho_k^*} < \frac{\frac{\sqrt{2}-1}{2} + \frac{\sqrt{2}/8}{4k^2-1} + \frac{(2-\sqrt{2})/2}{(4k^2-1)(2k+3)-\sqrt{2}/2}}{\frac{\sqrt{2}+1}{2} + \frac{\sqrt{2}/8}{4k^2-1} + \frac{(2-\sqrt{2})/2}{(4k^2-1)(2k+3)-\sqrt{2}/2}}.$$

The table below compares this error bound to the true truncation errors. In the last column we have replaced M_k by M_5 for $k > 5$. This bound is well suited to determine n in $S_n(\Gamma_n)$ to achieve a wanted accuracy. The true value of $\mathbf{K}(a_n/1)$ is π, and the approximants $S_n(\Gamma_n)$ converge very fast to this value.

n	$\lvert f - S_n(\Gamma_n)\rvert$	(2.6.1)	(2.6.1)modified
5	$2.6 \cdot 10^{-7}$	$7.4 \cdot 10^{-7}$	$7.4 \cdot 10^{-7}$
6	$2.7 \cdot 10^{-8}$	$7.6 \cdot 10^{-8}$	$7.6 \cdot 10^{-8}$
7	$3.0 \cdot 10^{-9}$	$8.5 \cdot 10^{-9}$	$8.5 \cdot 10^{-9}$
8	$3.5 \cdot 10^{-10}$	$1.0 \cdot 10^{-9}$	$1.0 \cdot 10^{-9}$
9	$4.3 \cdot 10^{-11}$	$1.2 \cdot 10^{-10}$	$1.2 \cdot 10^{-10}$
10	$5.4 \cdot 10^{-12}$	$1.6 \cdot 10^{-11}$	$1.6 \cdot 10^{-11}$
11	$7.1 \cdot 10^{-13}$	$2.0 \cdot 10^{-12}$	$2.1 \cdot 10^{-12}$
12	$9.5 \cdot 10^{-14}$	$2.7 \cdot 10^{-13}$	$2.8 \cdot 10^{-13}$
13	$1.3 \cdot 10^{-14}$	$3.7 \cdot 10^{-14}$	$3.9 \cdot 10^{-14}$
14	$1.8 \cdot 10^{-15}$	$5.2 \cdot 10^{-15}$	$5.4 \cdot 10^{-15}$
15	$2.5 \cdot 10^{-16}$	$7.2 \cdot 10^{-16}$	$7.6 \cdot 10^{-16}$

$\diamond$

Our next example demonstrates how difficult it can be to find useful value sets for limit periodic continued fractions of parabolic type.

Example 13. The continued fraction $\mathbf{K}(a_n/1)$ with

$$a_n := -\frac{1}{4} + \frac{(-1/2)^n}{4n+2} \quad \text{for } n = 1, 2, 3, \dots$$

is limit 1-periodic of parabolic type. Let all $\Gamma_n := -\frac{1}{2}$. Then $|1+\Gamma_n| = |\Gamma_{n-1}| = \frac{1}{2}$, and we can for instance use

$$\sigma_n := 2n+1 \quad \text{for } n \geq 0.$$

Then $\Psi_n := \sigma_n|1+\Gamma_n| - \sigma_{n-1}|\Gamma_{n-1}| = 1$ for all n. The choice (2.4.11) for $\{\rho_n\}$ then gives

$$\rho_{n-1} = \frac{1/2}{2n-1}\left(1 - \sqrt{1-4\delta_n}\right) \quad \text{where } \delta_n := \sup_{m\geq n}\left\{(4m^2-1)\frac{(1/2)^m}{4m+2}\right\}$$
$$\text{i.e., } \delta_n = \sup_{m\geq n}\{(\tfrac{1}{2})^{m+1}(2m-1)\} = (\tfrac{1}{2})^{n+1}(2n-1) \quad \text{for } n \geq 3$$

and $\delta_1 = \delta_2 = \frac{3}{8}$. In particular $4\delta_n < 1$ for $n \geq 4$, so $\mathbf{K}(a_n/1)$ has a sequence $\{V_n\}_{n=0}^{\infty}$ of value sets with $V_n := \mathfrak{B}(-\frac{1}{2}, \rho_n)$ for $n \geq 3$. Of course, this means that $\mathbf{K}(a_n/1)$ has a tail sequence $\{t_n\}$ with $t_n \in \mathfrak{B}(-\frac{1}{2}, \rho_n)$ for all n. But we do not know that this is a sequence of tail values. We can not even say whether $\mathbf{K}(a_n/1)$ converges generally or not at this point. $\diamond$

5.2.7 Error bounds based on idea 3

Idea 3 on page 239 was to combine the bound $|f^{(n)} - w_n| \leq \rho_n$ with known truncation error bounds $|f - S_n(\widehat{w}_n)| \leq \lambda_n$ to get

$$|f - S_n(w_n)| \leq \frac{|\widehat{w}_n - \zeta_n|}{|w_n - \zeta_n|}\,\frac{\lambda_n\rho_n}{|\widehat{w}_n - f^{(n)}|}. \tag{2.7.1}$$

For instance, if $\{\widehat{V}_n\}$ with $\widehat{V}_n := (-g_n + e^{i\alpha}\overline{\mathbb{H}})$ for $0 < g_n < 1$ and fixed $-\frac{\pi}{2} < \alpha < \frac{\pi}{2}$ is a sequence of value sets for $\mathbf{K}(a_n/1)$, then we know from Theorem 3.45 on page 155 that

$$\operatorname{diam} S_n(\widehat{V}_n) \leq \frac{|a_1|/((1-g_1)\cos\alpha)}{\displaystyle\prod_{j=2}^{n}\left(1 + \frac{\widehat{P}_{j-1}g_{j-1}(1-g_j)\cos^2\alpha}{|a_j|\widehat{\Sigma}_{j-2}}\right)} \tag{2.7.2}$$

where

$$\widehat{\Sigma}_n := \sum_{m=0}^{n}\widehat{P}_m \quad \text{and} \quad \widehat{P}_m := \prod_{k=1}^{m}\frac{1-g_k}{g_k}. \tag{2.7.3}$$

Another typical example is valid for S-fractions $\mathbf{K}(a_n z/1)$ with $a_n > 0$ and $z := r\,e^{2i\alpha}$ for some fixed $r > 0$ and $-\frac{\pi}{2} < \alpha < \frac{\pi}{2}$. Then the Thron-Gragg-Warner bound

$$|f - S_n(\widehat{w})| \leq \frac{2a_1 r}{\cos\alpha}\prod_{k=2}^{n}\frac{\sqrt{1+4a_k r/\cos^2\alpha} - 1}{\sqrt{1+4a_k r/\cos^2\alpha} + 1} =: \lambda_n \tag{2.7.4}$$

from page 128 gives a very good truncation error bound in this case. A natural idea is therefore to use this error in (2.7.1).

Example 14. Again we consider $\mathbf{K}(a_n/1)$ with $a_1 := 4$ and $a_{n+1} := n^2/(4n^2-1)$. A truncation error bound for

$$|f - S_n(w_n)| \quad \text{for } w_n := \Gamma_n := \frac{\sqrt{2}-1}{2} + \frac{\sqrt{2}/8}{4n^2-1}$$

was established by means of the Oval Sequence Theorem in Example 12. We shall now derive bounds for $|f - S_n(w_n)$ by means of (2.7.1). If we base our analysis on (2.7.2) with $\alpha := 0$ and all $g_n := \frac{1}{2}$, then

$$|f - S_n(\infty)| \le 8 \prod_{j=2}^{n} \left(1 + \frac{1/4}{(j-1)|a_j|}\right)^{-1} = 8 \prod_{j=1}^{n-1} \left(1 + \frac{4j^2-1}{4j^3}\right)^{-1}.$$

In Example 12 we found that

$$|f^{(n)} - w_n| \le \rho_n^* = \frac{(2-\sqrt{2})/2}{(4n^2-1)(2n+3) - \sqrt{2}/2} \quad \text{for } n \ge 1.$$

Since all $a_n > 0$, we definitely have

$$\zeta_n = -1 - \frac{a_n}{1} \, \frac{a_{n-1}}{+\,1} \, \frac{}{+\cdots+} \, \frac{a_2}{1} < -1.$$

Hence $|w_n - \zeta_n| > |1 + w_n| - \rho_n^*$, and thus it follows from (2.7.1) with $\widehat{w}_n := \infty$ that

$$|f - S_n(w_n)| < \frac{\rho_n^*}{|1+w_n| - \rho_n^*} \, 8 \prod_{j=1}^{n-1} \left(1 + \frac{4j^2-1}{4j^3}\right)^{-1}. \tag{2.7.5}$$

Now, $\mathbf{K}(a_n/1)$ can also be considered as an S-fractions with $z := 1$. The Thron-Gragg-Warner bound (2.7.4) can therefore also be used; i.e.,

$$|f - S_n(w_n)| < \frac{\rho_n^*}{|1+w_n| - \rho_n^*} \, 8 \prod_{k=1}^{n-1} \frac{\sqrt{1+4k^2/(4k^2-1)} - 1}{\sqrt{1+4k^2/(4k^2-1)} + 1}. \tag{2.7.6}$$

The table below shows the true truncation error $|f - S_n(w_n)|$ and the three truncation error bounds (2.6.1), (2.7.5) and (2.7.6).

n	$\lvert f - S_n(\Gamma_n)\rvert$	(2.6.1)	(2.7.5)	(2.7.6)
5	$2.6 \cdot 10^{-7}$	$7.4 \cdot 10^{-7}$	$3.6 \cdot 10^{-4}$	$1.7 \cdot 10^{-6}$
6	$2.7 \cdot 10^{-8}$	$7.6 \cdot 10^{-8}$	$1.8 \cdot 10^{-4}$	$1.8 \cdot 10^{-7}$
7	$3.0 \cdot 10^{-9}$	$8.5 \cdot 10^{-9}$	$9.9 \cdot 10^{-5}$	$2.0 \cdot 10^{-8}$
8	$3.5 \cdot 10^{-10}$	$1.0 \cdot 10^{-9}$	$5.9 \cdot 10^{-5}$	$2.3 \cdot 10^{-9}$
9	$4.3 \cdot 10^{-11}$	$1.2 \cdot 10^{-10}$	$3.8 \cdot 10^{-5}$	$2.9 \cdot 10^{-10}$
10	$5.4 \cdot 10^{-12}$	$1.6 \cdot 10^{-11}$	$2.5 \cdot 10^{-5}$	$3.7 \cdot 10^{-11}$
11	$7.1 \cdot 10^{-13}$	$2.0 \cdot 10^{-12}$	$1.7 \cdot 10^{-5}$	$4.8 \cdot 10^{-12}$
12	$9.5 \cdot 10^{-14}$	$2.7 \cdot 10^{-13}$	$1.2 \cdot 10^{-5}$	$6.4 \cdot 10^{-13}$
13	$1.3 \cdot 10^{-14}$	$3.7 \cdot 10^{-14}$	$9.0 \cdot 10^{-6}$	$8.7 \cdot 10^{-14}$
14	$1.8 \cdot 10^{-15}$	$5.2 \cdot 10^{-15}$	$6.8 \cdot 10^{-6}$	$1.2 \cdot 10^{-14}$
15	$2.5 \cdot 10^{-16}$	$7.2 \cdot 10^{-16}$	$5.2 \cdot 10^{-6}$	$1.7 \cdot 10^{-15}$

◇

5.3 Stable computation of approximants

5.3.1 Stability of the backward recurrence algorithm

In section 1.1.3 on page 10 we suggested three algorithms for computing approximants

$$S_n(w_n) = \frac{a_1}{b_1} + \frac{a_2}{b_2} + \cdots + \frac{a_n}{b_n + w_n} = \frac{A_{n-1}w_n + A_n}{B_{n-1}w_n + B_n}. \tag{3.1.1}$$

1. The forward recurrence algorithm where A_n and B_n are computed by means of the recurrence relation

$$A_k = b_k A_{k-1} + a_k A_{k-2}, \qquad B_k = b_k B_{k-1} + a_k B_{k-2}, \tag{3.1.2}$$

   starting with $A_{-1} = 1$, $A_0 = 0$ and $B_{-1} = 0$, $B_0 = 1$.

2. The backward recurrence algorithm where we use the recursion

$$q_{k-1} = s_k(q_k) := \frac{a_k}{b_k + q_k} \quad \text{for } k = n, n-1, \ldots, 1 \tag{3.1.3}$$

   starting with $q_n := w_n$.

3. The Euler-Minding summation where we compute $\{B_k\}_{k=1}^n$ by means of (3.1.2) and

$$S_n(0) = \frac{a_1}{B_0 B_1} - \frac{a_1 a_2}{B_1 B_2} + \frac{a_1 a_2 a_3}{B_2 B_3} - \frac{a_1 a_2 a_3 a_4}{B_3 B_4} + \cdots - \frac{\prod_{k=1}^n (-a_k)}{B_{n-1} B_n} \tag{3.1.4}$$

   which normally only is suited to find classical approximants for continued fractions $\mathbf{K}(1/b_n)$, even though it can be modified to give general approximants $S_n(w_n)$.

The emphasis in section 1.1.3 was on the complexity of the algorithm; i.e., then number of operations needed to reach $S_n(w_n)$. Even more important is the stability of the computation, in particular if we want to compute high order approximants $S_n(w_n)$. The forward algorithm is clearly unstable if for instance $b_k A_{k-1} \approx -a_k A_{k-2}$ or $b_k B_{k-1} \approx -a_k B_{k-2}$ for some indices. One can also run into very large or very small values for $|A_n|$ or $|B_n|$ fairly soon, which complicates the computation. The same problems may ruin the Euler-Minding summation.

In section 1.1.3 we claimed that the backward algorithm is usually more stable than the other two. So, what are we actually doing in this algorithm? Well, let us assume that $\mathbf{K}(a_n/b_n)$ converges generally to f, and that the purpose of computing $S_n(w)$ is to approximate f. We are therefore not worried if the computed value $\widehat{S}_n(w)$ of $S_n(w)$ is closer to f than it should be. What we do, is: we start with a given value $w \in \widehat{\mathbb{C}}$ and compute

$$w_{n,n} := w, \quad w_{n,k-1} := s_k(w_{n,k}) \quad \text{for } k = n, n-1, \ldots, 1. \tag{3.1.5}$$

That is, we compute the first $(n+1)$ terms of the tail sequence $\{t_{n,k}\}_k$ where

$$t_{n,k} = w_{n,k} = s_{k+1} \circ s_{k+2} \circ \cdots \circ s_n(w) = S_{n-k}^{(k)}(w). \tag{3.1.6}$$

We know a lot about tail sequences for a generally convergent continued fraction $\mathbf{K}(a_n/b_n)$:

- there are two kinds of tail sequences for $\mathbf{K}(a_n/b_n)$; the sequence $\{t_n\}$ of tail values where $t_0 = f$, and all the other tail sequences which in the literature often are called *the wrong tail sequences.*
- all the wrong tail sequences have the same asymptotic behavior in the sense that $\lim \mathfrak{m}(t_n, \tilde{t}_n) = 0$ when $t_0 \neq f$ and $\tilde{t}_0 \neq f$.
- if we start with a given value for t_0 and compute the tail sequence by the recurrence
$$t_{n+1} := s_n^{-1}(t_n) \quad \text{for } n = 0, 1, 2, \dots, \tag{3.1.7}$$
then the computed sequence $\{\widehat{t_n}\}$ will always behave like a wrong tail sequence asymptotically, even if we started with $t_0 = f$. The reason is that even minor inaccuracies makes $t_n \neq f^{(n)}$, at least if
$$\liminf_{n\to\infty} \mathfrak{m}(f^{(n)}, t_n) > 0 \quad \text{for } t_0 \neq f. \tag{3.1.8}$$

Therefore, the computation of a wrong tail sequence is stable when (3.1.8) holds, whereas the computation of the tail values is highly unstable with this algorithm. In the backward recurrence algorithm the recursion (3.1.7) is inverted. We start with t_n and compute $t_{n-1} = s_n(t_n)$. Therefore, the computation of the sequence of tail values by this algorithm is highly stable, whereas computation of wrong tail sequences are unstable: they will be pulled in towards f, exactly what we want.

We can also argue in terms of value sets. Let for instance V be a simple bounded closed value set for $\mathbf{K}(a_n/1)$ with $V^\circ \neq \emptyset$ and $V \cap (-1 - V) = \emptyset$. Let $w \in V_0$. Then the part $\{t_{n,k}\}_{k=0}^n$ of the tail sequence $\{t_{n,k}\}$ given by (3.1.6) is contained in V_0. As n increases, this tail sequence will be more and more like the sequence of tail values from V_0. The other tail sequences have all their limit points in $-1 - V_0$.

5.4 Remarks

1. **Convergence acceleration.** The idea of choosing $\{w_n\}$ carefully to make $\{S_n(w_n)\}$ converge faster is an old idea. Indeed, already in 1869 Sylvester ([Sylv69]) claimed: "I think a substantial difference does arise in favor of the continued fraction form, inasmuch as it indicates a certain obvious correction to be applied in order that the convergence may become more exact." The correction he had in mind was exactly the idea of using $S_n(w)$ instead of $S_n(0)$. He was comparing the continued fraction to the series $\sum(f_n - f_{n-1})$ where $f_n := S_n(0)$.

This idea has been applied by a number of authors, but the first systematic investigation for complex continued fractions is due to Thron and Waadeland ([ThWa80a]) who used the attracting fixed point of limit 1-periodic continued fractions $\mathbf{K}(a_n/1)$ to accelerate the convergence. Their investigations were inspired by previous work by Gill ([Gill75]). Later on, Gill and the authors of this book have extended the theory in a number of papers.

2. **Birkhoff-Trjzinski theory.** In order to accelerate the convergence of a continued fraction $\mathbf{K}(a_n/b_n)$, we use approximants $S_n(w_n)$ where w_n approximates its tail value $f^{(n)}$. Now, a tail sequence $\{t_n\}$ can always be written as $\{-P_n/P_{n-1}\}$ where $\{P_n\}$ is a non-trivial solution of the recurrence relation

$$P_n = b_n P_{n-1} + a_n P_{n-2} \quad \text{for } n = 1, 2, 3, \dots .$$

The Birkhoff-Trjzinski theory gives a method to approximate $\{P_n\}$ when $a_n = a(n)$ and $b_n = b(n)$ are functions of n with asymptotic expansions of the form

$$a(n) \sim n^{\lambda_1/\omega} \sum_{m=0}^{\infty} \alpha_m \, n^{-m/\omega}, \quad b(n) \sim n^{\lambda_2/\omega} \sum_{m=0}^{\infty} \beta_m \, n^{-m/\omega} \quad \text{as } n \to \infty,$$

where λ_i are integers, $\omega \in \mathbb{N}$ and $\alpha_0\beta_0 \neq 0$. They proved that every solution of the recurrence relation can be written as a linear combination of two solutions $P_n := P(n)$ which have asymptotic expansions of the form

$$\begin{aligned}
&e^{Q(\rho,n)} s(\rho, n) \quad \text{where} \\
&Q(\rho, n) := \delta_0 \, n \ln(n) + \sum_{j=1}^{\rho} \delta_j n^{(\rho+1-j)/\rho}, \\
&s(\rho, n) := n^{\theta} \sum_{j=0}^{t} (\ln n)^j n^{r_{t-1}/\rho} q_j(\rho, n) \\
&q_j(\rho, n) := \sum_{m=0}^{\infty} c_{j,m} n^{-m/\rho}
\end{aligned}$$

where the integers $\rho \geq 1$, r_j and $\delta_0\rho$ and the complex numbers δ_j, θ and $c_{j,m}$ with $c_{j,0} \neq 0$, $r_0 := 0$ and $-\pi \leq \operatorname{Im} \delta_1 < \pi$ can be determined by substituting this into the recurrence relation and comparing coefficients, ([Birk30], [BiTr32]). For more information we refer to [Wimp84].

5.5 Problems

1. **Limit periodic continued fraction.** Let $\mathbf{K}(1/b_n)$ be the continued fraction where $b_n = 4 + (0.9)^n$ for all n.
   (a) Find a value set $V := \mathcal{B}(\Gamma, \rho)$ for $\mathbf{K}(1/b_n)$. (Try to make V small.)
   (b) Does $\mathbf{K}(1/b_n)$ converge to a finite value f?
   (c) Are the classical approximants f_n of $\mathbf{K}(1/b_n)$ all distinct; i.e. $f_n \neq f_m$ if $n \neq m$?
   (d) Use the value set V found in (a) to derive upper bounds for the truncation error $|f - S_n(w)|$ for suitably chosen $w \in \mathbb{C}$.

2. **Simple element set.** Let $\mathbf{K}(a_n/1)$ have all elements $a_n \in E := \{w \in \mathbb{C};\ |w-3-i| < 0.4\}$. Find a $\Gamma \in \mathbb{C}$ and a $0 < \rho < |1+\Gamma|$ such that E is a subset of the cartesian oval in the Oval Sequence Theorem on page 243 with all $\Gamma_n := \Gamma$ and $\rho_n := \rho$. Use this to prove that $\mathbf{K}(a_n/1)$ converges to a finite value and to find truncation error bounds for suitably chosen approximants of $\mathbf{K}(a_n/1)$.

3. **The logarithm of $(1+i)$.** Use the Oval Sequence Theorem on page 243 with all $\Gamma_n := \Gamma$ and $\rho_n := \rho$ to find bounds for $|f - S_n(w)|$ for the continued fraction

$$\operatorname*{K}\frac{a_n i}{1} := \frac{i}{1+}\,\frac{i/2}{1}\,+\,\frac{i/6}{1}\,+\,\frac{2i/6}{1}\,+\,\frac{2i/10}{1}\,+\,\frac{3i/10}{1}\,+\cdots$$

when $w = (\sqrt{1+i}-1)/2$.

4. **Limit periodic continued fraction.** We want to improve the speed of convergence of the continued fraction

$$\operatorname*{K}\frac{a_n}{1} = \frac{6+(0.9)}{1}\,+\,\frac{6+(0.9)^2}{1}\,+\,\frac{6+(0.9)^3}{1}\,+\,\frac{6+(0.9)^4}{1}\,+\cdots.$$

   (a) Prove that $\mathbf{K}(a_n/1)$ converges to a finite value f.

   (b) Suggest a value $w_n = w$ for all n such that

$$\lim_{n\to\infty}\frac{f - S_n(w)}{f - S_n(0)} = 0.$$

   Does the sequence $\{S_n(w)\}$ have an oscillating character which can be used to obtain upper bounds for the truncation error $|f - S_n(w)|$?

   (c) Suggest values for w_n such that

$$\lim_{n\to\infty}\frac{f - S_n(w_n)}{f - S_n(w)} = 0$$

   where w is the value from a). Does $\{S_n(w_n)\}$ have such an oscillating character?

5. **The case $a_n \to \infty$.** Suggest expressions for w_n such that the approximants $S_n(w_n)$ of

$$\operatorname*{K}\frac{a_n}{1} = \frac{1^2}{1+}\,\frac{5\cdot 2^2}{1}\,+\frac{3^2}{1+}\,\frac{5\cdot 4^2}{1}\,+\frac{5^2}{1+}\,\frac{5\cdot 6^2}{1}\,+\frac{7^2}{1}+\cdots$$

(hopefully) converge faster to the value f of $\mathbf{K}(a_n/1)$ than $S_n(0)$. Compute the first 6 approximants of $S_n(0)$ and $S_n(w_n)$ and use the oscillating character to determine an error bound for $S_6(0)$. Find an upper bound for $|f - S_n(w_n)|$.

6. **Choice of approximations.**

   (a) Suggest some favorable approximants $S_n(w_n)$ to compute $\mathbf{K}((-0.2+(0.4)^n)/1)$.

   (b) Compute the value of $\mathbf{K}((-0.2+(0.4)^n)/1)$ with an absolute error less than 0.05.

7. **The improvement machine.** Let $\mathbf{K}(a_n/1)$ be given by

$$a_n = x(1+x) + r^n \quad \text{where } 0 < |x| < |1+x| \text{ and } 0 < |r| < 1\,.$$

Use the improvement machine on page 227 to derive approximants $S_n(w_n^{(m)})$ for $\mathbf{K}(a_n/1)$ such that

$$\lim_{n \to \infty} \frac{f - S_n(w_n^{(m)})}{f - S_n(w_n^{(m-1)})} = 0 \quad \text{for } m = 1, 2, 3, \dots\,.$$

8. ♠ **The Oval Theorem.** Set all $\Gamma_n := \Gamma$ and $\rho_n := \rho$ in the oval sequence theorem on page 243, where $\operatorname{Re}\Gamma > -\frac{1}{2}$ and $\rho < |1+\Gamma|$.

   (a) Prove that

   $$E := \left\{a \in \mathbb{C};\ \left|\frac{a(1+\overline{\Gamma})}{|1+\Gamma|^2 - \rho^2} - \Gamma\right| + \frac{|a|\rho}{|1+\Gamma|^2 - \rho^2} \le \rho\right\}$$

   is the simple element set for continued fractions $\mathbf{K}(a_n/1)$ corresponding to the simple value set $V := \mathfrak{B}(\Gamma, \rho)$.

   (b) Prove that every continued fraction $\mathbf{K}(a_n/1)$ from E converges to some value in V. (Hint: the Parabola Theorem on page 151 or the lemma on page 113 may be of help.)

   (c) Prove that

   $$|f - S_n(w_n)| \le 2\rho \frac{|\Gamma| + \rho}{|1+\Gamma| - \rho} M^{n-1} \quad \text{where } M := \max_{v \in V} \left|\frac{v}{1+v}\right|$$

   for every continued fraction $\mathbf{K}(a_n/1)$ from E and $w_n \in V$.

   (d) Prove that $M < 1$ if $\rho < \frac{1}{2}(|1+\Gamma| - |\Gamma|)$.

   (e) Prove that $M < 1$ if $\Gamma > 0$ and $\rho < \Gamma + \frac{1}{2}$.

9. ♠ **A bridge between the Worpitzky disk and the Parabola Theorem.** Let E be as in Problem 8 with $\rho := \operatorname{Re}\Gamma + \frac{1}{2} > 0$.

   (a) Prove that $-\frac{1}{4} \in \partial E$.

   (b) Prove that E reduces to the Worpitzky disk if $\Gamma := 0$.

   (c) Prove that the Worpitzky disk from (b) is contained in E for all Γ with $\operatorname{Re}\Gamma > 0$.

   (d) Prove that E is contained in the parabolic region E_α from the Parabola Theorem on page 151 when $\arg \Gamma(1+\Gamma) = 2\alpha$.

   (e) Prove that $\lim E = E_\alpha$ from the Parabola Theorem when $\arg \Gamma(1+\Gamma) = 2\alpha$ is kept fixed and $\Gamma \to \infty$.

Appendix A

Some continued fraction expansions

This is a catalogue of some of the known continued fraction expansions. The list is in no way complete. Still it can be useful, both to find a continued fraction expansion of some given function and to "sum" a given continued fraction.

We have not attempted to find the origin of each result. The references we give are therefore just pointing to books or papers where the expansion also can be found.

A.1 Introduction

A.1.1 Notation

We write

$$f(\cdots) = \mathbf{K}(a_n(\cdots)/b_n(\cdots)); \quad (\cdots) \in D \tag{1.1.1}$$

to say that the continued fraction converges in the classical sense to $f(\cdots)$ for the parameters $(\cdots)$ in the set D. In the literature the set D is often far too restrictive, if it is given at all. We have determined a (possibly larger) set D_c where the continued fraction converges. This is done by methods presented in this book. However, it may well happen that the equality (1.1.1) fails in a subset of D_c, even if $f(\cdots)$ is interpreted as an analytic continuation of the expression in question. In some cases we therefore give expressions for sets D_c and D_f such that $\mathbf{K}(a_n(\cdots)/b_n(\cdots))$ converges in D_c and the equality holds in $D_f \subseteq D_c$. What happens outside these sets is not checked. The identities normally holds also at points where $a_n(\cdots) = 0$ unless otherwise stated.

The elements $a_n(\cdots)$ and $b_n(\cdots)$ of almost all continued fractions in this appendix are polynomials in the parameters. The classical approximants are then rational functions of the parameters, and can therefore not converge to multivalued functions. If the left hand side of (1.1.1) is a multivalued function, we always take the principal part unless otherwise stated. The principal part is often written with a capital first letter, such as Ln z, Arctan z etc.

A.1.2 Transformations

It is evident that not every continued fraction expansion can find room in a book like this. On the other hand, quite a number of the known continued fraction expansions can be derived from one another by simple transformations. We have for instance

$$f = b_0 + \mathbf{K}(a_n/b_n) \Longleftrightarrow c \cdot \frac{1}{f} = \frac{c}{b_0 +} \frac{a_1}{b_1 +} \frac{a_2}{b_2 + \cdots}; \quad c \neq 0. \tag{1.2.1}$$

Similarly, if $f = b_0 + \mathbf{K}(a_n/b_n)$, then $g = (f-1)/(f+1) = 1 - 2/(1+f)$; i.e.,

$$f = b_0 + \mathbf{K}(a_n/b_n) \Longleftrightarrow \frac{f-1}{f+1} = 1 - \frac{2}{1 + b_0 +} \frac{a_1}{b_1 +} \frac{a_2}{b_2 + \cdots}. \tag{1.2.2}$$

Another simple transformation is maybe most easily described for S-fractions. Assume that $f(z) = b_0 + \mathbf{K}(a_n z/1)$. Then $f(z^{-1}) = b_0 + \mathbf{K}(a_n z^{-1}/1)$. Equivalence transformations lead to

$$\begin{aligned} f\left(\tfrac{1}{z}\right) &= b_0 + \frac{a_1}{z +} \frac{a_2}{1 +} \frac{a_3}{z +} \frac{a_4}{1 +} \frac{a_5}{z + \cdots} \\ &= b_0 + \frac{a_1/z}{1} \frac{a_2}{+ z +} \frac{a_3}{1 +} \frac{a_4}{z +} \frac{a_5}{1 + \cdots} \\ &= b_0 + \frac{a_1/\xi}{\xi} \frac{a_2}{+ \xi +} \frac{a_3}{\xi +} \frac{a_4}{\xi +} \frac{a_5}{\xi + \cdots}, \end{aligned}$$

where $\xi^2 = z$. We shall normally not list equivalent continued fractions like this separately.

Another situation that often arises is the following: We have

$$f(z) = b_0 + \frac{a_1 z^2}{1} \frac{a_2 z^2}{+ 1} \frac{a_3 z^2}{+ 1} + \cdots. \tag{1.2.3a}$$

Then

$$f(iz) = b_0 - \frac{a_1 z^2}{1} \frac{a_2 z^2}{- 1} \frac{a_3 z^2}{- 1} - \cdots. \tag{1.2.4b}$$

Of course, every time we have a continued fraction expansion $f = b_0 + \mathbf{K}(a_n/b_n)$ with all $a_n, b_n \neq 0$, we can take its even or odd part and obtain a "new" continued fraction converging to the same value f. Some of these variations will be listed, in particular if they turn out to be nice and simple.

A.2 Elementary functions

A.2.1 Mathematical constants

$$\pi = 3 + \frac{1}{7+} \frac{1}{15+} \frac{1}{1+} \frac{1}{292+} \frac{1}{1+} \frac{1}{1+} \frac{1}{1+} \frac{1}{2+} \frac{1}{1+} \frac{1}{3+} \frac{1}{1+} \frac{1}{14+\cdots}, \tag{2.1.1}$$

([JoTh80], p 23). This is the regular continued fraction expansion of π.

$$\pi = \frac{4}{1+}\frac{1^2}{3+}\frac{2^2}{5+}\frac{3^2}{7+}\frac{4^2}{9+\cdots}, \tag{2.1.2}$$

([JoTh80], p 25), (see also (3.6.1)).

$$\frac{\pi}{2} = 1 + \frac{1}{1+}\frac{1\cdot 2}{1+}\frac{2\cdot 3}{1+}\frac{3\cdot 4}{1+\cdots}, \quad \text{([Khru06a])}. \tag{2.1.3}$$

For the **Riemann zeta function** we have

$$\tfrac{1}{2}\zeta(2) = \frac{\pi^2}{12} = \frac{1}{1+}\frac{1^4}{3+}\frac{2^4}{5+}\frac{3^4}{7+\cdots}, \tag{2.1.4}$$

([Bern89], p 150).

$$\zeta(2) = \frac{\pi^2}{6} = 1 + \frac{1}{1+}\frac{1^2}{1+}\frac{1\cdot 2}{1+}\frac{2^2}{1+}\frac{2\cdot 3}{1+}\frac{3^2}{1+}\frac{3\cdot 4}{1+}\frac{4^2}{1+\cdots}, \tag{2.1.5}$$

([Bern89], p 153).

Apery's constant:

$$\zeta(3) = 1 + \frac{1}{4+}\frac{1^3}{1+}\frac{1^3}{12+}\frac{2^3}{1+}\frac{2^3}{20+}\frac{3^3}{1+}\frac{3^3}{28+}\frac{4^3}{1+}\frac{4^3}{36+\cdots}, \tag{2.1.6}$$

([Bern89], p 155), (see also (4.7.37)).

Euler's number:

$$e = \frac{1}{1-}\frac{1}{1+}\frac{1}{2-}\frac{1}{3+}\frac{1}{2-}\frac{1}{5+}\frac{1}{2-}\frac{1}{7+\cdots}, \tag{2.1.7}$$

([JoTh80], p 25), (see also (3.2.1)).

$$e = 2 + \frac{1}{1+}\frac{1}{2+}\frac{1}{1+}\frac{1}{1+}\frac{1}{4+}\frac{1}{1+}\frac{1}{1+}\frac{1}{6+\cdots}, \tag{2.1.8}$$

([JoTh80], p 23).

$$e = 1 + \frac{2}{1+}\frac{1}{6+}\frac{1}{10+}\frac{1}{14+}\frac{1}{18+\cdots}, \tag{2.1.9}$$

([Khov63], p 114). (See also (3.2.2)).

$$e = 2 + \frac{2}{2+}\frac{3}{3+}\frac{4}{4+}\frac{5}{5+\cdots}, \tag{2.1.10}$$

([Perr57], p 57).

$$e = \frac{1}{1-}\frac{2}{3+}\frac{1}{6+}\frac{1}{10+}\frac{1}{14+}\frac{1}{18+\cdots}, \tag{2.1.11}$$

([Khov63], p 114). (See also (3.2.2) for $z = -1$).

$$\sqrt{e} = 1 + \frac{1}{1+}\frac{1}{1+}\frac{1}{5+}\frac{1}{1+}\frac{1}{1+}\frac{1}{9+}\frac{1}{1+}\frac{1}{1+}\frac{1}{13+\cdots}, \tag{2.1.12}$$

([Euler37]).

$$\sqrt[3]{e} = 1 + \frac{1}{2+}\frac{1}{1+}\frac{1}{1+}\frac{1}{8+}\frac{1}{1+}\frac{1}{1+}\frac{1}{14+}\frac{1}{1+}\frac{1}{1+}\frac{1}{20+}\cdots, \tag{2.1.13}$$

([Euler37]). (See also (2.2.4).)

$$\coth\tfrac{1}{2} = \frac{e+1}{e-1} = 2 + \frac{1}{6+}\frac{1}{10+}\frac{1}{14+}\frac{1}{18+}\cdots, \tag{2.1.14}$$

([Khru06b]).

The golden ratio:

$$\frac{\sqrt{5}-1}{2} = \frac{1}{1+}\frac{1}{1+}\frac{1}{1+}\cdots, \tag{2.1.15}$$

([JoTh80], p 23).

Catalan's constant: $G := \sum_{k=0}^{\infty}(-1)^k/(2k+1)^2$:

$$2G = 2 - \frac{1^2}{3\ +}\frac{2^2}{1\ +}\frac{2^2}{3\ +}\frac{4^2}{1\ +}\frac{4^2}{3\ +}\frac{6^2}{1\ +}\frac{6^2}{3\ +}\cdots, \tag{2.1.16}$$

([Bern89], p 151). (See also (4.7.30) with $z := 2$.)

$$2G = 1 + \frac{1}{1/2+}\frac{1^2}{1/2+}\frac{1\cdot 2}{1/2\ +}\frac{2^2}{1/2+}\frac{2\cdot 3}{1/2\ +}\frac{3^2}{1/2+}\frac{3\cdot 4}{1/2\ +}\cdots, \tag{2.1.17}$$

([Bern89], p 153). (See also (4.7.32) with $z := \frac{1}{2}$.)

A.2.2 The exponential function

$$\begin{aligned} e^z &= {}_1F_1(1;1;z) = \frac{1}{1-}\frac{z}{1+}\frac{z}{2-}\frac{z}{3+}\frac{z}{2-}\frac{z}{5+}\frac{z}{2-}\frac{z}{7+}\cdots \\ &= \frac{1}{1-}\frac{z}{1+}\frac{1z}{2\ -}\frac{1z}{3\ +}\frac{2z}{4\ -}\frac{2z}{5\ +}\frac{3z}{6\ -}\frac{3z}{7\ +}\frac{4z}{8\ -}\frac{4z}{9\ +}\cdots; \quad z \in \mathbb{C}, \end{aligned}$$

([JoTh80], p 207). (See also (4.1.4).) The odd part of this continued fraction is

$$e^z = 1 + \frac{2z}{2-z+}\frac{z^2}{6\ +}\frac{z^2}{10+}\frac{z^2}{14+}\frac{z^2}{18+}\cdots; \quad z \in \mathbb{C}, \tag{2.2.1}$$

([Khov63], p 114).

$$e^z = 1 + \frac{z}{1-z+}\frac{1z}{2-z+}\frac{2z}{3-z+}\frac{3z}{4-z+}\cdots; \ z \in \mathbb{C}, \tag{2.2.2}$$

([JoTh80], p 272).

Since $e^z = 1/e^{-z}$, we can find three more expansions from (2.2.1)–(2.2.2) by use of (1.2.1). For instance, (2.2.2) transforms into

$$e^z = \frac{1}{1-}\frac{z}{1+z-}\frac{1z}{2+z-}\frac{2z}{3+z-}\frac{3z}{4+z+}\cdots; \ z \in \mathbb{C}, \tag{2.2.3}$$

([Khov63], p 113).

$$e^{1/z} = 1 + \frac{1}{z-1+}\frac{1}{1+}\frac{1}{1+}\frac{1}{3z-1+}\frac{1}{1+\cdots}\frac{1}{1+}\frac{1}{5z-1+}\frac{1}{1+\cdots} \quad z \in \mathbb{C}, \tag{2.2.4}$$

([Khru06b]).

Lambert's continued fraction

$$\frac{e^z - e^{-z}}{e^z + e^{-z}} = \frac{z}{1+}\frac{z^2}{3+}\frac{z^2}{5+}\frac{z^2}{7+\cdots}; \; z \in \mathbb{C}, \tag{2.2.5}$$

([Wall48], p 349), is easily obtained from (2.2.1) by use of (1.2.2).

A.2.3 The general binomial function

The general binomial function $(1+z)^\alpha$ is a multivalued function. As already mentioned in the introduction, we shall always let $(1+z)^\alpha$ mean the principal part of this function; i.e., as always,

$$(1+z)^\alpha := \exp(\alpha \operatorname{Ln}(1+z)) \text{ where } -\pi < \operatorname{Im}(\operatorname{Ln}(1+z)) \le \pi.$$

We then have the following expansions:

$$\begin{aligned}(1+z)^\alpha &= {}_2F_1(-\alpha,1;1;-z)\\ &= \frac{1}{1-}\frac{\alpha z}{1+}\frac{(1+\alpha)z}{2+}\frac{(1-\alpha)z}{3+}\frac{(2+\alpha)z}{2+}\frac{(2-\alpha)z}{5+}\frac{(3+\alpha)z}{2+}\frac{(3-\alpha)z}{7+\cdots}\end{aligned} \tag{2.3.1}$$

for $\alpha \in \mathbb{C}$ and $|\arg(z+1)| < \pi$, ([JoTh80], p 202). (See also (3.1.6).) The odd part of this continued fraction is

$$\begin{aligned}&(1+z)^\alpha =\\ &1 + \frac{2\alpha z}{2+(1-\alpha)z-}\frac{(1^2-\alpha^2)z^2}{3(z+2)-}\frac{(2^2-\alpha^2)z^2}{5(z+2)-}\frac{(3^2-\alpha^2)z^2}{7(z+2)-\cdots}\end{aligned} \tag{2.3.2}$$

for $\alpha \in \mathbb{C}$ and $|\arg(z+1)| < \pi$, ([Khov63], p 105). (2.3.2) is also the odd part of

$$\begin{aligned}&(1+z)^\alpha =\\ &\frac{1}{1-}\frac{\alpha z}{1(1+z)-}\frac{(1-\alpha)z}{2-}\frac{(1+\alpha)z}{3(1+z)-}\frac{(2-\alpha)z}{2-}\frac{(2+\alpha)z}{5(1+z)-}\frac{(3-\alpha)z}{2-\cdots}\end{aligned} \tag{2.3.3}$$

for $\alpha \in \mathbb{C}$ and $|\arg(z+1)| < \pi$, ([Khov63], p 101).

$$(1+z)^\alpha = \frac{1}{1-}\frac{\alpha z}{1+(1+\alpha)z-}\frac{1(1+\alpha)z(1+z)}{2+(3+\alpha)z-}\frac{2(2+\alpha)z(1+z)}{3+(5+\alpha)z-\cdots} \tag{2.3.4}$$

([Khov63], p 101). $D_c := \{(\alpha,z) \in \mathbb{C}^2;\, \operatorname{Re}(z) \neq -\frac{1}{2},\, z \neq -1\}$, $D_f := \{(\alpha,z) \in \mathbb{C}^2;\, \operatorname{Re}(z) > -\frac{1}{2}\}$.

$$(1+z)^\alpha = \frac{1}{1-}\frac{\alpha z}{1+\alpha z+}\frac{1(1-\alpha)z}{2-(1-\alpha)z+}\frac{2(2-\alpha)z}{3-(2-\alpha)z+}\frac{3(3-\alpha)z}{4-(3-\alpha)z+\cdots}, \tag{2.3.5}$$

([Khov63], p 102). $D_c := \{(\alpha, z) \in \mathbb{C}^2;\ |z| \neq 1\}$, $D_c := \{(\alpha, z) \in \mathbb{C}^2;\ |z| < 1\}$.

The general binomial function satisfies $(1+z)^\alpha = 1/(1+z)^{-\alpha}$. Hence the equality (1.2.1) applied to these 5 expansions gives us 5 new ones. To find a continued fraction expansion for

$$\left(\frac{z+1}{z-1}\right)^\alpha = \left(1 + \frac{2}{z-1}\right)^\alpha \tag{2.3.6}$$

we can use any of the 5 expansions (2.3.1)–(2.3.5) with z replaced by $2/(z-1)$.

Laguerre's continued fraction

$$\left(\frac{z+1}{z-1}\right)^\alpha = 1 + \frac{2\alpha}{z-\alpha+}\,\frac{\alpha^2-1^2}{3z}\,\underset{+}{}\,\frac{\alpha^2-2^2}{5z}\,\underset{+}{}\,\frac{\alpha^2-3^2}{7z}\,\underset{+}{}\cdots, \tag{2.3.7}$$

for $\alpha \in \mathbb{C}$ and $z \in \mathbb{C} \setminus [-1,1]$, ([Perr57], p 153).

$$\begin{aligned} z^{1/z} = 1 + \frac{z-1}{1z}\,\underset{+}{}\,\frac{(1z-1)(z-1)}{2}\,\underset{+}{}\,\frac{(1z+1)(z-1)}{3z}\,\underset{+}{} \\ \frac{(2z-1)(z-1)}{2}\,\underset{+}{}\,\frac{(2z+1)(z-1)}{5z}\,\underset{+}{}\,\frac{(3z-1)(z-1)}{2}\,\underset{+}{}\,\frac{(3z+1)(z-1)}{7z}\,\underset{+}{}\cdots \end{aligned} \tag{2.3.8}$$

for $|\arg z| < \pi$, ([Khov63], p 109). The even part of (2.3.8) is

$$\begin{aligned} z^{1/z} = 1 + \frac{2(z-1)}{z^2+1}\,\underset{-}{}\,\frac{(1^2z^2-1)(z-1)^2}{3z(z+1)}\,\underset{-}{} \\ \frac{(2^2z^2-1)(z-1)^2}{5z(z+1)}\,\underset{-}{}\,\frac{(3^2z^2-1)(z-1)^2}{7z(z+1)}\,\underset{-}{}\cdots \end{aligned} \tag{2.3.9}$$

for $|\arg z| < \pi$, ([Khov63], p 110).

$$\begin{aligned} \left(\frac{1+az}{1+bz}\right)^\alpha = 1 + \frac{2\alpha(a-b)z}{2+(a+b-\alpha(a-b))z-}\,\frac{(a-b)^2(1^2-\alpha^2)z^2}{3(2+(a+b)z)}\,\underset{-}{} \\ \frac{(a-b)^2(2^2-\alpha^2)z^2}{5(2+(a+b)z)}\,\underset{-}{}\,\frac{(a-b)^2(3^2-\alpha^2)z^2}{7(2+(a+b)z)}\,\underset{-}{}\cdots \end{aligned} \tag{2.3.10}$$

for $\alpha \in \mathbb{C}$ and $\frac{2+(a+b)z}{(a-b)z} \in \mathbb{C} \setminus [-1,1]$, ([Perr57], p 264). (Note that $\frac{2+(a+b)z}{(a-b)z} = \frac{x+y}{x-y}$ for $x := 1+az$, $y := 1+bz$.)

A.2.4 The natural logarithm

$$\begin{aligned} \operatorname{Ln}(1+z) &= z\,{}_2F_1(1,1;2;-z) = z\int_0^1 \frac{dt}{1+zt} \\ &= \frac{z}{1+}\,\frac{1z}{2}\,\underset{+}{}\,\frac{1z}{3}\,\underset{+}{}\,\frac{2z}{2}\,\underset{+}{}\,\frac{2z}{5}\,\underset{+}{}\,\frac{3z}{2}\,\underset{+}{}\,\frac{3z}{7}\,\underset{+}{}\,\frac{4z}{2}\,\underset{+}{}\,\frac{4z}{9}\,\underset{+}{}\cdots \\ &= \frac{z}{1+}\,\frac{1^2z}{2}\,\underset{+}{}\,\frac{1^2z}{3}\,\underset{+}{}\,\frac{2^2z}{4}\,\underset{+}{}\,\frac{2^2z}{5}\,\underset{+}{}\,\frac{3^2z}{6}\,\underset{+}{}\,\frac{3^2z}{7}\,\underset{+}{}\,\frac{4^2z}{8}\,\underset{+}{}\,\frac{4^2z}{9}\,\underset{+}{}\cdots \end{aligned} \tag{2.4.1}$$

for $|\arg(1+z)| < \pi$, ([JoTh80], p 203). (See also (3.1.6).)

$$\mathrm{Ln}(1+z) = \frac{z}{1+z} \frac{}{-} \frac{z}{1} \frac{}{+} \frac{1}{1+z} \frac{}{-} \frac{z}{1} \frac{}{+} \frac{1/2}{1+z} \frac{}{-} \frac{z}{1} \frac{}{+} \frac{1}{1+z} \frac{}{-} \frac{z}{1} \frac{}{+} \frac{2/3}{1+z} \frac{}{-} \cdots \tag{2.4.2}$$

for $|\arg(1+z)| < \pi$, ([JoTh80], p 319). Here the continued fraction has the form $\mathbf{K}(a_n(z)/b_n(z))$ where all $a_{2n}(z) = -z$, $a_{4n-1}(z) = 1$ and $a_{4n+1}(z) = n/(n+1)$. (2.4.1) is the even part of (2.4.2). The odd part of (2.4.2) can be written

$$\begin{aligned}\mathrm{Ln}(1+z) &= \frac{z}{1+z}\left\{1 + \frac{z}{2} \frac{}{+} \frac{2z}{3} \frac{}{+} \frac{z}{2} \frac{}{+} \frac{3z}{5} \frac{}{+} \frac{2z}{2} \frac{}{+} \frac{4z}{7} \frac{}{+} \frac{3z}{2} \frac{}{+} \frac{5z}{9} \frac{}{+} \frac{4z}{2} \frac{}{+} \cdots\right\} \\ &= \frac{z}{1+z}\left\{1 + \frac{z}{2} \frac{}{+} \frac{1\cdot 2z}{3} \frac{}{+} \frac{1\cdot 2z}{4} \frac{}{+} \frac{2\cdot 3z}{5} \frac{}{+} \frac{2\cdot 3z}{6} \frac{}{+} \frac{3\cdot 4z}{7} \frac{}{+} \frac{3\cdot 4z}{8} \frac{}{+} \cdots\right\}\end{aligned} \tag{2.4.3}$$

for $|\arg(1+z)| < \pi$. The even part of (2.4.1) is

$$\mathrm{Ln}(1+z) = \frac{2z}{1(2+z)} \frac{}{-} \frac{1^2z^2}{3(2+z)} \frac{}{-} \frac{2^2z^2}{5(2+z)} \frac{}{-} \frac{3^2z^2}{7(2+z)} \frac{}{-} \cdots \tag{2.4.4}$$

for $|\arg(1+z)| < \pi$, ([Khov63], p 111).

$$\mathrm{Ln}(1+z) = \frac{z}{1} \frac{}{+} \frac{1^2z}{2-z} \frac{}{+} \frac{2^2z}{3-2z} \frac{}{+} \frac{3^2z}{4-3z} \frac{}{+} \frac{4^2z}{5-4z} \frac{}{+} \frac{5^2z}{6-5z} \frac{}{+} \cdots, \tag{2.4.5}$$

([Khov63], p 111). $D_c := \{z \in \mathbb{C};\ |z| \neq 1\}$, $D_f := \{z \in \mathbb{C};\ |z| < 1\}$.

The connection

$$\mathrm{Ln}(1+z) = -\mathrm{Ln}\left(\frac{1}{1+z}\right) = -\mathrm{Ln}\left(1 - \frac{z}{1+z}\right) \tag{2.4.6}$$

can be applied to (2.4.1)–(2.4.5) to get 5 new continued fraction expansions. For instance, from (2.4.1) we get

$$\mathrm{Ln}(1+z) = \frac{z}{1+z} \frac{}{-} \frac{1z}{2} \frac{}{-} \frac{1z}{3(1+z)} \frac{}{-} \frac{2z}{2} \frac{}{-} \frac{2z}{5(1+z)} \frac{}{-} \frac{3z}{2} \frac{}{-} \frac{3z}{7(1+z)} \frac{}{-} \cdots \tag{2.4.7}$$

for $|\arg(1+z)| < \pi$, ([Khov63], p 110), and from (2.4.5)

$$\mathrm{Ln}(1+z) = \frac{z}{1+z} \frac{}{-} \frac{1^2z(1+z)}{2+3z} \frac{}{-} \frac{2^2z(1+z)}{3+5z} \frac{}{-} \frac{3^2z(1+z)}{4+7z} \frac{}{-} \cdots, \tag{2.4.8}$$

([Khov63], p 111). $D_c := \{z \in \mathbb{C};\ \mathrm{Re}(z) \neq -\frac{1}{2}\}$, $D_f := \{z \in \mathbb{C};\ \mathrm{Re}(z) > -\frac{1}{2}\}$.

$$\begin{aligned}\mathrm{Ln}\Big(\frac{1+z}{1-z}\Big) &= 2z\,{}_2F_1(\tfrac{1}{2}, 1; \tfrac{3}{2}; z^2) = \mathrm{Ln}\Big(1 + \frac{2z}{1-z}\Big) = z\int_{-1}^{1} \frac{dt}{1+tz} \\ &= \frac{2z}{1} \frac{}{-} \frac{1^2z^2}{3} \frac{}{-} \frac{2^2z^2}{5} \frac{}{-} \frac{3^2z^2}{7} \frac{}{-} \frac{4^2z^2}{9} \frac{}{-} \cdots\end{aligned} \tag{2.4.9}$$

for $|\arg(1-z^2)| < \pi$, ([JoTh80], p 203). (See also (3.1.6).) From this we also get

$$\mathrm{Ln}\Big(\frac{z+1}{z-1}\Big) = \mathrm{Ln}\Big(\frac{1+1/z}{1-1/z}\Big) = \frac{2}{z} \frac{}{-} \frac{1^2}{3z} \frac{}{-} \frac{2^2}{5z} \frac{}{-} \frac{3^2}{7z} \frac{}{-} \frac{4^2}{9z} \frac{}{-} \cdots \tag{2.4.10}$$

for $z \in \mathbb{C} \setminus [-1, 1]$, ([Perr57], p 155). Of course, also other continued fraction expansions for $\mathrm{Ln}(1+z)$ can be used to derive expressions for $\mathrm{Ln}((1+z)/(1-z))$.

A.2.5 Trigonometric and hyperbolic functions

$$\tan z := \frac{\sin z}{\cos z} = z\frac{{}_0F_1(3/2;-z^2/4)}{{}_0F_1(1/2;-z^2/4)} = \frac{z}{1-}\frac{z^2}{3-}\frac{z^2}{5-}\frac{z^2}{7-}\frac{z^2}{9-\cdots}; \; z \in \mathbb{C} \tag{2.5.1}$$

([JoTh80], p 211), (See also (3.1.1).) The odd part of (2.5.1) is

$$\begin{aligned}\tan z = z + \frac{5z^3}{1\cdot 3\cdot 5 - 6z^2 -}\frac{1\cdot 9z^4}{5\cdot 7\cdot 9 - 14z^2 -}\\ \frac{5\cdot 13z^4}{9\cdot 11\cdot 13 - 22z^2 -}\frac{9\cdot 17z^4}{13\cdot 15\cdot 17 - 30z^2 -\cdots}; \; z\in\mathbb{C},\end{aligned} \tag{2.5.2}$$

Another type of expansion is

$$\tan\frac{z\pi}{4} = \frac{z}{1+}\frac{1^2-z^2}{2+}\frac{3^2-z^2}{2+}\frac{5^2-z^2}{2+}\frac{7^2-z^2}{2+\cdots}; \; z\in\mathbb{C}, \tag{2.5.3}$$

([Perr57], p 35). From these expansions one also gets continued fractions for $\cot z = 1/\tan z$, $\tanh z = -i\tan(iz)$ and $\coth z = i/\tan(iz)$.

Quite another type of expansion for $\tan z$ follows from the identity

$$\tan\alpha z = -i\frac{(1+i\tan z)^\alpha - (1-i\tan z)^\alpha}{(1+i\tan z)^\alpha + (1-i\tan z)^\alpha} = -i\frac{y-1}{y+1}, \tag{2.5.4}$$

where $y := ((1+i\tan z)/(1-i\tan z))^\alpha$ can be expanded according to (2.3.7). Combined with (1.2.2) we get

$$\tan\alpha z = \frac{\alpha\tan z}{1-}\frac{(\alpha^2-1^2)\tan^2 z}{3-}\frac{(\alpha^2-2^2)\tan^2 z}{5-}\frac{(\alpha^2-3^2)\tan^2 z}{7-\cdots}, \tag{2.5.5}$$

([Khov63], p 108). $D_c := \{(\alpha,z)\in\mathbb{C}^2; \mathrm{Re}(\cos z)\neq 0\}$, $D_f := \{(\alpha,z)\in\mathbb{C}^2; |\mathrm{Re}(z)| < \pi/2\}$.

$$\coth\tfrac{1}{z} = 1 + \frac{1}{3z+}\frac{1}{5z+}\frac{1}{7z+}\frac{1}{9z+\cdots}; z\in\mathbb{C}, \tag{2.5.6}$$

([Khru06b]).

$$\frac{\pi z}{2}\coth\frac{\pi z}{2} = 1 + \frac{z^2}{1+}\frac{1^2(z^2+1^2)}{3+}\frac{2^2(z^2+2^2)}{5+}\frac{3^2(z^2+3^2)}{7+\cdots} \tag{2.5.7}$$

for all $z\in\mathbb{C}$, ([ABJL92], entry 44).

$$\frac{a\tanh(\pi b/2) - b\tanh(\pi a/2)}{a\tanh(\pi a/2) - b\tanh(\pi b/2)} = \frac{ab}{1+}\frac{(a^2+1^2)(b^2+1^2)}{3+}\frac{(a^2+2^2)(b^2+2^2)}{5+\cdots} \tag{2.5.8}$$

for all $a, b\in\mathbb{C}$, ([ABJL92], entry 47).

$$\frac{\sinh(\pi z) - \sin(\pi z)}{\sinh(\pi z) + \sin(\pi z)} = \frac{2z^2}{1+}\frac{4z^4+1^4}{3+}\frac{4z^4+2^4}{5+}\frac{4z^4+3^4}{7+\cdots} \tag{2.5.9}$$

for all $z\in\mathbb{C}$, ([ABJL92], entry 49).

A.2.6 Inverse trigonometric and hyperbolic functions

$$\begin{aligned}\operatorname{Arctan} z &= z\,{}_2F_1(\tfrac{1}{2},1;\tfrac{3}{2};-z^2) = -\frac{i}{2}\operatorname{Ln}\Big(\frac{1+iz}{1-iz}\Big)\\ &= \frac{z}{1+}\,\frac{1^2z^2}{3}\,{+}\,\frac{2^2z^2}{5}\,{+}\,\frac{3^2z^2}{7}\,{+}\,\frac{4^2z^2}{9}\,{+}\cdots\end{aligned} \tag{2.6.1}$$

for $|\arg(1+z^2)|<\pi$; i.e. $z\in\mathbb{C}\setminus i((-\infty,-1]\cup[1,\infty))$, ([JoTh80], p 202). (See also (3.1.6).) This continued fraction can also be written

$$\operatorname{Arctan} z = z - \frac{z^3}{3+}\,\frac{3^2z^2}{5}\,{+}\,\frac{2^2z^2}{7}\,{+}\,\frac{5^2z^2}{9}\,{+}\,\frac{4^2z^2}{11}\,{+}\,\frac{7^2z^2}{13}\,{+}\,\frac{6^2z^2}{15}\,{+}\cdots \tag{2.6.2}$$

for $z\in\mathbb{C}\setminus i((-\infty,-1]\cup[1,\infty))$, ([Khov63], p 117).

$$\operatorname{Arctan} z = \frac{z}{1(1+z^2)-}\,\frac{1\cdot 2z^2}{3}\,\frac{1\cdot 2z^2}{-5(1+z^2)-}\,\frac{3\cdot 4z^2}{7}\,\frac{3\cdot 4z^2}{-9(1+z^2)-}\,\frac{5\cdot 6z^2}{11}\,-\cdots \tag{2.6.3}$$

for $z\in\mathbb{C}\setminus i((-\infty,-1]\cup[1,\infty))$, ([Khov63], p 121). (This follows from (2.6.6) with z replaced by $z(1+z^2)^{-1/2}$.) Since $\operatorname{Artanh} z = i\operatorname{Arctan}(-iz)$, we also get continued fraction expansions for $\operatorname{Artanh} z$ from (2.6.1)– (2.6.3).

Also expressions for

$$\operatorname{Arcsin} z = \operatorname{Arctan}\Big(\frac{z}{\sqrt{1-z^2}}\Big),\quad \operatorname{Arccos} z = \operatorname{Arctan}\Big(\frac{\sqrt{1-z^2}}{z}\Big)$$

can be obtained. For instance, from (2.6.1) we get

$$\frac{\operatorname{Arcsin} z}{\sqrt{1-z^2}} = \frac{z}{1(1-z^2)+}\,\frac{1^2z^2}{3}\,\frac{2^2z^2}{+5(1-z^2)+}\,\frac{3^2z^2}{7}\,\frac{4^2z^2}{+9(1-z^2)+\cdots} \tag{2.6.4}$$

for $|\arg(1-z^2)|<\pi$, ([Khov63], p 118) and

$$\frac{\operatorname{Arccos} z}{\sqrt{1-z^2}} = \frac{1}{z+}\,\frac{1^2(1-z^2)}{3z}\,{+}\,\frac{2^2(1-z^2)}{5z}\,{+}\,\frac{3^2(1-z^2)}{7z}\,{+}\cdots, \tag{2.6.5}$$

([Khov63], p 119). Here $D_c := \{z\in\mathbb{C};\ \operatorname{Re} z\neq 0\}$ and $D_f := \{z\in\mathbb{C};\ \operatorname{Re} z>0\}$. We also have

$$\begin{aligned}\frac{\operatorname{Arcsin} z}{\sqrt{1-z^2}} &= z\frac{{}_2F_1(\frac{1}{2},\frac{1}{2};\frac{3}{2};z^2)}{{}_2F_1(\frac{1}{2},-\frac{1}{2};\frac{1}{2};z^2)}\\ &= \frac{z}{1-}\,\frac{1\cdot 2z^2}{3}\,{-}\,\frac{1\cdot 2z^2}{5}\,{-}\,\frac{3\cdot 4z^2}{7}\,{-}\,\frac{3\cdot 4z^2}{9}\,{-}\,\frac{5\cdot 6z^2}{11}\,{-}\,\frac{5\cdot 6z^2}{13}\,{-}\cdots\end{aligned} \tag{2.6.6}$$

for $|\arg(1-z^2)|<\pi$, ([JoTh80], p 203), and thus, since $\operatorname{Arccos} z = \operatorname{Arcsin}\sqrt{1-z^2}$ for $0\le z\le\frac{\pi}{2}$

$$\frac{\operatorname{Arccos z}}{\sqrt{1-z^2}} = \frac{z}{1-}\,\frac{1\cdot 2(1-z^2)}{3}\,{-}\,\frac{1\cdot 2(1-z^2)}{5}\,{-}\,\frac{3\cdot 4(1-z^2)}{7}\,{-}\,\frac{3\cdot 4(1-z^2)}{9}\,{-}\cdots, \tag{2.6.7}$$

([Khov63], p 121) where $D_c := \{z \in \mathbb{C};\ \mathrm{Re}(z) \neq 0\}$, $D_f := \{z \in \mathbb{C};\ \mathrm{Re}(z) > 0\}$.

Similar expressions for inverse hyperbolic functions can be derived, since $\mathrm{Arsinh}\, z = i\mathrm{Arcsin}(-iz)$ and $(\mathrm{Arcosh}\, z)/\sqrt{z^2-1} = (\mathrm{Arccos}\, z)/\sqrt{1-z^2}$ for $0 \leq z \leq \frac{\pi}{2}$.

A neat formula can be obtained from (3.2.6) in the following way

$$\begin{aligned}\left(\frac{iz+1}{iz-1}\right)^{i\alpha} &= \exp\left(i\alpha \mathrm{Ln}\Big(\frac{iz+1}{iz-1}\Big)\right) = \exp(2\alpha \mathrm{Arctan}(1/z)) \\ &= 1 + \frac{2\alpha}{z-\alpha +}\,\frac{\alpha^2+1^2}{3z}\,\frac{}{+}\,\frac{\alpha^2+2^2}{5z}\,\frac{}{+}\,\frac{\alpha^2+3^2}{7z}\,\frac{}{+\cdots}\end{aligned} \tag{2.6.8}$$

for $|\arg(1+1/z^2)| < \pi$, i.e. $z \notin i[-1,1]$, ([Wall57], p 346).

$$\frac{\mathrm{Arsinh}\, z}{(1+z^2)^{1/2}} = z\,{}_2F_1(1,1;\tfrac{3}{2};-z^2) = \frac{z}{1+}\,\frac{2z^2}{1\ +}\,\frac{2(1+z^2)}{1}\,\frac{}{+}\,\frac{4z^2}{1\ +}\,\frac{4(1+z^2)}{1}\,\frac{}{+\cdots} \tag{2.6.9}$$

for $z \in \mathbb{R}$, ([ABJL92], entry 37).

$$\mathrm{Arctan}\, z = z\,{}_2F_1(\tfrac{1}{2},1;\tfrac{3}{2};-z^2) = \frac{z}{1+}\,\frac{1z^2}{1\ +}\,\frac{2(1+z^2)}{1}\,\frac{}{+}\,\frac{3z^2}{1\ +}\,\frac{4(1+z^2)}{1}\,\frac{}{+\cdots} \tag{2.6.10}$$

for $z \in \mathbb{R}$, ([ABJL92], entry 38).

A.2.7 Continued fractions with simple values

The continued fractions in this section all have easy to find tail sequences ([Lore08c]). The first one is taken from ([Perr57], p 279). It converges for all pairs $(a,z) \in \mathbb{C}^2$, but the equality

$$0 = -a - z + \frac{z}{1-a-z+}\,\frac{2z}{2-a-z+}\,\frac{3z}{3-a-z+\cdots} \tag{2.7.1}$$

given by Perron is only claimed for $z \neq 0$ and $a \in \mathbb{N} \cup \{0\}$. In our next continued fraction the constant sequence $\{1\}$ is always a tail sequence, so

$$1 = \frac{z+1}{z}\,\frac{z+2}{+z+1+}\,\frac{z+3}{z+2+}\,\frac{z+4}{z+3+\cdots}\,; \quad z \in \mathbb{C} \setminus (-\mathbb{N}), \tag{2.7.2}$$

([Bern89], p 112). (See also (3.1.5) with $z := 1$, $a := z+1$ and $c := z+1$.) The sequence of tail values for the next continued fraction is also known ([Lore08c]); it is in fact $t_0 := 1$, $t_{n+1} := z + na$ for $n \geq 0$. Hence

$$1 = \frac{z+a}{a\ \ +}\,\frac{(z+a)^2-a^2}{a}\,\frac{}{+}\,\frac{(z+2a)^2-a^2}{a}\,\frac{}{+}\,\frac{(z+3a)^2-a^2}{a}\,\frac{}{+\cdots} \tag{2.7.3}$$

for $a \neq 0$ and $z/a \notin (-\mathbb{N})$, ([Bern89], p 118). (See also (3.1.8) with $z := -1$, $a := 0$, $b := (z/a) - 2$ and $c := z/a$.)

$$a = \frac{ab}{a+b+d-}\,\frac{(a+d)(b+d)}{a+b+3d}\;-\;\frac{(a+2d)(b+2d)}{a+b+5d}\;-\cdots, \tag{2.7.4}$$

([Bern89], p 119). Here $D_c := \{(a,b,d) \in \mathbb{C}^3;\ \mathrm{Re}((a-b)/d) \neq 0 \text{ or } a = b\}$. Both $\{-a-nd\}$ and $\{-b-nd\}$ are tail sequences, so the value of the continued fraction is either a or b. The value is a in $D_f := \{(a,b,d) \in \mathbb{C}^3;\ \mathrm{Re}((a-b)/d) < 0 \text{ or } a = b\}$. (See also (3.1.6) with $z = 1$, a replaced by $(a+d)/2d$, b replaced by $a/2d$ and c replaced by $(a+b+d)/2d$.) For $b = a$ replaced by $a+1$ and $d := 1$, (3.7.4) can be transformed into

$$a = 2a+1-\frac{(a+1)^2}{2a+3}\;-\;\frac{(a+2)^2}{2a+5}\;-\;\frac{(a+3)^2}{2a+7}\;-\cdots; \quad a \in \mathbb{C}, \tag{2.7.5}$$

([Perr57], p 105).

$$az = \frac{abz}{b-(a+1)z+}\,\frac{(a+1)(b+1)z}{b+1-(a+2)z+}\,\frac{(a+2)(b+2)z}{b+2-(a+3)z+\cdots}, \tag{2.7.6}$$

([Perr57], p 290). $D_c := \{(a,b,z) \in \mathbb{C}^3;\ |z| \neq 1 \text{ and } b \neq 0,-1,-2,\dots\}$, $D_f := \{(a,b,z) \in D_c;\ |z| < 1\}$. (See also (3.1.8) with z replaced by $-z$, a replaced by $b-a$, and $b = c$ replaced by $b-1$.)

$$\frac{z+a+1}{z+1} = \frac{z+a}{z-1+}\,\frac{z+2a}{z+a-1+}\,\frac{z+3a}{z+2a-1+\cdots}, \tag{2.7.7}$$

([Bern89], p 115). $D_c := \{(a,z) \in \mathbb{C}^2;\ a \neq 0 \text{ and } z/a \neq 0,-1,-2,\dots\} \cup \{(a,z) \in \mathbb{C}^2;\ a = 0 \text{ and } |z| \neq 1\}$, $D_f := \{(a,z) \in D_c;\ \text{if } a = 0, \text{ then } |z| > 1\}$. (See also (3.1.5) with z replaced by $1/a$, a replaced by $z/a+1$ and c replaced by z/a.) If we instead let $z = 1$ and replace a by $z-1$, c by $z-3$ in (3.1.5) we get

$$\frac{z^2+z+1}{z^2-z+1} = \frac{z}{z-3+}\,\frac{z+1}{z-2+}\,\frac{z+2}{z-1+}\,\frac{z+3}{z}\,\frac{z+4}{+z+1+\cdots}, \tag{2.7.8}$$

([Bern89], p 118). $D_c := \mathbb{C}$, $D_f := \{z \in \mathbb{C};\ z \neq 0,-1,-2,\dots\}$. For $z := 1, a := z-1$ and $c := z-4$ in (3.1.5) we get

$$\frac{z^3+2z+1}{(z-1)^3+2(z-1)+1} = \frac{z}{z-4+}\,\frac{z+1}{z-3+}\,\frac{z+2}{z-2+}\,\frac{z+3}{z-1+}\,\frac{z+4}{z}\;+\cdots, \tag{2.7.9}$$

([Bern89], p 118). $D_c := \mathbb{C}$, $D_f := \{z \in \mathbb{C};\ z \neq 0,-1,-2,\dots\}$.

A.3 Hypergeometric functions

A.3.1 General expressions

$$c\frac{{}_0F_1(c;z)}{{}_0F_1(c+1;z)} = c+\frac{z}{c+1+}\,\frac{z}{c+2+}\,\frac{z}{c+3+\cdots}; \quad (c,z) \in \mathbb{C}^2, \tag{3.1.1}$$

([JoTh80], p 210).

$$\frac{{}_2F_0(a,b;z)}{{}_2F_0(a,b+1;z)} = 1 - \frac{az}{1} \, - \, \frac{(b+1)z}{1} \, - \, \frac{(a+1)z}{1} \, - \, \frac{(b+2)z}{1} \, - \, \frac{(a+2)z}{1} \, - \cdots \tag{3.1.2}$$

for $(a, b, z) \in \mathbb{C}^3$ with $|\arg(-z)| < \pi$, ([JoTh80], p 213). The even part of this one is

$$\frac{{}_2F_0(a,b;z)}{{}_2F_0(a,b+1;z)} = 1 + \frac{az}{(b+1)z-1} - \frac{(a+1)(b+1)z^2}{(a+b+3)z-1} - \frac{(a+2)(b+2)z^2}{(a+b+5)z-1} - \cdots \tag{3.1.3}$$

for $(a, b, z) \in \mathbb{C}^3$ with $|\arg(-z)| < \pi$.

$$\begin{aligned} c \cdot \frac{{}_1F_1(a;c;z)}{{}_1F_1(a+1;c+1;z)} &= c - \frac{(c-a)z}{c+1} + \frac{(a+1)z}{c+2} \\ &- \frac{(c-a+1)z}{c+3} + \frac{(a+2)z}{c+4} - \frac{(c-a+2)z}{c+5} + \cdots; \quad (a,c,z) \in \mathbb{C}^3, \end{aligned} \tag{3.1.4}$$

([JoTh80], p 206).

$$\frac{{}_1F_1(a+1;c+1;z)}{{}_1F_1(a;c;z)} = \frac{c}{c-z} + \frac{(a+1)z}{c+1-z} + \frac{(a+2)z}{c+2-z} + \frac{(a+3)z}{c+3-z} + \cdots \tag{3.1.5}$$

for $(a, c, z) \in \mathbb{C}^3$, ([JoTh80], p 278).

$$\begin{aligned} c \cdot \frac{{}_2F_1(a,b;c;z)}{{}_2F_1(a,b+1;c+1;z)} &= c - \frac{a(c-b)z}{c+1} - \frac{(b+1)(c-a+1)z}{c+2} \\ &- \frac{(a+1)(c-b+1)z}{c+3} - \frac{(b+2)(c-a+2)z}{c+4} - \frac{(a+2)(c-b+2)z}{c+5} - \cdots \end{aligned} \tag{3.1.6}$$

for $(a, b, c, z) \in \mathbb{C}^4$ with $|\arg(1-z)| < \pi$, ([JoTh80], p 199). The Nörlund fraction has the form

$$\begin{aligned} c \cdot \frac{{}_2F_1(a,b;c;z)}{{}_2F_1(a+1,b+1;c+1;z)} &= c - (a+b+1)z + \frac{(a+1)(b+1)(z-z^2)}{c+1-(a+b+3)z} + \\ &\frac{(a+2)(b+2)(z-z^2)}{c+2-(a+b+5)z} + \frac{(a+3)(b+3)(z-z^2)}{c+3-(a+b+7)z} + \cdots \end{aligned} \tag{3.1.7}$$

([LoWa92], p 304). $D_c := \{(a,b,c,z) \in \mathbb{C}^4;\ \mathrm{Re}(z) \neq \frac{1}{2}\}$, $D_f := \{(a,b,c,z) \in \mathbb{C}^4;\ \mathrm{Re}(z) < \frac{1}{2}\}$. The Euler fraction has the form

$$\begin{aligned} c \cdot \frac{{}_2F_1(a,b;c;z)}{{}_2F_1(a,b+1;c+1;z)} &= c + (b-a+1)z - \frac{(c-a+1)(b+1)z}{c+1+(b-a+2)z} - \\ &\frac{(c-a+2)(b+2)z}{c+2+(b-a+3)z} - \frac{(c-a+3)(b+3)z}{c+3+(b-a+4)z} - \cdots, \end{aligned} \tag{3.1.8}$$

([LoWa92], p 308). $D_c := \{(a,b,c,z) \in \mathbb{C}^4;\ |z| \neq 1\}$, $D_c := \{(a,b,c,z) \in \mathbb{C}^4;\ |z| < 1,\ (c-a) \neq -1,-2,-3,\dots\}$.

By setting $b := 0$ in (3.1.2), (3.1.5), (3.1.6) or (3.1.7) and using (1.2.1) we get continued fraction expansions for ${}_2F_0(a,1;z)$ and ${}_2F_1(a,1;c+1;z)$. Similarly, $a := 0$ in (3.1.3) or (3.1.4) gives continued fraction expansions for ${}_1F_1(1;c+1;z)$. A different expansion is

$$
{}_2F_1(a,1;c+1;z) = \frac{\Gamma(1-a)\Gamma(c+1)}{\Gamma(c-a+1)} \frac{(1-z)^{c-a}}{(-z)^c} - \\
\frac{c}{1-c+(a-1)z+}\frac{1(1-c)(z-1)}{3-c+(a-2)z+}\frac{2(2-c)(z-1)}{5-c+(a-3)z+\cdots}, \tag{3.1.9}
$$

([Bern89], p 164). $D_c := \{(a,c,z) \in \mathbb{C}^3;\ |z-1| \neq 1\}$, $D_f := \{(a,c,z) \in \mathbb{C}^3;\ |z-1| < 1$ and $c \neq 0,1,2,\dots\}$. From this follows after some computation, ([Bern89], p 165) that

$$
\begin{aligned}
{}_1F_1(1;c+1;z) &= \frac{e^z\Gamma(c+1)}{z^c} - \frac{c}{z+}\frac{1-c}{1}\frac{1}{+z+}\frac{2-c}{1}\frac{2}{+z+}\frac{3-c}{1}+\cdots \\
&= \frac{e^z\Gamma(c+1)}{z^c} - \frac{c}{z+1-c-}\frac{1(1-c)}{z+3-c-}\frac{2(2-c)}{z+5-c-}\frac{3(3-c)}{z+7-c-\cdots}
\end{aligned} \tag{3.1.10}
$$

for $|\arg z| < \pi$, ([Bern89], p 165) (the second continued fraction is the even part of the first one).

A.3.2 Special examples with ${}_0F_1$

The **Bessel function** of the first kind and order ν is

$$
J_\nu(z) := \left(\frac{z}{2}\right)^\nu \sum_{k=0}^{\infty} \frac{(-1)^k (z/2)^{2k}}{k!\,\Gamma(\nu+k+1)} = \frac{(z/2)^\nu}{\Gamma(\nu+1)}\,{}_0F_1(\nu+1;-\tfrac{z^2}{4}), \tag{3.2.1}
$$

so that by (3.1.1)

$$
\begin{aligned}
\frac{J_{\nu+1}(z)}{J_\nu(z)} &= \frac{z}{2(\nu+1)} \cdot \frac{{}_0F_1(\nu+2;-z^2/4)}{{}_0F_1(\nu+1;-z^2/4)} \\
&= \frac{z}{2(\nu+1)-}\frac{z^2}{2(\nu+2)-}\frac{z^2}{2(\nu+3)-}\frac{z^2}{2(\nu+4)-\cdots}
\end{aligned} \tag{3.2.2}
$$

for $z \in \mathbb{C}$, $\nu \neq -1,-2,-3,\dots$, ([JoTh80], p 211).

A.3.3 Special examples with ${}_2F_0$

The connection (see for instance ([Wall48], p 352, p 355))

$$
{}_2F_0(a,b;-z) \sim \frac{1}{\Gamma(a)} \int_0^\infty \frac{e^{-t}t^{a-1}}{(1+tz)^b}dt = \frac{1}{\Gamma(b)} \int_0^\infty \frac{e^{-t}t^{b-1}}{(1+tz)^a}dt \tag{3.3.1}
$$

implies that (3.1.2) – (3.1.3) lead to continued fraction expansions for ratios of such integrals. In particular, the **incomplete gamma function** $\Gamma(a,z)$ satisfies

$$
\Gamma(a,z) := \int_z^\infty e^{-t}t^{a-1}dt \sim e^{-z}z^{a-1}{}_2F_0(1-a,1;-1/z), \tag{3.3.2}
$$

([EMOT53], p 266). Hence, by (3.1.2)

$$\begin{aligned}\Gamma(a,z) &= \frac{e^{-z}z^a}{z} \, \frac{1-a}{+\;\;1} \, \frac{1}{+z+} \, \frac{2-a}{1} \, \frac{2}{+z+} \, \frac{3-a}{1} \, \frac{3}{+z+\cdots} \\ &= \frac{e^{-z}z^a}{1+z-a-} \, \frac{1(1-a)}{3+z-a-} \, \frac{2(2-a)}{5+z-a-} \, \frac{3(3-a)}{7+z-a-\cdots}\end{aligned} \tag{3.3.3}$$

for $(a,z) \in \mathbb{C}^2$ with $|\arg z| < \pi$, ([AbSt65], p 260, p 263), ([Khov63], p 144), where the second continued fraction is the even part of the first one.

This (and the expressions to come) are to be interpreted in the following way: The integral in (3.3.2) is taken for real z. Then $\Gamma(a,z)$ is the analytic continuation of this function to the given domain.

The **complementary error function** erfc z satisfies

$$\operatorname{erfc} z := \frac{2}{\sqrt{\pi}} \int_z^\infty e^{-t^2}\,dt = \frac{1}{\sqrt{\pi}}\Gamma(\tfrac{1}{2}, z^2) \sim \frac{1}{\sqrt{\pi}} e^{-z^2} z^{-1} {}_2F_0(\tfrac{1}{2}, 1; -1/z^2) \tag{3.3.4}$$

([EMOT53], p 266), which means that by (3.1.2)

$$\begin{aligned}\operatorname{erfc} z &= \frac{2}{\sqrt{\pi}} e^{-z^2} \left\{ \frac{1}{2z+} \, \frac{2}{2z+} \, \frac{4}{2z+} \, \frac{6}{2z+} \, \frac{8}{2z+\cdots} \right\} \\ &= \frac{2}{\sqrt{\pi}} e^{-z^2} \left\{ \frac{z}{1+2z^2-} \, \frac{1\cdot 2}{5+2z^2-} \, \frac{3\cdot 4}{9+2z^2-} \, \frac{5\cdot 6}{13+2z^2-} \, \frac{7\cdot 8}{17+2z^2-\cdots} \right\},\end{aligned} \tag{3.3.5}$$

([JoTh80], p 219). (There is a slightly different notation in ([JoTh80]).) $D_c := \{z \in \mathbb{C};\ \operatorname{Re} z \neq 0\}$, $D_f := \{z \in \mathbb{C};\ \operatorname{Re} z > 0\}$. Again the second continued fraction is the even part of the first one. If we integrate this complementary error function we get similar expressions:

$$i^{-1}\operatorname{erfc} z = \frac{2}{\sqrt{\pi}} e^{-z^2}, \; i^0\operatorname{erfc} z = \operatorname{erfc} z, \; i^n\operatorname{erfc} z = \int_z^\infty i^{n-1}\operatorname{erfc} t\,dt \tag{3.3.6}$$

for $n = 1, 2, 3, \ldots$. Therefore

$$\begin{aligned}\frac{i^{n-1}\operatorname{erfc} z}{i^n\operatorname{erfc} z} &= 2z \frac{{}_2F_0(\frac{n+1}{2}, \frac{n}{2}; -1/z^2)}{{}_2F_0(\frac{n+1}{2}, \frac{n}{2}+1; -1/z^2)} \\ &= 2z + \frac{2(n+1)}{2z} \, \frac{2(n+2)}{+\;\;2z} \, \frac{2(n+3)}{+\;\;2z} \, {+\cdots}\end{aligned} \tag{3.3.7}$$

([JoTh80], p 219). $D_c := \{(n,z) \in \mathbb{C}^2;\ \operatorname{Re} z \neq 0\}$, $D_f := \{(n,z) \in \mathbb{C}^2;\ \operatorname{Re} z > 0\}$.

For the **exponential integral**

$$-\operatorname{Ei}(-z) := \int_z^\infty \frac{e^{-t}}{t}\,dt \sim \frac{e^{-z}}{z} {}_2F_0(1, 1; -\tfrac{1}{z}), \tag{3.3.8}$$

([EMOT53], p 267), we get by (3.1.2) and its even part

$$\begin{aligned}\operatorname{Ei}(-z) &= -\frac{e^{-z}}{z} \, \frac{1}{+1+} \, \frac{1}{z+} \, \frac{2}{1+} \, \frac{2}{z+} \, \frac{3}{1+} \, \frac{3}{z+} \, \frac{4}{1+\cdots} \\ &= -\frac{e^{-z}}{1+z-} \, \frac{1^2}{3+z-} \, \frac{2^2}{5+z-} \, \frac{3^2}{7+z-} \, \frac{4^2}{9+z-\cdots}\end{aligned} \tag{3.3.9}$$

for $|\arg z| < \pi$, ([Khov63], p 145).

Similarly, for the **logarithmic integral**

$$\begin{aligned}\operatorname{li} z &:= \int_0^z \frac{dt}{\operatorname{Ln} t} = \operatorname{Ei}(\operatorname{Ln} z) = \frac{z}{\operatorname{Ln} z - 1 -}\,\frac{1}{\operatorname{Ln} z - 1 -}\,\frac{1}{\operatorname{Ln} z - 1 -}\,\frac{2}{\operatorname{Ln} z -}\,\frac{2}{\operatorname{Ln} z - \cdots}\\ &= -\frac{z}{1 - \operatorname{Ln} z -}\,\frac{1^2}{3 - \operatorname{Ln} z -}\,\frac{2^2}{5 - \operatorname{Ln} z -}\,\frac{3^2}{7 - \operatorname{Ln} z -}\,\frac{4^2}{9 - \operatorname{Ln} z - \cdots}\end{aligned} \tag{3.3.10}$$

for $|\arg(-\operatorname{Ln} z)| < \pi$.

The **plasma dispersion function** is

$$\begin{aligned}P(z) &:= \frac{1}{\sqrt{\pi}} \int_{-\infty}^{\infty} \frac{e^{-t^2}}{t-z}\,dt = i\sqrt{\pi}\, e^{-z^2} \operatorname{erfc}(-iz)\\ &= \frac{2z}{1 - 2z^2 -}\,\frac{1\cdot 2}{5 - 2z^2 -}\,\frac{3\cdot 4}{9 - 2z^2 -}\,\frac{5\cdot 6}{13 - 2z^2 -}\,\frac{7\cdot 8}{17 - 2z^2 - \cdots},\end{aligned} \tag{3.3.11}$$

([JoTh80], p 219). $D_c := \{z \in \mathbb{C};\ \operatorname{Im}(z) \neq 0\}$, $D_f := \{z \in \mathbb{C};\ \operatorname{Im}(z) > 0\}$.

$$\frac{\int_0^\infty t^a e^{-bt-t^2/2}\,dt}{\int_0^\infty t^{a-1} e^{-bt-t^2/2}\,dt} = \frac{a}{b} \cdot \frac{{}_2F_0\left(\frac{a}{2}, \frac{a+1}{2}; -\frac{1}{b^2}\right)}{{}_2F_0\left(\frac{a}{2}, \frac{a-1}{2}; -\frac{1}{b^2}\right)} = \frac{a}{b +}\,\frac{a+1}{b +}\,\frac{a+2}{b +}\,\frac{a+3}{b + \cdots}, \tag{3.3.12}$$

([Perr57], p 297). $D_c := \{(a,b) \in \mathbb{C}^2;\ \operatorname{Re} b \neq 0\}$, $D_f := \{(a,b) \in \mathbb{C}^2;\ \operatorname{Re} b > 0\}$.

A.3.4 Special examples with ${}_1F_1$

From ([EMOT53], p 255) it follows that

$${}_1F_1(a;c;z) = \frac{\Gamma(c)}{\Gamma(a)\Gamma(c-a)} \int_0^1 e^{tz} t^{a-1} (1-t)^{c-a-1}\,dt \tag{3.4.1}$$

for $\operatorname{Re}(c) > 0$, $\operatorname{Re}(a) > 0$. Hence (3.1.4), (3.1.5) and (3.1.10) lead to continued fraction expansions of ratios of such integrals.

The **error function** is given by

$$\begin{aligned}\operatorname{erf}(z) &:= \frac{2}{\sqrt{\pi}} \int_0^z e^{-t^2}\,dt = \frac{2}{\sqrt{\pi}} z\, {}_1F_1(\tfrac{1}{2}; \tfrac{3}{2}; -z^2), \quad \text{([EMOT53], p 266)}\\ &= \frac{2}{\sqrt{\pi}} z e^{-z^2}\, {}_1F_1(1; \tfrac{3}{2}; z^2), \quad \text{([JoTh80], p 282)}.\end{aligned} \tag{3.4.2}$$

Hence,

$$\begin{aligned}\operatorname{erf}(z) &= \frac{2e^{-z^2}}{\sqrt{\pi}}\,\frac{z}{1 -}\,\frac{2z^2}{3 +}\,\frac{4z^2}{5 -}\,\frac{6z^2}{7 +}\,\frac{8z^2}{9 - \cdots}\\ &= \frac{2e^{-z^2}}{\sqrt{\pi}}\,\frac{z}{1 - 2z^2 +}\,\frac{4z^2}{3 - 2z^2 +}\,\frac{8z^2}{5 - 2z^2 +}\,\frac{12z^2}{7 - 2z^2 + \cdots}\end{aligned} \tag{3.4.3}$$

for $z \in \mathbb{C}$, ([JoTh80], p 208 and 282).

The error function is related to **Dawson's integral**

$$\int_0^z e^{t^2}\,dt = \frac{i\sqrt{\pi}}{2} \operatorname{erf}(-iz), \quad \text{([JoTh80], p 208)} \tag{3.4.4}$$

and to the **Fresnel integrals**

$$C(z) := \int_0^z \cos\left(\frac{\pi}{2}t^2\right) dt, \qquad S(z) := \int_0^z \sin\left(\frac{\pi}{2}t^2\right) dt \tag{3.4.5}$$

by

$$\begin{aligned} C(z) + iS(z) &= \int_0^z e^{it^2\pi/2}\, dt = \sqrt{\frac{-2}{i\pi}} \int_0^{\sqrt{-i\pi/2}\,\cdot\, z} e^{-u^2} du \\ &= \frac{1+i}{2} \operatorname{erf}\left(\frac{\sqrt{\pi}}{2}(1-i)z\right). \end{aligned} \tag{3.4.6}$$

The **incomplete gamma function**

$$\begin{aligned} \gamma(a,z) &:= \int_0^z e^{-t} t^{a-1} dt = \frac{z^a}{a} e^{-z}\, {}_1F_1(1; a+1; z) \\ &= \frac{z^a e^{-z}}{a} \frac{az}{-a+1} \frac{1z}{+a+2-} \frac{(a+1)z}{a+3} \frac{2z}{+a+4-} \frac{(a+2)z}{a+5} \frac{}{+\cdots} \\ &= \frac{z^a e^{-z}}{a} \frac{az}{-1+a+z-} \frac{(1+a)z}{2+a+z-} \frac{(2+a)z}{3+a+z-} \frac{(3+a)z}{4+a+z-\cdots} \end{aligned} \tag{3.4.7}$$

for all $(a,z) \in \mathbb{C}^2$, ([JoTh80], p 209), ([Khov63], p 149–150).

The **Coulomb wave function**

$$F_L(\eta,\rho) = \rho^{L+1} e^{-i\rho} C_L(\eta)\, {}_1F_1(L+1-i\eta; 2L+2; 2i\rho) \tag{3.4.8}$$

where $C_L(\eta) = 2^L \exp(-\pi\eta/2)|\Gamma(L+1+i\eta)|/(2L+1)!$ for $\eta \in \mathbb{R}$, $\rho > 0$ and $L \in \mathbb{N} \cup \{0\}$ satisfies

$$\begin{aligned} \frac{F_L(\eta,\rho)}{F_{L-1}(\eta,\rho)} &= \frac{(L+1)(L^2+\eta^2)^{1/2}}{(2L+1)(\eta+L(L+1)/\rho)-} \frac{L(L+2)((L+1)^2+\eta^2)}{(2L+3)(\eta+(L+1)(L+2)/\rho)-} \\ &\quad \frac{(L+1)(L+3)((L+2)^2+\eta^2)}{(2L+5)(\eta+(L+2)(L+3)/\rho)-\cdots}, \end{aligned} \tag{3.4.9}$$

([JoTh80], p 216). $D_c := \{(L,\eta,\rho) \in \mathbb{C}^3;\ \rho \neq 0\}$, $D_f := \{(L,\eta,\rho) \in (\mathbb{N}\cup\{0\}) \times \mathbb{C}^2;\ \rho \neq 0\}$.

It is well known that

$$\sum_{k=0}^{\infty} \frac{(-z)^k}{k!\,(a+k)} = e^{-z} \sum_{k=0}^{\infty} \frac{z^k}{(a)_{k+1}} = \frac{e^{-z}}{a}\, {}_1F_1(1; a+1; z), \tag{3.4.10}$$

([Bern89], p 166). This means for instance that

$$\sum_{k=0}^{\infty} \frac{(-z)^k}{k!\,(a+k)} = \frac{\Gamma(a)}{z^a} - \frac{e^{-z}}{z+1-a-} \frac{1(1-a)}{z+3-a-} \frac{2(2-a)}{z+5-a-\cdots} \tag{3.4.11}$$

for $(a,z) \in \mathbb{C}^2$ with $|\arg z| < \pi$, and

$$\sum_{k=0}^{\infty} \frac{z^k}{1 \cdot 3 \cdots (2k+1)} = \sqrt{\frac{\pi}{2z}} e^{z/2} - \frac{1}{z+1-} \frac{1 \cdot 2}{z+5-} \frac{3 \cdot 4}{z+9-} \frac{5 \cdot 6}{z+13-\cdots} \tag{3.4.12}$$

for $|\arg z| < \pi$, ([Bern89], p 166), and

$$\begin{aligned} ze^{-z^2}\sum_{k=0}^{\infty}\frac{(2z^2)^k}{1\cdot 3\cdots(2k+1)} &= \int_0^z e^{-t^2}\,dt \\ &= \frac{\sqrt{\pi}}{2} - \frac{e^{-z^2}}{2z^2+1-}\,\frac{1\cdot 2}{2z^2+5-}\,\frac{3\cdot 4}{2z^2+9-}\,\frac{5\cdot 6}{2z^2+13-}\cdots, \end{aligned} \tag{3.4.13}$$

([Bern89], p 166). $D_c := \{z\in\mathbb{C};\ \operatorname{Re} z \neq 0\}$, $D_f := \{z\in\mathbb{C};\ \operatorname{Re} z > 0\}$.

A.3.5 Special examples with $_2F_1$

$$\begin{aligned} \int_0^z \frac{t^p\,dt}{1+t^q} &= \frac{z^{p+1}}{q}\,{}_2F_1\left(\frac{p+1}{q},1;1+\frac{p+1}{q};-z^q\right) \\ &= \frac{z^{p+1}}{0q+p+1+}\,\frac{(0q+p+1)^2z^q}{1q+p+1}\,\frac{(1q)^2z^q}{+2q+p+1+} \\ &\quad \frac{(1q+p+1)^2z^q}{3q+p+1}\,\frac{(2q)^2z^q}{+4q+p+1+}\cdots \end{aligned} \tag{3.5.1}$$

for $p,q>0$ with $|\arg(1+z^q)|<\pi$, ([Khov63], p 126).

Incomplete beta functions are given by

$$B_x(p,q) := \int_0^x t^{p-1}(1-t)^{q-1}\,dt = \frac{x^p}{p}\,{}_2F_1(p,1-q;p+1;x) \tag{3.5.2}$$

for $p>0, q>0$ and $0\le x\le 1$, ([EMOT53], p 87). Hence, by (3.1.6) and (3.1.8)

$$\begin{aligned} \frac{B_x(p+1,q)}{B_x(p,q)} &= \frac{px}{p+1}\,\frac{{}_2F_1(p+1,1-q;p+2;x)}{{}_2F_1(p,1-q;p+1;x)} \\ &= \frac{px}{p+1-}\,\frac{1(1-q)x}{p+2}\,-\,\frac{(p+1)(p+q+1)x}{p+3}\,- \\ &\quad \frac{2(2-q)x}{p+4}\,-\,\frac{(p+2)(p+q+2)x}{p+5}\,-\cdots;\quad |\arg(1-x)|<\pi \end{aligned} \tag{3.5.3}$$

for $p>0$, $q>0$, ([JoTh80], p 217), and

$$\begin{aligned} \frac{B_x(p+1,q)}{B_x(p,q)} &= \frac{px}{p+1+(p+q)x-}\,\frac{(p+q+1)(p+1)x}{p+2+(p+q+1)x} \\ &\quad \frac{(p+q+2)(p+2)x}{-p+3+(p+q+2)x-}\cdots, \end{aligned} \tag{3.5.4}$$

([JoTh80], p 217). $D_c := \{p>0,\ q>0,\ x\in\mathbb{C};\ |x|\neq 1\}$, $D_f := \{(p,q,x)\in D_c;\ |x|<1\}$.

Legendre functions of the first kind of degree $\alpha\in\mathbb{R}$ and order $m\in\mathbb{N}\cup\{0\}$ are given by

$$\begin{aligned} P_\alpha^m(z) &:= \frac{1}{\pi}\frac{\Gamma(\alpha+m+1)}{\Gamma(\alpha+1)}\int_0^\pi \left(z+(z^2-1)^{1/2}\cos t\right)^\alpha \cos mt\,dt \\ &= \frac{1}{\Gamma(1-m)}\left(\frac{z+1}{z-1}\right)^{m/2}\,{}_2F_1\left(-\alpha,\alpha+1;1-m;\frac{1-z}{2}\right). \end{aligned} \tag{3.5.5}$$

From ([Gaut67], formula (6.1) on p 55) it follows that

$$\frac{P_\alpha^m(z)}{P_\alpha^{m-1}(z)} = \cfrac{(m+\alpha)(m-\alpha-1)}{-\cfrac{2mz}{(z^2-1)^{1/2}}} - \cfrac{(m+1+\alpha)(m-\alpha)}{-\cfrac{2(m+1)z}{(z^2-1)^{1/2}}} - \cfrac{(m+2+\alpha)(m+1-\alpha)}{-\cfrac{2(m+2)z}{(z^2-1)^{1/2}}} - \cdots$$
$$\sim -\frac{(m+\alpha)(m-\alpha-1)\sqrt{z^2-1}}{2mz} - \frac{(m+1+\alpha)(m-\alpha)(z^2-1)}{2(m+1)z} - \frac{(m+2+\alpha)(m+1-\alpha)(z^2-1)}{2(m+2)z} - \frac{(m+3+\alpha)(m+2-\alpha)(z^2-1)}{2(m+3)z} - \cdots \tag{3.5.6}$$

$D_c := \{(\alpha, m, z) \in \mathbb{C}^3;\ \mathrm{Re}(z) \neq 0\}$, $D_f := \{(\alpha, m, z) \in \mathbb{R} \times (\mathbb{N} \cup \{0\}) \times \mathbb{C};\ \mathrm{Re}(z) > 0\}$.

Legendre functions of the second kind of degree $\alpha \in \mathbb{R}$ and order $m \in \mathbb{N} \cup \{0\}$ are given by

$$Q_\alpha^m(z) := (-1)^m \frac{\Gamma(\alpha+1)}{\Gamma(\alpha-m+1)} \int_0^\infty \frac{\cosh mt}{(z+(z^2-1)^{1/2}\cosh t)^{\alpha+1}} dt$$
$$= \frac{\sqrt{\pi}\, e^{im\pi}}{(2z)^{\alpha+1}} \left(1 - \frac{1}{z^2}\right)^{m/2} \frac{\Gamma(\alpha+m+1)}{\Gamma(\alpha+m+\frac{3}{2})} \cdot {}_2F_1\left(\tfrac{\alpha+m+2}{2}, \tfrac{\alpha+m+1}{2}; \alpha+\tfrac{3}{2}; 1/z^2\right). \tag{3.5.7}$$

In ([JoTh80], p 205) it is proved that

$$\frac{Q_\alpha^m(z)}{Q_{\alpha+1}^m(z)} = \frac{1}{\alpha+m+1}\left\{(2\alpha+3)z - \frac{(\alpha+m+2)^2}{(2\alpha+5)z} - \frac{(\alpha+m+3)^2}{(2\alpha+7)z} - \frac{(\alpha+m+4)^2}{(2\alpha+9)z} - \frac{(\alpha+m+5)^2}{(2\alpha+11)z} - \frac{(\alpha+m+6)^2}{(2\alpha+13)z} - \cdots\right\}. \tag{3.5.8}$$

$D_c := \{(\alpha, m, z) \in \mathbb{C}^3;\ z \notin [-1, 1]\}$, $D_f := \{(\alpha, m, z) \in \mathbb{R} \times (\mathbb{N} \cup \{0\}) \times \mathbb{C};\ z \notin [-1, 1]\}$.

A.3.6 Some integrals

Hypergeometric functions can be written in terms of integrals. This has already been used to some extent in the preceding subsections, and we refer to ([AbSt65]) and ([EMOT53]) for further details. Here we shall just list some simple examples without bringing in the hypergeometric functions themselves.

$$\int_0^1 x^s e^{1-x} dx = \frac{1}{s+}\frac{1}{s+1+}\frac{1}{s+1+}\frac{1}{s+1+}\cdots = \sum_{n=1}^\infty \frac{1}{(s+1)_n};\ s \in \mathbb{C}, \tag{3.6.1}$$

([Khru06b]).

$$\int_0^1 \frac{x^s}{1+x^2} dx = \frac{1}{s+}\frac{1^2}{s+}\frac{2^2}{s+}\frac{3^2}{s+}\cdots = \sum_{k=0}^\infty \frac{2\cdot(-1)^k}{s+2k+1} = \sum_{k=1}^\infty \frac{4}{(s+2k)^2-1}, \tag{3.6.2}$$

([Khru06b]). $D_c := \{s \in \mathbb{C};\ \operatorname{Re} s \neq 0\}$, $D_f := \{s \in \mathbb{C};\ \operatorname{Re} s > 0\}$.

$$\int_0^\infty \frac{e^{-t}dt}{t+z} = \frac{1}{z+1-}\frac{1^2}{z+3-}\frac{2^2}{z+5-}\frac{3^2}{z+7-}\cdots; \quad |\arg z| < \pi, \tag{3.6.3}$$

([BoSh89], p 20).

$$\begin{aligned}\int_0^\infty \frac{e^{-t/z}}{(1+t)^n}dt &= \frac{z}{1+}\frac{nz}{1}\frac{1z}{+1+}\frac{(n+1)z}{1}\frac{2z}{+1+}\frac{(n+2)z}{1}\frac{3z}{+1}+\cdots \\ &= \frac{z}{1+nz-}\frac{nz^2}{1+(n+2)z-}\frac{2(n+1)z^2}{1+(n+4)z-}\frac{3(n+2)z^2}{1+(n+6)z-}\cdots\end{aligned} \tag{3.6.4}$$

for $(n,z) \in \mathbb{R}\times\mathbb{C}$ with $|\arg z| < \pi$, ([BoSh89], p 157). The second continued fraction is the even part of the first one.

$$\int_0^\infty \frac{e^{-tz}}{\cosh^2 t}dt = \frac{1}{z+}\frac{1\cdot 2}{z}\frac{2\cdot 3}{+z}\frac{3\cdot 4}{+z}+\cdots = 2z\sum_{k=0}^\infty \frac{(-1)^k}{(z+2k)(z+2k+2)}, \tag{3.6.5}$$

([Khru06b]). $D_c := \{z \in \mathbb{C};\ \operatorname{Re} z \neq 0\}$, $D_f := \{z \in \mathbb{C};\ \operatorname{Re} z > 0\}$.

For **Jacobi's elliptic functions** $\operatorname{sn} t$, $\operatorname{cn} t$ and $\operatorname{dn} t$ with modulus k we have

$$\int_0^\infty e^{-tz}\operatorname{sn} t\,dt = \frac{1}{1^2(1+k^2)+z^2-}\frac{1\cdot 2^2\cdot 3k^2}{3^2(1+k^2)+z^2-}\frac{3\cdot 4^2\cdot 5k^2}{5^2(1+k^2)+z^2-}\cdots, \tag{3.6.6}$$

([Wall48], p 374). $D_c := \{(k,z) \in \mathbb{C}^2;\ |k| \neq 1\}$, $D_f := \{(k,z) \in \mathbb{C}^2;\ |k| < 1\}$,

$$\int_0^\infty e^{-tz}\operatorname{sn}^2 t\,dt = \frac{2}{2^2(1+k^2)+z^2-}\frac{2\cdot 3^2\cdot 4k^2}{4^2(1+k^2)+z^2-}\frac{4\cdot 5^2\cdot 6k^2}{6^2(1+k^2)+z^2-}\cdots, \tag{3.6.7}$$

([Wall48], p 375). $D_c := \{(k,z) \in \mathbb{C}^2;\ |k| \neq 1\}$, $D_f := \{(k,z) \in \mathbb{C}^2;\ |k| < 1\}$,

$$\int_0^\infty e^{-tz}\operatorname{cn} t\,dt = \frac{1}{z+}\frac{1^2}{z+}\frac{2^2k^2}{z}\frac{3^2}{+z+}\frac{4^2k^2}{z}\frac{5^2}{+z+}\cdots, \tag{3.6.8}$$

([Perr57], p 220). $D_c := \{(k,z) \in \mathbb{R}\times\mathbb{C};\ \operatorname{Re} z \neq 0\}$, $D_f := \{(k,z) \in \mathbb{R}\times\mathbb{C};\ \operatorname{Re} z > 0\}$,

$$\int_0^\infty e^{-tz}\operatorname{dn} t\,dt = \frac{1}{z+}\frac{1^2k^2}{z}\frac{2^2}{+z+}\frac{3^2k^2}{z}\frac{4^2}{+z+}\frac{5^2k^2}{z}\cdots, \tag{3.6.9}$$

([Wall48], p 374). $D_c := \{(k,z) \in \mathbb{R}\times\mathbb{C};\ \operatorname{Re} z \neq 0\}$, $D_f := \{(k,z) \in \mathbb{R}\times\mathbb{C};\ \operatorname{Re} z > 0\}$, and

$$\begin{aligned}&\int_0^\infty \frac{\operatorname{sn} t\operatorname{cn} t}{\operatorname{dn} t}e^{-tz}dt = \\ &\frac{1}{2\cdot 1^2(2-k^2)+z^2-}\frac{1\cdot 2^2\cdot 3k^4}{2\cdot 3^2(2-k^2)+z^2-}\frac{3\cdot 4^2\cdot 5k^4}{2\cdot 5^2(2-k^2)+z^2-}\cdots\end{aligned} \tag{3.6.10}$$

for $(k,z) \in \mathbb{C}^2$ with $|1-k^2|<1$, ([Wall48], p 375).

$$\int_0^\infty \left(\frac{1-c}{e^{t(1-c)}-c^b}\right)^a e^{-tz}dt = \frac{r^a}{z+}\,\frac{ar}{1+}\,\frac{rc^b}{z+}\,\frac{(a+1)r}{1}\,\frac{2rc^b}{+\ z}\,\frac{(a+2)r}{+\ 1}+\cdots \tag{3.6.11}$$

where $r := (1-c)/(1-c^b)$, for $(a,b,c,z) \in \mathbb{C}^4$ with $a>0$, $c^b>0$ and $|\arg(r/z)|<\pi$, ([Wall48], p 359).

$$\int_0^\infty \frac{te^{-tz}}{\sinh t}dt = \frac{1}{z+}\,\frac{1^4}{3z+}\,\frac{2^4}{5z+}\,\frac{3^4}{7z+}\cdots, \tag{3.6.12}$$

([Wall48], p 371). $D_c := \{z \in \mathbb{C};\ \mathrm{Re}\, z \neq 0\}$, $D_f := \{z \in \mathbb{C};\ \mathrm{Re}\, z > 0\}$.

$$\begin{aligned}\int_0^\infty \frac{2te^{-tz}}{e^t+e^{-t}}dt &= \int_0^\infty \frac{te^{-tz}}{\cosh t}dt = 2\sum_{n=0}^\infty \frac{(-1)^n}{(z+1+2n)^2}\\ &= \frac{1}{z^2-1+}\,\frac{4\cdot 1^2}{1}\,\frac{4\cdot 1^2}{+z^2-1+}\,\frac{4\cdot 2^2}{1}\,\frac{4\cdot 2^2}{+z^2-1+}\,\frac{4\cdot 3^2}{1}+\cdots,\end{aligned} \tag{3.6.13}$$

([Perr57], p 30). $D_c := \{z \in \mathbb{C};\ |\arg(z^2-1)|<\pi\}$, $D_f := \{z \in \mathbb{C};\ \mathrm{Re}\, z > 0$ and $z \notin (0,1]\}$. For instance, for $z =: \sqrt{5}$ we get

$$\int_0^\infty \frac{4te^{-\sqrt{5}t}}{\cosh t}dt = \frac{1}{1+}\,\frac{1^2}{1+}\,\frac{1^2}{1+}\,\frac{2^2}{1+}\,\frac{2^2}{1+}\,\frac{3^2}{1+}\,\frac{3^2}{1+}\cdots, \text{ ([Perr57], p 30).} \tag{3.6.14}$$

A.3.7 Gamma function expressions by Ramanujan

Ramanujan produced quite a number of continued fraction expansions of ratios of gamma functions. These ratios have all proved to be connected to hypergeometric functions, ([Rama57], [Bern89]). We use Ramanujan's notation

$$\begin{aligned}\prod_\varepsilon \Gamma(a+\varepsilon b+c) &:= \Gamma(a+b+c)\Gamma(a-b+c),\\ \prod_\varepsilon \Gamma(a+\varepsilon b+\varepsilon c+d) &:= \Gamma(a+b+c+d)\Gamma(a-b+c+d)\times\\ &\quad \times \Gamma(a+b-c+d)\Gamma(a-b-c+d)\end{aligned} \tag{3.7.1}$$

and so on. That is, $\varepsilon = \pm 1$, and the product is taken over all different combinations of the εs.

$$\begin{aligned}&\frac{1-R}{1+R} = \frac{p}{z+}\,\frac{1^2-q^2}{z}\,\frac{2^2-p^2}{+\ z}\,\frac{3^2-q^2}{+\ z}\,\frac{4^2-p^2}{+\ z}+\cdots\\ &\text{where } R = \prod_\varepsilon \frac{\Gamma\left(\frac{z+p+\varepsilon q+1}{4}\right)}{\Gamma\left(\frac{z+p+\varepsilon q+3}{4}\right)}\cdot \prod_\varepsilon \frac{\Gamma\left(\frac{z-p+\varepsilon q+3}{4}\right)}{\Gamma\left(\frac{z-p+\varepsilon q+1}{4}\right)},\end{aligned} \tag{3.7.2}$$

([Bern89], p 156). $D_c := \{(p,q,z) \in \mathbb{C}^3;\ \mathrm{Re}\, z \neq 0\}$, $D_f := \{(p,q,z) \in \mathbb{C}^3;\ \mathrm{Re}\, z > 0\}$.

From this it follows that

$$\sum_{k=1}^{\infty}\left\{\frac{(-1)^{k+1}}{z+q+2k-1}+\frac{(-1)^{k+1}}{z-q+2k-1}\right\} = \int_0^{\infty}\frac{\cosh(qt)e^{-tz}}{\cosh t}dt = \frac{1}{z+}\,\frac{1^2-q^2}{z}\,\frac{2^2}{+\,z+}\,\frac{3^2-q^2}{z}\,\frac{4^2}{+\,z\,+}\cdots, \tag{3.7.3}$$

([Bern89], p 148), $D_c := \{(q,z)\in\mathbb{C}^2;\ \operatorname{Re} z\neq 0\}$, $D_f := \{(q,z)\in\mathbb{C}^2;\ \operatorname{Re} z>0\}$, and

$$\tanh\left\{\int_0^{\infty}\frac{\sinh(at)e^{-tz}}{t\cosh t}dt\right\} = \frac{a}{z+}\,\frac{1^2}{z+}\,\frac{2^2-a^2}{z}\,\frac{3^2}{+\,z+}\,\frac{4^2-a^2}{z}\,+\cdots, \tag{3.7.4}$$

([Wall48], p 372), $D_c := \{(a,z)\in\mathbb{C}^2;\ \operatorname{Re} z\neq 0\}$, $D_f := \{(a,z)\in\mathbb{C}^2;\ \operatorname{Re} z>0\}$, and

$$\tanh\left\{\frac{1}{2}\int_0^{\infty}\frac{\sinh(2at)e^{-tz}}{t\cosh t}dt\right\} = \frac{a}{z+}\,\frac{1^2-a^2}{z}\,+\,\frac{2^2-a^2}{z}\,+\,\frac{3^2-a^2}{z}\,+\,\frac{4^2-a^2}{z}\,+\cdots, \tag{3.7.5}$$

([Wall48], p 371), $D_c := \{(a,z)\in\mathbb{C}^2;\ \operatorname{Re} z\neq 0\}$, $D_f := \{(a,z)\in\mathbb{C}^2;\ \operatorname{Re} z>0\}$.

Solving (3.7.2) for $1/R$ gives

$$\frac{1}{R} = 1+\frac{2p}{z-p+}\,\frac{1^2-q^2}{z}\,+\,\frac{2^2-p^2}{z}\,+\,\frac{3^2-q^2}{z}\,+\,\frac{4^2-p^2}{z}\,+\cdots, \tag{3.7.6}$$

([Perr57], p 34). $D_c := \{(p,z)\in\mathbb{C}^2;\ \operatorname{Re} z\neq 0\}$, $D_f := \{(p,z)\in\mathbb{C}^2;\ \operatorname{Re} z>0\}$. The values $p := q := 1/2$ lead to

$$\frac{z}{4}\frac{\Gamma^2\left(\frac{z}{4}\right)}{\Gamma^2\left(\frac{z+2}{4}\right)} = 1+\frac{2}{2z-1+}\,\frac{1\cdot 3}{2z}\,+\,\frac{3\cdot 5}{2z}\,+\,\frac{5\cdot 7}{2z}\,+\cdots \quad \text{for } \operatorname{Re}(z)>0 \tag{3.7.7}$$

and thus, for $z := 4n$ or $z := 4n-2$ where $n\in\mathbb{N}$, we have

$$\frac{1}{n\pi}\left(\frac{2\cdot 4\cdot\cdots\cdot(2n)}{1\cdot 3\cdot\cdots\cdot(2n-1)}\right)^2 = 1+\frac{2}{8n-1+}\,\frac{1\cdot 3}{8n}\,+\,\frac{3\cdot 5}{8n}\,+\,\frac{5\cdot 7}{8n}\,+\cdots, \tag{3.7.8}$$

([Perr57], p 34),

$$\frac{2n^2\pi}{2n-1}\left(\frac{1\cdot 3\cdot\cdots\cdot(2n-1)}{2\cdot 4\cdot\cdots\cdot(2n)}\right)^2 = 1+\frac{2}{8n-5+}\,\frac{1\cdot 3}{8n-4+}\,\frac{3\cdot 5}{8n-4+}\,\frac{5\cdot 7}{8n-4+}\cdots \tag{3.7.9}$$

for $n\in\mathbb{N}$, ([Perr57], p 34).

$$\frac{a+1}{a}\,\frac{\displaystyle\int_0^1 t^a\left(\frac{1-t}{1+t}\right)^b dt}{\displaystyle\int_0^1 t^{a-1}\left(\frac{1-t}{1+t}\right)^b dt} = \frac{a+1}{2b}\,+\,\frac{(a+1)(a+2)}{2b}\,+\,\frac{(a+2)(a+3)}{2b}\,+\cdots, \tag{3.7.10}$$

([Perr57], p 299). $D_c := \{(a,b) \in \mathbb{C}^2;\ \mathrm{Re}(b) \neq 0\}$, $D_f := \{(a,b) \in \mathbb{C}^2;\ \mathrm{Re}(b) > 0\}$. From this follows directly that also

$$\frac{\displaystyle\int_0^1 t^a \left(\frac{1-t}{1+t}\right)^b \frac{dt}{1-t}}{\displaystyle\int_0^1 t^a \left(\frac{1-t}{1+t}\right)^b \frac{dt}{1-t^2}} = 1 + \frac{a+1}{2b} \genfrac{}{}{0pt}{}{}{+} \frac{(a+1)(a+2)}{2b} \genfrac{}{}{0pt}{}{}{+} \frac{(a+2)(a+3)}{2b} \genfrac{}{}{0pt}{}{}{+\cdots}, \tag{3.7.11}$$

([Perr57], p 300). $D_c := \{(a,b) \in \mathbb{C}^2;\ \mathrm{Re}(b) \neq 0\}$, $D_f := \{(a,b) \in \mathbb{C}^2;\ \mathrm{Re}(b) > 0\}$. A formula of the same character as (3.7.2) is

$$\prod_\varepsilon \left(\Gamma\Big(\frac{z+\varepsilon p+\varepsilon q+1}{4}\Big)\right) \Big/ \left(\Gamma\Big(\frac{z+\varepsilon p+\varepsilon q+3}{4}\Big)\right) = \frac{8}{\frac{1}{2}(z^2-p^2+q^2-1)+} \frac{1^2-q^2}{1} \genfrac{}{}{0pt}{}{}{+} \frac{1^2-p^2}{z^2-1} \genfrac{}{}{0pt}{}{}{+} \frac{3^2-q^2}{1} \genfrac{}{}{0pt}{}{}{+} \frac{3^2-p^2}{z^2-1} \genfrac{}{}{0pt}{}{}{+\cdots}, \tag{3.7.12}$$

([Bern89], p 159). $D_c := \{(p,q,z) \in \mathbb{C}^3;\ |\arg(z^2-1)| < \pi\}$, $D_f := \{(p,q,z) \in \mathbb{C}^3;\ \mathrm{Re}\, z > 0\} \setminus \{(p,q,z) \in \mathbb{C}^3;\ 0 < z \leq 1\}$.

$$\prod_\varepsilon \frac{\Gamma\left(\dfrac{z+\varepsilon q+1}{4}\right)}{\Gamma\left(\dfrac{z+\varepsilon q+3}{4}\right)} = \frac{4}{z+} \frac{1^2-q^2}{2z} \genfrac{}{}{0pt}{}{}{+} \frac{3^2-q^2}{2z} \genfrac{}{}{0pt}{}{}{+} \frac{5^2-q^2}{2z} \genfrac{}{}{0pt}{}{}{+\cdots}, \tag{3.7.13}$$

([Bern89], p 140). $D_c := \{(q,z) \in \mathbb{C}^2;\ \mathrm{Re}\, z \neq 0\}$, $D_f := \{(q,z) \in \mathbb{C}^2;\ \mathrm{Re}\, z > 0\}$. For $q := 0$ and $z := 4n-1$ or $z := 4n+1$ for an $n \in \mathbb{N}$, this reduces to

$$4\pi n^2 \left(\frac{1\cdot 3 \cdot \cdots \cdot (2n-1)}{2\cdot 4 \cdot \cdots \cdot (2n)}\right)^2 = 4n-1+\frac{1^2}{8n-2+} \frac{3^2}{8n-2+} \frac{5^2}{8n-2+\cdots}, \tag{3.7.14}$$

([Perr57], p 36), or

$$\frac{1}{\pi}\left(\frac{2n+1}{n+1}\right)^2 \left(\frac{2\cdot 4 \cdot \cdots \cdot (2n+2)}{1\cdot 3 \cdot \cdots \cdot (2n+1)}\right)^2 = 4n+1+\frac{1^2}{8n+2+} \frac{3^2}{8n+2+} \frac{5^2}{8n+2+\cdots}, \tag{3.7.15}$$

([Perr57], p 36).

A formula closely related to (3.7.13) is

$$\exp\left\{\int_0^\infty \left(1-\frac{\cosh 2at}{\cosh 2t}\right) e^{-tz} \frac{dt}{t}\right\} = 1+\frac{2(1^2-a^2)}{z^2} \genfrac{}{}{0pt}{}{}{+} \frac{3^2-a^2}{1} \genfrac{}{}{0pt}{}{}{+} \frac{5^2-a^2}{z^2} \genfrac{}{}{0pt}{}{}{+\cdots}, \tag{3.7.16}$$

([Wall48], p 371). $D_c := \{(a,z) \in \mathbb{C}^2;\ \mathrm{Re}\, z \neq 0\}$, $D_f := \{(a,z) \in \mathbb{C}^2;\ \mathrm{Re}\, z > 0\}$.

The most involved of Ramanujan's formulas of this type is

$$\frac{R-Q}{R+Q} = \frac{8abcdh}{1\{2S_4 - (S_2 - 2\cdot 0\cdot 1)^2 - 4(0^2+0+1)^2\}+} \frac{64(a^2-1^2)(b^2-1^2)(c^2-1^2)(d^2-1^2)(h^2-1^2)}{3\{2S_4 - (S_2 - 2\cdot 1\cdot 2)^2 - 4(1^2+1+1)^2\}} + \frac{64(a^2-2^2)(b^2-2^2)(c^2-2^2)(d^2-2^2)(h^2-2^2)}{5\{2S_4 - (S_2 - 2\cdot 2\cdot 3)^2 - 4(2^2+2+1)^2\}} + \cdots, \tag{3.7.17}$$

where $S_4 := a^4 + b^4 + c^4 + d^4 + h^4 + 1$, $S_2 := a^2 + b^2 + c^2 + d^2 + h^2 - 1$, and

$$R := \prod_\varepsilon \Gamma\left(\frac{a+\varepsilon(b+c)+\varepsilon(d+h)+1}{2}\right)\cdot \prod_\varepsilon \Gamma\left(\frac{a+\varepsilon(b+d)+\varepsilon(c+h)+1}{2}\right),$$
$$Q := \prod_\varepsilon \Gamma\left(\frac{a+\varepsilon(b-c)+\varepsilon(d+h)+1}{2}\right)\cdot \prod_\varepsilon \Gamma\left(\frac{a+\varepsilon(b+c)+\varepsilon(d-h)+1}{2}\right), \tag{3.7.18}$$

([Bern89], p 163). The expansion (3.7.17) only holds if the continued fraction terminates.

$$\frac{1-R}{1+R} = \frac{2abc}{z^2-a^2-b^2-c^2+1+} \frac{4(a^2-1^2)(b^2-1^2)(c^2-1^2)}{3(z^2-a^2-b^2-c^2+5)} + \frac{4(a^2-2^2)(b^2-2^2)(c^2-2^2)}{5(z^2-a^2-b^2-c^2+13)} + \cdots \quad \text{for } (a,b,c,z)\in\mathbb{C}^4 \quad \text{where} \tag{3.7.19}$$
$$R := \prod_\varepsilon \frac{\Gamma\left(\frac{z+a+\varepsilon(b+c)+1}{2}\right)}{\Gamma\left(\frac{z-a+\varepsilon(b+c)+1}{2}\right)} \cdot \prod_\varepsilon \frac{\Gamma\left(\frac{z-a+\varepsilon(b-c)+1}{2}\right)}{\Gamma\left(\frac{z+a+\varepsilon(b-c)+1}{2}\right)},$$

([Bern89], p 157). The last number in each partial denominator of the continued fraction (i.e., 1, 5, 13, ...) is the number $2n^2+2n+1$ for $n = 0, 1, 2, \ldots$.)

$$\frac{1-R}{1+R} = \frac{ab}{z+} \frac{(a^2-1^2)(b^2-1^2)}{3z} + \frac{(a^2-2^2)(b^2-2^2)}{5z} + \frac{(a^2-3^2)(b^2-3^2)}{7z} + \cdots \tag{3.7.20}$$
$$\text{where } R = \prod_\varepsilon \Gamma\left(\frac{z+\varepsilon(a-b)+1}{2}\right) \Big/ \prod_\varepsilon \Gamma\left(\frac{z+\varepsilon(a+b)+1}{2}\right),$$

([Bern89], p 155). $D_c := \{(a,b,z)\in\mathbb{C}^3;\ \mathrm{Re}\, z \neq 0\}$, $D_f := \{(a,b,z)\in\mathbb{C}^3;\ \mathrm{Re}\, z > 0\}$. In particular

$$\sum_{k=0}^\infty \left\{\frac{1}{z-a+2k+1} - \frac{1}{z+a+2k+1}\right\} = \lim_{b\to 0}\frac{1}{b}\frac{1-R}{1+R} = \frac{a}{z+} \frac{1^2(1^2-a^2)}{3z} + \frac{2^2(2^2-a^2)}{5z} + \frac{3^2(3^2-a^2)}{7z} + \cdots \tag{3.7.21}$$

for $\mathrm{Re}(z) > 0$, ([Bern89], p 149).

$$\sum_{k=1}^\infty \frac{(-1)^{k+1}}{(a+k)(b+k)} = \frac{1}{(a+1)(b+1)+} \frac{(a+1)^2(b+1)^2}{a+b+3} + \frac{(a+2)^2(b+2)^2}{a+b+5} + \cdots \tag{3.7.22}$$

for $(a,b) \in \mathbb{C}^2$ with $b \neq -1$ if $a \in (-\mathbb{N})$ and $a \neq -1$ if $b \in (-\mathbb{N})$, ([Bern89], p 123).

$$\begin{aligned}\frac{1-R}{1+R} &= \frac{ab}{z^2-1-a^2+}\,\frac{2^2-b^2}{1}\,\frac{2^2-a^2}{+\,z^2-1+}\,\frac{4^2-b^2}{1}\,\frac{4^2-a^2}{+\,z^2-1+\cdots};\\ R &:= \prod_{\varepsilon} \frac{\Gamma\Bigl(\frac{z+\varepsilon(a+b)+3}{4}\Bigr)}{\Gamma\Bigl(\frac{z+\varepsilon(a+b)+1}{4}\Bigr)} \Big/ \prod_{\varepsilon} \frac{\Gamma\Bigl(\frac{z+\varepsilon(a-b)+3}{4}\Bigr)}{\Gamma\Bigl(\frac{z+\varepsilon(a-b)+1}{4}\Bigr)},\end{aligned} \tag{3.7.23}$$

([Bern89], p 158). $D_c := \{(a,b,z) \in \mathbb{C}^3;\ |\arg(z^2-1)| < \pi\}$, $D_f := \{(a,b,z) \in \mathbb{C}^3;\ \operatorname{Re} z > 0\} \setminus \{(a,b,z) \in \mathbb{C}^3;\ 0 < z \leq 1\}$. Dividing (3.7.23) by a and letting $a \to 0$ in this equality gives

$$\begin{aligned}\sum_{k=0}^{\infty}\left\{\frac{(-1)^k}{z-b+2k+1} - \frac{(-1)^k}{z+b+2k+1}\right\} &= \int_0^{\infty} e^{-tz}\frac{\sinh(bt)}{\cosh t}\,dt\\ &= \frac{b}{z^2-1+}\,\frac{2^2-b^2}{1}\,\frac{2^2}{+\,z^2-1+}\,\frac{4^2-b^2}{1}\,{+\cdots},\end{aligned} \tag{3.7.24}$$

([Bern89], p 150). $D_c := \{(b,z) \in \mathbb{C}^2;\ |\arg(z^2-1)| < \pi\}$, $D_f := \{(b,z) \in \mathbb{C}^2;\ \operatorname{Re} z > 0\} \setminus \{(b,z) \in \mathbb{C}^2;\ 0 < z \leq 1\}$. Of course, dividing this again by b and letting $b \to 0$ gives

$$2\sum_{k=0}^{\infty} \frac{(-1)^k}{(z+2k+1)^2} = \frac{1}{z^2-1+}\,\frac{2^2}{1}\,\frac{2^2}{+\,z^2-1+}\,\frac{4^2}{1}\,\frac{4^2}{+\,z^2-1+\cdots}, \tag{3.7.25}$$

([Bern89], p 151). $D_c := \{z \in \mathbb{C};\ |\arg(z^2-1)| < \pi\}$, $D_f := \{z \in \mathbb{C};\ \operatorname{Re} z > 0\} \setminus \{z \in \mathbb{C};\ 0 < z \leq 1\}$.

$$1 + 2z\sum_{k=1}^{\infty} \frac{(-1)^k}{z+2k} = \frac{1}{z+}\,\frac{1\cdot 2}{z}\,\frac{2\cdot 3}{+\,z}\,\frac{3\cdot 4}{+\,z}\,{+\cdots}, \tag{3.7.26}$$

([Bern89], p 151). $D_c := \{z \in \mathbb{C};\ \operatorname{Re} z \neq 0\}$, $D_f := \{z \in \mathbb{C};\ \operatorname{Re} z > 0\}$.

$$1 + 2z^2\sum_{k=1}^{\infty} \frac{(-1)^k}{(z+k)^2} = \frac{1}{z+}\,\frac{1^2}{z+}\,\frac{1\cdot 2}{z}\,\frac{2^2}{+\,z+}\,\frac{2\cdot 3}{z}\,\frac{3^2}{+\,z+\cdots}, \tag{3.7.27}$$

([Bern89], p 152). $D_c := \{z \in \mathbb{C};\ \operatorname{Re} z \neq 0\}$, $D_f := \{z \in \mathbb{C};\ \operatorname{Re} z > 0\}$.

$$\begin{aligned}c\int_0^{\infty} \frac{\sinh at \sinh bt}{\sinh ct} e^{-tz}\,dt &=\\ &\frac{ab}{1(z^2+c^2-a^2-b^2)-}\,\frac{4\cdot 1^2(1^2c^2-a^2)(1^2c^2-b^2)}{3(z^2+5c^2-a^2-b^2)}\,-\\ &\qquad\frac{4\cdot 2^2(2^2c^2-a^2)(2^2c^2-b^2)}{5(z^2+13c^2-a^2-b^2)}\,{-\cdots},\end{aligned} \tag{3.7.28}$$

where the coefficients for c^2 are $2k^2+2k+1$ in the denominators ([Wall48], p 370). $D_c := \{(a,b,c,z) \in \mathbb{C}^4;\ \operatorname{Re}\frac{z}{c} \neq 0\}$, $D_f := \{(a,b,c,z) \in \mathbb{R}^4;\ \operatorname{Re}\frac{z}{c} > 0$ and $\operatorname{Re}(z+c-a-b) > 0\}$.

$$c\int_0^{\infty} \frac{\sinh at}{\sinh ct} e^{-tz}\,dt = \frac{a}{z+}\,\frac{1^2(1^2c^2-a^2)}{3z}\,\frac{2^2(2^2c^2-a^2)}{+\,5z}\,{+\cdots}, \tag{3.7.29}$$

([Wall48], p 370). $D_c := \{(a,c,z) \in \mathbb{C}^3;\ \mathrm{Re}(z/c) \neq 0\}$, $D_f := \{(a,c,z) \in \mathbb{C}^3;\ \mathrm{Re}(z/c) > 0\}$.

$$\int_0^\infty \frac{e^{-tz}dt}{(\cosh t + a\sinh t)^b} = \frac{1}{z+ab+}\,\frac{1\cdot b(1-a^2)}{z+a(b+2)+}\,\frac{2(b+1)(1-a^2)}{z+a(b+4)}\,+\,\frac{3(b+2)(1-a^2)}{z+a(b+6)}\,+\cdots, \tag{3.7.30}$$

([Wall48], p 369). $D_c := \{(a,b,z) \in \mathbb{C}^3;\ \mathrm{Re}\,a \neq 0\}$, $D_f := \{(a,b,z) \in \mathbb{C}^3;\ \mathrm{Re}\,a > 0$ and $\mathrm{Re}(b+z) > 0\}$.

$$\int_0^\infty {}_2F_1\Big(a,b;\frac{a+b+1}{2};-\sinh^2 t\Big)e^{-tz}dt = \frac{1}{z+}\,\frac{4\cdot 1ab}{(a+b+1)z+}\,+\,\frac{4\cdot 2(a+1)(b+1)(a+b)}{(a+b+3)z}\,+\,\frac{4\cdot 3(a+2)(b+2)(a+b+1)}{(a+b+5)z}\,+\cdots \tag{3.7.31}$$

([Wall48], p 370). $D_c := \{(a,b,z) \in \mathbb{C}^3;\ \mathrm{Re}\,z \neq 0\}$, $D_f := \{(a,b,z) \in \mathbb{C}^3;\ \mathrm{Re}\,z > 0\}$.

$$\zeta(3,z+1) := \sum_{k=1}^\infty \frac{1}{(z+k)^3} = \frac{1}{2(z^2+z)+}\,\frac{1^3}{1+}\,\frac{1^3}{6(z^2+z)+}\,\frac{2^3}{1+}\,\frac{2^3}{10(z^2+z)+}\cdots, \tag{3.7.32}$$

([Bern89], p. 153). $D_c := \{z \in \mathbb{C};\ |\arg(z^2+z)| < \pi\} = \{z \in \mathbb{C};\ \mathrm{Re}\,z \neq -\frac{1}{2}\} \setminus [-1,0]$, $D_f := \{z \in \mathbb{C};\ \mathrm{Re}\,z > -\frac{1}{2}\} \setminus [-\frac{1}{2},0]$. The even part of this continued fraction is

$$\zeta(3,z+1) = \frac{1}{1(2z^2+2z+1)-}\,\frac{1^6}{3(2z^2+2z+3)-}\,\frac{2^6}{5(2z^2+2z+7)-}\,\frac{3^6}{7(2z^2+2z+13)-}\cdots. \tag{3.7.33}$$

$D_c := \{z \in \mathbb{C};\ \mathrm{Re}(z^2+\frac{1}{2}) \neq 0\}$, $D_f := \{z \in \mathbb{C};\ \mathrm{Re}(z^2+\frac{1}{2}) > 0\}$. (The numbers 1, 3, 7, 13, ... in the denominators are n^2+n+1 for $n = 0,1,2,\ldots$.)

$$\begin{aligned}&\sum_{k=0}^\infty\Big\{\frac{1}{z+a+b+2k+1}+\frac{1}{z-a-b+2k+1}-\frac{1}{z+a-b+2k+1}-\frac{1}{z-a+b+2k+1}\Big\}\\&=\sum_{k=0}^\infty\frac{8ab(z+2k+1)}{\{(z+2k+1)^2-a^2-b^2\}^2-4a^2b^2}\\&=\frac{2ab}{1(z^2-1)+b^2-a^2+}\,\frac{2(1^2-b^2)}{1}\,+\,\frac{2(1^2-a^2)}{3(z^2-1)+b^2-a^2+}\,\frac{4(2^2-b^2)}{1}\,+\,\frac{4(2^2-a^2)}{5(z^2-1)+b^2-a^2+}\cdots,\end{aligned} \tag{3.7.34}$$

([Bern89], p 158). $D_c := \{(a,b,z) \in \mathbb{C}^3;\ |\arg(z^2-1)| < \pi\} = \{(a,b,z) \in \mathbb{C}^3;\ \mathrm{Re}\,z \neq 0$ and $z \notin (-1,1)\}$, $D_f := \{(a,b,z) \in \mathbb{C}^3;\ \mathrm{Re}\,z > 0$ and $z \notin (0,1)\}$. Dividing by $2a$ and letting $a \to 0$ in (3.7.34) leads to

$$\sum_{k=0}^{\infty}\left\{\frac{1}{(z-b+2k+1)^2}-\frac{1}{(z+b+2k+1)^2}\right\}=\sum_{k=0}^{\infty}\frac{4b(z+2k+1)}{\{(z+2k+1)^2-b^2\}^2}$$
$$=\frac{b}{1(z^2-1)+b^2+}\,\frac{2(1^2-b^2)}{1}\,\frac{2\cdot 1^2}{+3(z^2-1)+b^2+}\,\frac{4(2^2-b^2)}{1}\,\frac{4\cdot 2^2}{+5(z^2-1)+b^2+\cdots}, \tag{3.7.35}$$

([Bern89], p 158). $D_c := \{(b,z)\in\mathbb{C}^2;\ \operatorname{Re} z\neq 0 \text{ and } z\notin(-1,1)\}$, $D_f := \{(b,z)\in\mathbb{C}^2;\ \operatorname{Re} z>0 \text{ and } z\notin(0,1)\}$. The even part of this continued fraction is

$$\sum_{k=0}^{\infty}\left\{\frac{1}{(z-b+2k+1)^2}-\frac{1}{(z+b+2k+1)^2}\right\}=\sum_{k=0}^{\infty}\frac{4b(z+2k+1)}{\{(z+2k+1)^2-b^2\}^2}$$
$$=\frac{b}{1(z^2-b^2+1)-}\,\frac{4(1^2-b^2)1^4}{3(z^2-b^2+5)-}\,\frac{4(2^2-b^2)2^4}{5(z^2-b^2+13)-}\,\frac{4(3^2-b^2)3^4}{7(z^2-b^2+25)-\cdots}. \tag{3.7.36}$$

$D_c := \{(b,z)\in\mathbb{C}^2;\ \operatorname{Re} z\neq 0\}$, $D_f := \{(b,z)\in\mathbb{C}^2;\ \operatorname{Re} z>0\}$. (The numbers 1, 5, 13, 25, ... in the denominators have the form $2n^2+2n+1$ for $n=0,1,2,3,\dots$.)

$$\frac{u-v}{u+v}=\frac{2a^2}{1z+}\,\frac{4a^4+1^4}{3z}\,+\,\frac{4a^4+2^4}{5z}\,+\,\frac{4a^4+3^4}{7z}\,+\cdots \quad\text{where}$$
$$u:=\prod_{k=0}^{\infty}\left\{1+\left(\frac{2a}{z+2k+1}\right)^2\right\},\quad v:=\frac{\Gamma^2\left(\frac{z+1}{2}\right)}{\Gamma\left(\frac{z+2a+1}{2}\right)\Gamma\left(\frac{z-2a+1}{2}\right)} \tag{3.7.37}$$

([ABJL92], entry 48). $D_c := \{(a,z)\in\mathbb{C};\ \operatorname{Re} z\neq 0\}$, $D_f := \{(a,z)\in\mathbb{C};\ \operatorname{Re} z>0\}$.

$$\frac{u-v}{u+v}=\frac{a^3}{1(2z^2+2z+1)+}\,\frac{a^6-1^6}{3(2z^2+2z+3)+}\,\frac{a^6-2^6}{5(2z^2+2z+7)+\cdots}$$
where
$$u:=\prod_{k=1}^{\infty}\left\{1+\left(\frac{a}{z+k}\right)^3\right\},\qquad v:=\prod_{k=1}^{\infty}\left\{1-\left(\frac{a}{z+k}\right)^3\right\}, \tag{3.7.38}$$

([ABJL92], entry 50). $D_c := \{(a,z)\in\mathbb{C}^2;\ \operatorname{Re} z\neq -\frac{1}{2}\}$, $D_f := \{(a,z)\in\mathbb{C}^2;\ \operatorname{Re} z> -\frac{1}{2}\}$. (The numbers $1,3,7,\dots$ in the denominators are the numbers n^2+n+1 for $n=0,1,2,\dots$.)

$$2\sum_{k=0}^{\infty}\frac{(-1)^k y^{2k+1}}{r+2k+1}=\frac{z}{1+a+}\,\frac{1^2z^2}{3+a+}\,\frac{2^2z^2}{5+a+}\,\frac{3^2z^2}{7+a+\cdots}$$
$$\text{where}\quad y:=(\sqrt{1+z^2}-1)/z \quad\text{and}\quad r:=a/\sqrt{1+z^2} \tag{3.7.39}$$

for $(a,z)\in\mathbb{C}^2$ with $|\arg(z^2+1)|<\pi$; i.e., $z\in\mathbb{C}\setminus i((-\infty,-1]\cup[1,\infty))$, ([ABJL92], entry 14). With the same notation and same region for (a,z), also

$$y+r\left(y+\frac{1}{y}\right)\sum_{k=1}^{\infty}\frac{(-1)^k y^{2k}}{r+2k}=\frac{z}{2+a+}\,\frac{1\cdot 2z^2}{4+a}\,+\,\frac{2\cdot 3z^2}{6+a}\,+\,\frac{3\cdot 4z^2}{8+a}\,+\cdots, \tag{3.7.40}$$

([ABJL92], entry 15) and

$$\left(1+\frac{1}{z^2}\right)^{(b-1)/2}(2y)^b\sum_{k=0}^{\infty}\frac{(-1)^k(b)_k y^{2k}}{k!(r+b+2k)}= \\ \frac{z}{a+b+}\frac{1\cdot bz^2}{a+b+2+}\frac{2(b+1)z^2}{a+b+4+}\frac{3(b+2)z^2}{a+b+6+}\cdots, \tag{3.7.41}$$

([ABJL92], entry 17), where $b \in \mathbb{C}$.

A.4 Basic hypergeometric functions

In this chapter we use the standard notation

$$ {}_2\varphi_1(a,b;c;q;z) := \sum_{n=0}^{\infty}\frac{(a;q)_n(b;q)_n}{(c;q)_n(q;q)_n}z^n$$
$$\text{whcro } (d;q)_0 := 1,\ (d;q)_n := (1-d)(1-dq)\cdots(1-dq^{n-1}) \text{ for } n\in\mathbb{N}.$$

For convenience we always assume that $q \in \mathbb{C}$ with $|q| < 1$, although the continued fraction may well converge, even to the right value, for other values of $q \in \mathbb{C}$.

A.4.1 General expressions

$$(1-c)\frac{{}_2\varphi_1(a,b;c;q;z)}{{}_2\varphi_1(a,bq;cq;q;z)}= \\ 1-c+\frac{(1-a)(c-b)z}{1-cq\quad+}\frac{(1-bq)(cq-a)z}{1-cq^2\quad+}\frac{(1-aq)(cq-b)qz}{1-cq^3\quad+} \\ \frac{(1-bq^2)(cq^2-a)qz}{1-cq^4\quad+}\frac{(1-aq^2)(cq^2-b)q^2z}{1-cq^5\quad+\cdots} \tag{4.1.1}$$

for $(a,b,c,z)\in\mathbb{C}^4$, ([ABBW85], p 14).

$$(1-c)\frac{{}_2\varphi_1(a,b;c;q;z)}{{}_2\varphi_1(aq,bq;cq;q;z)} = b_0 + \mathbf{K}(a_n/b_n) \\ \text{where } a_n := (1-aq^n)(1-bq^n)cq^{n-1}(1-zabq^n/c)z \\ b_n := 1-cq^n-(a+b-abq^n-abq^{n+1})q^n z \tag{4.1.2}$$

for $(a,b,c,z)\in\mathbb{C}^4$.

$$q(1-c)\frac{{}_2\varphi_1(a,b;c;q;z)}{{}_2\varphi_1(a,bq;cq;q;z)} = (1-c)q+(a-bq)z-\frac{(a-cq)(1-bq)qz}{(1-cq)q+(a-bq^2)z-} \\ \frac{(a-cq^2)(1-bq^2)qz}{(1-cq^2)q+(a-bq^3)z}-\frac{(a-cq^3)(1-bq^3)qz}{(1-cq^3)q+(a-bq^4)z}-\cdots \tag{4.1.3}$$

for $(a,b,c,z)\in\mathbb{C}^4$, ([ABBW85], p 18).

If we choose $b = 1$ in (4.1.1), (4.1.2) or (4.1.3) we obtain continued fraction expansions for ${}_2\varphi_1(a, q; cq; q; z)$ or ${}_2\varphi_1(aq, q; cq; q; z)$.

A.4.2 Two general results by Andrews

$$\frac{G(a,b,c;q)}{G(aq,b,cq;q)} = 1 + \frac{aq+cq}{1} {}_{+} \frac{bq+cq^2}{1} {}_{+} \frac{aq^2+cq^3}{1} {}_{+} \frac{bq^2+cq^4}{1} {}_{+} \cdots$$
$$\text{where} \quad G(a,b,c;q) := \sum_{k=0}^{\infty} \frac{(-\frac{c}{a};q)_k q^{k(k+1)/2} a^k}{(q;q)_k (-bq;q)_k} \tag{4.2.1}$$

for $(a, b, c) \in \mathbb{C}^3$, ([ABJL89], p 80).

$$\frac{H(a_1,a_2;z;q)}{H(a_1,a_2;qz;q)} = 1 + bqz + \frac{(1+aq^2z)qz}{1+bq^2z} {}_{+} \frac{(1+aq^3z)q^2z}{1+bq^3z} {}_{+} \cdots$$
where $a := -1/a_1a_2$ and $b := -1/a_1 - 1/a_2$ and
$$H(a_1,a_2;z;q) := \frac{\left(\frac{qz}{a_1};q\right)_\infty \left(\frac{qz}{a_2};q\right)_\infty}{(qz;q)_\infty (1-z)} \times \tag{4.2.2}$$
$$\times \sum_{k=0}^{\infty} \frac{(1-zq^{2k})(z;q)_k (a_1;q)_k (a_2;q)_k q^{k(3k+1)/2} (az^2)^k}{(q;q)_k \left(\frac{qz}{a_1};q\right)_k \left(\frac{qz}{a_2};q\right)_k}$$

for $(1/a_1,\ 1/a_2, z) \in \mathbb{C}^3$, ([ABJL89], p 79).

A.4.3 q-expressions by Ramanujan

The formula (4.2.1) can also be found in Ramanujan's lost notebook ([Andr79], p 90). Quite a number of Ramanujan's expressions are special cases of (4.2.1) and (4.2.2). We refer in particular to ([ABJL92]) for more details. From (4.1.1) we find that

$$\frac{(-a;q)_\infty(b;q)_\infty - (a;q)_\infty(-b;q)_\infty}{(-a;q)_\infty(b;q)_\infty + (a;q)_\infty(-b;q)_\infty} = \frac{a-b}{1-q} \cdot \frac{{}_2\varphi_1\left(\frac{bq}{a}, \frac{bq^2}{a}; q^3; q^2; a^2\right)}{{}_2\varphi_1\left(\frac{bq}{a}, \frac{b}{a}; q; q^2; a^2\right)} \tag{4.3.1}$$
$$= \frac{a-b}{1-q} {}_{+} \frac{(a-bq)(aq-b)}{1-q^3} {}_{+} \frac{(a-bq^2)(aq^2-b)q}{1-q^5} {}_{+} \frac{(a-bq^3)(aq^3-b)q^2}{1-q^7} {}_{+} \cdots$$

for $(a, b) \in \mathbb{C}^2$, ([ABBW85], p 14).

$$\frac{(a^2q^3;q^4)_\infty(b^2q^3;q^4)_\infty}{(a^2q;q^4)_\infty(b^2q;q^4)_\infty} = \frac{1}{1-ab} {}_{+} \frac{(a-bq)(b-aq)}{(1-ab)(q^2+1)} {}_{+} \frac{(a-bq^3)(b-aq^3)}{(1-ab)(q^4+1)} {}_{+} \cdots \tag{4.3.2}$$

for $(a,b) \in \mathbb{C}^2$, ([ABBW85], entry 12).

$$\begin{aligned} \frac{F(b;a)}{F(b;aq)} &= 1 + \frac{aq}{1+bq+}\frac{aq^2}{1+bq^2+}\frac{aq^3}{1+bq^3+}\cdots \\ \text{where } F(b;a) &:= \sum_{k=0}^{\infty} \frac{a^k q^{k^2}}{(-bq;q)_k (q;q)_k} \end{aligned} \tag{4.3.3}$$

for $(a,b) \in \mathbb{C}^2$, ([ABBW85], entry 15).

If we set $a := 0$ in (4.2.1) we get

$$\begin{aligned} \frac{\varphi(c)}{\varphi(cq)} &= 1 + \frac{cq}{1+}\frac{bq+cq^2}{1}\,+\frac{cq^3}{1+}\frac{bq^2+cq^4}{1}\,+\frac{cq^5}{1}+\cdots \\ \text{where } \varphi(c) &:= \sum_{k=0}^{\infty} \frac{q^{k^2} c^k}{(q;q)_k(-bq;q)_k)}, \end{aligned} \tag{4.3.4}$$

([ABJL92], entry 56). If we moreover set $b := -c$, this reduces to

$$\sum_{k=0}^{\infty} (-c)^k q^{k(k+1)/2} = \frac{1}{1+}\frac{cq}{1+}\frac{c(q^2-q)}{1}\,+\frac{cq^3}{1+}\frac{c(q^4-q^2)}{1}\,+\cdots, \tag{4.3.5}$$

([ABBW85], p 22).

$$\begin{aligned} \frac{G(z)}{G(qz)} &= 1 - \frac{qz}{1+q+}\frac{q^3 z}{1+q^2-}\frac{q^2 z}{1+q^3+}\frac{q^6 z}{1+q^4-}\frac{q^3 z}{1+q^5+}\frac{q^9 z}{1+q^6-}\cdots \\ \text{where } G(z) &:= \sum_{k=0}^{\infty} \frac{(-z)^k q^{k(k+1)/2}}{(q^2;q^2)_k} \end{aligned} \tag{4.3.6}$$

for $z \in \mathbb{C}$, ([ABJL92], formula 9.1).

$$\frac{(q^2;q^3)_\infty}{(q;q^3)_\infty} = \frac{1}{1-}\frac{q}{1+q-}\frac{q^3}{1+q^2-}\frac{q^5}{1+q^3-}\frac{q^7}{1+q^4-}\cdots, \tag{4.3.7}$$

([ABJL92], entry 10).

$$\frac{(q^3;q^4)_\infty}{(q;q^4)_\infty} = \frac{1}{1-}\frac{q}{1+q^2-}\frac{q^3}{1+q^4-}\frac{q^5}{1+q^6-}\cdots, \tag{4.3.8}$$

([ABJL92], entry 11).

$$\frac{(-q^2;q^2)_\infty}{(-q;q^2)_\infty} = \frac{1}{1+}\frac{q}{1+}\frac{q^2+q}{1}\,+\frac{q^3}{1+}\frac{q^4+q^2}{1}\,+\frac{q^5}{1}+\cdots, \tag{4.3.9}$$

([ABJL92], entry 12).

$$\frac{(q;q^2)_\infty}{\{(q^3;q^6)_\infty\}^3} = \frac{1}{1+}\frac{q+q^2}{1}\,+\frac{q^2+q^4}{1}\,+\frac{q^3+q^6}{1}\,+\cdots, \tag{4.3.10}$$

([ABJL89], thm 7).

$$\frac{(q;q^5)_\infty(q^4;q^5)_\infty}{(q^2;q^5)_\infty(q^3;q^5)_\infty}=\frac{1}{1+}\frac{q}{1+}\frac{q^2}{1+}\frac{q^3}{1+}\frac{q^4}{1+}\cdots, \tag{4.3.11}$$

([ABJL89], (5)).

$$\frac{(q;q^8)_\infty(q^7;q^8)_\infty}{(q^3;q^8)_\infty(q^5;q^8)_\infty}=\frac{1}{1+}\frac{q+q^2}{1}\,+\frac{q^4}{1+}\frac{q^3+q^6}{1}\,+\frac{q^8}{1+}\frac{q^5+q^{10}}{1}\,+\cdots, \tag{4.3.12}$$

([ABJL89], thm 6).

$$\sum_{k=1}^{\infty}\frac{(a;q)_\infty a^k}{(q;q)_k(1+q^kz)}=\frac{a}{1+}\frac{(1-a)qz}{1}\,+\frac{(1-q)aqz}{1}\,+\frac{(1-aq)q^2z}{1}\,+\frac{(1-q^2)aq^2z}{1}\,+\frac{(1-aq^2)q^3z}{1}\,+\cdots \tag{4.3.13}$$

for $(a,z)\in\mathbb{C}^2$, ([Wall48], p 376).

Bibliography

[ABBW85] C. Adiga, B. C. Berndt, S. Bhargava and G. N. Watson, "Chapter 16 of Ramanujan's Second Notebook: Theta–Functions and q–Series", Mem. of the Amer. Math. Soc., no. **315**, Providence, 1985.

[ABJL89] G.E. Andrews, B.C. Berndt, L. Jacobsen and R.L. Lamphere, *Variations on the Rogers–Ramanujan Continued Fraction in Ramanujan's Notebooks*, In: "Number Theory, Madras 1987" (K. Alladi ed.) Lecture Notes in Math. Springer Verlag, **1395** (1989), 73–83.

[ABJL92] G. E. Andrews, B. C. Berndt, L. Jacobsen and R. L. Lamphere, "The Continued Fractions Found in the Unorganized Portions of Ramanujan's Notebooks". Memoirs of the Amer. Math. Soc., Providencs R.I., Vol **99**, No 477, 1992.

[AbSt65] M. Abramowitz and I. A. Stegun, "Handbook of Mathematical Functions", Dover, New York 1965.

[Ahlf53] L. Ahlfors, "Complex Analysis", McGraw-Hill, New York (1953).

[AlBe88] G. Almkvist and B. Berndt, *Gauss, Landen, Ramanujan, the Arithmetic-Geometric Mean, Ellipses, π, and the Ladies Diary.* Amer. Math. Monthly (1988), 585–608.

[Andr79] G.E. Andrews, *An Introduction to Ramanujan's "Lost" Notebook*, Amer. Math. Monthly, **86**, (1979), 89–108.

[AnHi05] D. Angell and M. D. Hirschhorn, *A remarkable continued fraction*, Bull, Austral. Math. Soc. vol **72** (2005), 45–52.

[BaGM96] G. A. Baker and P. Graves-Morris, "Padé Approximants", Cambridge University Press (1996).

[BaJo85] C. Baltus and W. B. Jones, *Truncation error bounds for limit-periodic continued fractions* $\mathbf{K}(a_n/1)$ *with* $\lim a_n = 0$, Numer. Math. **46** (1985), 541–569.

[Bauer72] G. Bauer, *Von einem Kettenbruch Eulers und einem Theorem von Wallis*, Abh. München **11** (1872).

[BBCM07] D. Borwein, J. Borwein, R. Crandall and R. Mayer, *On the dynamics of certain recurrence relations*, Ramanujan J. **13** No 1-3 (2007), 63–101.

[Bear01a] A. F. Beardon, *Worpitzky theorem on continued fractions*, Jour. Comput. Appl. Math. **131** (2001), 143–148.

[Bear01b] A. F. Beardon, *The Worpitzky-Pringsheim theorem on continued fractions*, Rocky Mountain J. Math. **31** (2001), 389–399.

[Bear04] A. F. Beardon, Oral communication.

[BeLo01] A. F. Beardon and L. Lorentzen, *Approximants of Śleszyński-Pringsheim continued fractions*, J. Comput. and Appl. Math. **132** (2001), 467–477.

[Bern78] B. C. Berndt, *Ramanujan's Notebooks*, Math. Mag. **51** (1978), 147–164.

[Bern89] B. C. Berndt, "Ramanujan's Notebooks, Part II", Springer Verlag, New York, 1989.

[Berno75] D. Bernoulli, *Disquisitiones ulteriores de indole fractionum continarum*, Novi Comm. Acad. Imp. St Petersbourg **20** (1775), 24–47.

[Berno78] D. Bernoulli, *Observationes de seribus quae formantur ex additione vel subtractione quacunque terminorum se multuo consequentium*, Comm. Acad. Sci. Imp. Petropolilanae, Vol IV (1778), 85–100.

[BeSh07] A. F. Beardon and I. Short, *Van Vleck's theorem on continued fractions*, Comp. Methods and Funct. Theory **7** (2007), 185–203.

[Birk30] G. D. Birkhoff, *Formal theory of irregular difference equations*, Acta Math. **54** (1930), 205–246.

[BiTr32] G. D. Birkhoff and W. J. Trjitzinsky, *Analytic theory of singular difference equations*, Acta Math. **60** (1932), 1–89.

[BoCF04] J. M. Borwein, R. E. Crandall and G. J. Fee, *On the Ramanujan AGM fraction. I*, Experiment. Math. **13** No3 (2004), 275–285.

[BoCR04] J. M. Borwein, R. E. Crandall and E. Richard, *On the Ramanujan AGM fraction. II*, Experiment. Math. **13** No3 (2004), 287–295.

[Bomb72] R. Bombelli, "L'Algebra", Venezia 1572.

[BoML04] D. Bowman and J. Mc Laughlin, *A theorem on divergence in the general sense for continued fractions*, J. Comput. Appl. Math. **172**(2) (2004), 363–373.

[BoML06] D. Bowman and J. Mc Laughlin, *Continued fractions and generalizations with many limits: a survey*, In: Proceedings from Conference on Diphantine Analysis and Related Fields, Seminar on Mathematical Sciences **35**, Keio University, Dept. of Math., Tokohama, Japan (2006), 19–38.

[BoML07] D. Bowman and J. Mc Laughlin, *Continued fractions with multiple limits*, Advances in Math. **210**, no 2 (2007), 578–606.

[BoML08] D. Bowman and J. Mc Laughlin, *Asymptotics and limit sets for continued fractions, infinite matrix products, and recurrence relations.* Submitted.

[BoSh89] K. O. Bowman and L. R. Shenton, "Continued Fractions in Statistical Applications", Marcel Dekker, Inc., New York and Basel, 1989.

[Brez91] C. Brezinski, "History of Continued Fractions and Padé Approximants", Springer Series in Computational Mathematics, **12**, Springer-Verlag, Berlin (1991).

[Brom77] K. E. Broman, *Om konvergensen och divergensen af Kedjebråk*, Diss. Universitetet i Uppsala 1877.

[Cata13] P. Cataldi, *Trattato del modo brevissimo di trocare la radice quadro delli numeri*, Bologna 1613.

[Chih78] T. S. Chihara, "Orthogonal Polynomials",Gordon and Breach Science Publ. 1978.

[CJPVW7] A. Cuyt, W. B. Jones, V. B. Petersen, B. Verdonk and H. Waadeland, "Handbook of Continued Fractions for Special Functions", Klüwer Academic Publishers (2007).

[Cord92] A. Y. Córdova, *A convergence theorem for continued fractions*, Thesis. Bayrischen Julius-Maximilians-Universität Würzburg. 1992.

[CrJT94] C. M. Craviotto, W. B. Jones and W. T. Thron, *Best truncation error bounds for continued fractions* $\mathbf{K}(1/b_n)$, $\lim_{n\to\infty} b_n = \infty$, In: Continued fractions and orthogonal functions (eds.: S. Clement Cooper and W. J. Thron) Lecture Notes in Pure and Applied Math. **154**, Marcel Dekker (1994), 115–128.

[CuWu87] A. Cuyt and L. Wuytack, "Nonlinear Methods in Numerical Analysis", North-Holland Mathematics Studies **136**, Amsterdam (1987).

[Diri63] G. P. L. Dirichlet, *Vorlesungen über Zahlentheorie*, Braunschweig, 1863.

[EMOT53] A. Erdélyi, W. Magnus, F. Oberhettinger and F. G. Tricomi, "Higher Transcendental Functions", Vol. **1**, McGraw-Hill, New York, 1953.

[Eucl56] Euclid, *The thirteen books of Euclid's Elements*, Dover Publ. Inc. 2. ed, 1956.

[Euler37] L. Euler, *De fractionibus continuus*, Dissertatio, Comment. Acad. Sci. Imp. Petrop., XI (1737), 98–137.

[Euler48] L. Euler, "Introductio in analysin infitorum", chapter 18, Lausanne (1748).

[Euler67] L. Euler, *De fractionibus continuis observationes*, Comm. Acad. Sci. Imp. St. Pétersbourg, 11(1767), 32–81 Opera Omnia, Ser. **1**, Vol. **14**, B. G. Teubner, Lipsiae, 1925, 291–349.

[Fibo02] L. Fibonacci, "Liber Abaci", 1202.

[FiJo72] D. A. Field and W. B. Jones, *A priori estimates for truncation error of continued fractions*, Numer. math. **19** (1972), 283–302.

[Galo28] E. Galois, *Théoréme sur les fractions continues périodiques*, Ann. de Math. (gergonne) **19** (1828/29), 294; Oevres mathematiques, Gauthier Villars, Paris 1951.

[Gauss13] C. F. Gauss, *Disquisitiones generales circa seriem infinitam* $1 + \frac{\alpha\beta}{1\cdot\gamma}x + \frac{\alpha(\alpha+1)\beta(\beta+1)}{1\cdot2\cdot\gamma\cdot(\gamma+1)}xx + \frac{\alpha(\alpha+1)(\alpha+2)\beta(\beta+1)(\beta+2)}{1\cdot2\cdot3\cdot\gamma\cdot(\gamma+1)(\gamma+2)}x^3$ *etc*, Commentationes Societatis Regiae Scientiarium Gottingensis Recentiores **2** (1813), 1–46; Werke, Vol. **3** Göttingen (1876), 123–162.

[Gaut67] W. Gautschi, *Computational Aspects of Tree-Term Recurrence Relations*, SIAM Review **9** (1967), 24–82.

[Gaut70] W. Gautschi, *Efficient Computation of the Complex Error Function*, SIAM J. Numer. Anal. **7** (1970), 187–198.

[Gill73] J. Gill, *Infinite Compositions of Möbius Transformations*, Trans. Amer. Math. Soc. **176** (1973), 479–487.

[Gill75] J. Gill, *The Use of Attractive Fixed Points in Accelerating the Convergence of Limit-Periodic Continued Fractions*, Proc. Amer. Math. Soc. **47** (1975), 119–126.

[Gill80] J. Gill, *Convergence Acceleration for Continued Fractions* $\mathbf{K}(a_n/1)$ *with* $\lim a_n = 0$, "Analytic Theory of Continued Fractions", (W.B.Jones, W.J.Thron, H.Waadeland, eds), Lecture Notes in Mathematics **932**, Springer–Verlag, Berlin (1980), 67–70.

[Glai74] J. W. L. Glaisher, *On the transformation of continued products into continued fractions*, Proc. London Math. Soc. **5** (1874), 78–88.

[GrWa83] W. B. Gragg and D. D. Warner, *Two Constructive Results in Continued Fractions*, SIAM J. Numer. Anal. **20** (1983), 1187–1197.

[Hamel18] G. Hamel, *Über einen limitärperiodischen Kettenbruch*, Archiv der Math. und Phys. **27** (1918), 37–43.

[Henr77] P. Henrici, "Applied and Computational Complex Analysis", I, II and III, J. Wiley & Sons, New York (1974, 1977, 1986).

[Hens06] D. Hensley, "Continued fractions", World Scientific Publ. Co., 2006.

[HePf66] P. Henrici and P. Pfluger, *Truncation Error Estimates for Stieltjes Fractions*, Numer. Math. **9** (1966), 120–138.

[Hille62] E. Hille, "Analytic Function Theory", Vol **2**, Ginn, Boston (1962).

[HiTh65] K. L. Hillam and W. J. Thron, *A General Convergence Criterion for Continued Fractions* $\mathbf{K}(a_n/b_n)$, Proc. Amer. Math. Soc. **16** (1965), 1256–1262.

[Hurw95] A. Hurwitz, *Über die Bedingungen unter welchen eine Gleichung nur Wurzeln mit negativen reellen Teilen besitzt*, Math. Annalen **46** (1895), 273–284.

[Huyg95] C. Huygens, "Descriptio automati plenetarii", Haag (1695).

[Hütte55] Hütte, "Des Ingenieurs Taschenbuch", 28. Aufl. **1**, Wilhelm Ernst & Sohn, Berlin (1955), Seite 139.

[Ince26] E. L. Ince, "Ordinary Differential Equations", Longmans, Green & Co (1926).

[Jaco82] L. Jacobsen, *Some Periodic Sequences of Circular Convergence Regions*, "Analytic Theory of Continued Fractions", Lecture Notes in Mathematics **932** (W. B. Jones, W. J. Thron and H. Waadeland eds.), Springer-Verlag, Berlin (1982), 87–98.

[Jaco83] L. Jacobsen, *Convergence Acceleration and Analytic Continuation by Means of Modification of Continued Fractions*, Det Kgl. Norske Vid. Selsk. Skr. No **1** (1983), 19–33.

[Jaco84] L. Jacobsen, *Further results on convergence acceleration for continued fractions* $\mathbf{K}(a_n/1)$, Trans. Amer. math. Soc. **281**(1) (1984), 129–146.

[Jaco86] L. Jacobsen, *General Convergence of Continued Fractions*, Trans. Amer. Math. Soc. **294**(2) (1986), 477–485.

[Jaco86a] L. Jacobsen, *Composition of linear fractional transformations in terms of tail sequences*, Proc. Amer. math. Soc **97** (1986), 97–104.

[Jaco86b] L. Jacobsen, *A theorem on simple convergence regions for continued fractions* $\mathbf{K}(a_n/1)$, In: Analytic Theory for Continued Fractions II, Lecture Notes in Math **1199**, Springer-verlag (1986), 59–66.

[Jaco87] L. Jacobsen, *Nearness of continued fractions*, Math. Scand. **60** (1987), 129–147.

[Jaco88] L. Jacobsen, *Meromorphic continuation of functions given by limit k-periodic continued fractions*, Appl. Numer. Math. **4** (1988), 323–336.

[Jaco90] L. Jacobsen, *On the Bauer-Muir transformation for continued fractions and its applications*, J. Math. Ana. Appl. **152**(2) (1990), 496–514.

[JaJW87] L. Jacobsen, W. B. Jones and H. Waadeland, *Convergence Acceleration for Continued Fractions* $\mathbf{K}(a_n/1)$ *where* $a_n \to \infty$, "Rational Approximation and Its Applications in Mathematics and Physics", Lecture Notes in Mathematics **1237** (J. Gilewicz, M. Pindor and W. Siemaszko eds.) Springer-Verlag, Berlin (1987), 177–187.

[JaMa90] L. Jacobsen and D. R. Masson, *On the Convergence of Limit Periodic Continued Fractions* $\mathbf{K}(a_n/1)$, *where* $a_n \to -1/4$. Part III., Constr. Approx. **6** (1990), p.363–374.

[JaTh86] L. Jacobsen and W. J. Thron, *Oval Convergence Regions and Circular Limit Regions for Continued Fractions* $\mathbf{K}(a_n/1)$, "Analytic Theory of Continued Fractions" II, Lecture Notes in Mathematics **1199** (W. J. Thron ed.), Springer-Verlag, Berlin (1986), 90–126.

[JaTh87] L. Jacobsen and W. J. Thron, *Limiting structures for sequences of linear fractional transformations*, Proc. Amer. Math. Soc. **99** (1987), 141–146.

[JaTW89] L. Jacobsen, W. J. Thron and H. Waadeland, *Julius Worpitzky, his contributions to the analytic theory of continued fractions and his times*, In: Analytic Theory of Continued Fractions III (ed L. Jacobsen) Springer Lecture Notes in Math. **1406** (1989), 25–47.

[JaWa82] L. Jacobsen and H. Waadeland, *Some useful formulas involving tails of continued fractions*, In: Analytic Theory of Continued Fractions (eds. W. B. Jones, W. J. Thron and H. Waadeland) Springer lecture Notes in Math. **932**, (1982), 99–105.

[JaWa85] L. Jacobsen and H. Waadeland, *Glimt fra analytisk teori for kjedebrøker. Del 2.* (in Norwegian) Normat **33**(4) (1985), 168–175.

[JaWa86] L. Jacobsen and H. Waadeland, *Even and Odd Parts of Limit Periodic Continued Fractions*, J. Comp. Appl. Math. **15** (1986), 225–233.

[JaWa88] L. Jacobsen and H. Waadeland, *Convergence Acceleration of Limit Periodic Continued Fractions under Asymptotic Side Conditions*, Numer. Math. **53** (1988), 285–298.

[JaWa89] L. Jacobsen and H. Waadeland, *When does* $f(z)$ *have a Regular C-Fraction or a Normal Padé Table?*, Journ. Comp. and Appl. Math. **28** (1989), 199–206.

[JaWa90] L. Jacobsen and H. Waadeland, *An Asymptotic Property for Tails of Limit Periodic Continued Fractions*, Rocky Mountain J. of Math. **20**(1) (1990), 151–163.

[Jens09] J. L. W. V. Jensen, *Bidrag til Kædebrøkenes teori.* Festskrift til H. G. Zuhten, 1909.

[JoTh68] W. B. Jones and W. J. Thron, *Convergence of Continued Fractions*, Canad. J. of Math. **20** (1968), 1037–1055.

[JoTh70] W. B. Jones and W. J. Thron, *Twin-Convergence Regions for Continued Fractions* $\mathbf{K}(a_n/1)$, Trans. Amer. Math. Soc. **150** (1970), 93–119.

[JoTh76] W. B. Jones and W. J. Thron, *Truncation Error Analysis by Means of Approximant Systems and Inclusion Regions*, Numer. Math. **26** (1976), 117–154.

[JoTh80] W. B. Jones and W. J. Thron, "Continued Fractions: Analytic Theory and Applications", Encyclopedia of Mathematics and its Applications, **11**, Addison-Wesley Publishing Company, Reading, Mass. (1980). Now distributed by Cambridge University Press, New York.

[JoTW82] W. B. Jones, W. J. Thron and H. Waadeland, *Modifications of continued fractions*, Lecture Notes in Math., Springer–Verlag **932** (1982), 38–66.

[JTWa89] L. Jacobsen, W. J. Thron and H. Waadeland, *Julius Worpitzky, his contributions to analytic theory of continued fractions and his times*, Lecture Notes in Math., Springer–Verlag **1406** (1989), 25–47.

[Khov63] A. N. Khovanskii, "The Application of Continued Fractions and Their Generalizations to Problems in Approximation Theory", P. Noordhoff, Groningen (1963).

[Khru06a] S. Khrushchev, *A recovery of Brouncker's proof of the quadrature continued fraction*, Publications Mathematiques **50**(1) (2006), 3–42.

[Khru06b] S. Khrushchev, *On Euler's differential methods for continued fractions*, Electr. Trans. Numer. Anal. Kent State Univ. **25** (2006), 178–200.

[Klein95] F. Klein, *Über eine geometrische Auffassung der gewöhnlichen Kettenbruchentwicklung*, Nachr. der königl. Gesell. der Wiss., Göttingen, Math. Phys. Klasse (1895), 357–359.

[Koch95a] H. von Koch, *Quelques théorémes concernant la théorie générale des fractions continues*, Öfversigt af Kgl. Vetenskaps-Akad. Förh. **52** (1893).

[Koch95b] H. von Koch, *Sur une théor'eme de Stieltjes et sur les fonctions définies par des fractions continues*, Bull. S. M. F. **23** (1895).

[Lagr70] J. L. Lagrange, *Additions au mémoire sur la résolution des équations numériques*, Mém. Berl. **24** (1770).

[Lagr76] J. L. Lagrange, *Sur l'usage des fractions continues dans le calcul intégral*, Nouveaux Mém. Acad. Sci. Berlin **7** (1776), 236–264; Oeuvres, **4** (J. A. Serret, ed.), Gauthier Villars, Paris (1869), 301–322.

[Lagr98] J. L. Lagrange, *Additions aux Eléments d'Algèbre d'Euler*, Lyon (1798). Oeuvres VII.

[Lamb61] J. H. Lambert, *Mémoire sur quelques propriétés remarquables de quantités transcendantes circulaires et logarithmiques*, Histoire de l'Académie, Berlin (1761), 265–322.

[Lane45] R. E. Lane, *The Convergence and Values of Periodic Continued Fractions*, Bull. Amer. Math. Soc. **51** (1945), 246–250.

[Lange66] L. J. Lange, *On a family of twin convergence regions for continued fractions*, Illinois J. Math. vol **10** (1966), 97–108.

[Lange94] L. J. Lange, *Strip convergence regions for continued fractions*, In: Continued fractions and orthogonal functions (eds.: S. Clement Cooper and W. J. Thron), Lecture Notes in Pure and Applied Mathematics **154**, Marcel Dekker (1994), 211–232.

[Lange95] L. J. Lange, *Convergence region inclusion theorems for continued fractions* $\mathbf{K}(a_n/1)$, Constr. Approx. **11** (1995), 321–329.

[Lange99a] L. J. Lange, *A generalization of Van Vleck's Theorem and More on Complex Continued Fractions*, Contemporary Math. **236** (1999), 179–192.

[Lange99b] L. J. Lange, *Convergence regions with bounded convex complements for continued fractions* $\mathbf{K}(1/b_n)$, J. Comput. Appl. Math. vol **105** (1999), 355–366.

[LaTh60] L. J. Lange and W. J. Thron, *A two-parameter family of best twin convergence regions for continued fractions*, Math. Zeitschr. vol **73** (1960), 295–311.

[LaWa49] R. E. Lane and H. S. Wall, *Continued Fractions with Absolutely Convergent Even and Odd Parts*, Trans. Amer. Math. Soc. **67** (1949), 368–380.

[Lawd89] D. F. Lawden, "Elliptic Functions and Applications", Springer-Verlag, Applied Mathematical Sciences Vol. **80**, New York, 1989.

[LeTh42] W. Leighton and W. J. Thron, *Continued Fractions with Complex Elements*, Duke. Math. J. **9** (1942), 763–772.

[Levr89] P. Levrie, *Improving a Method for Computing Non-dominant Solutions of Certain Second-Order Recurrence Relations of Poincaré-Type*, Numer. Math. **56** (1989), 501–512.

[Lore92] L. Lorentzen, *Bestness of the parabola theorem for continued fractions*, J. Comp. Appl. Math. **40** (1992), 297–304.

[Lore93] L. Lorentzen, *Analytic Continuation of Functions Represented by Continued Fractions, Revisited*, Rocky Mountain J. of Math. **23**(2) (1993), 683–706.

[Lore94a] L. Lorentzen, *A convergence property for sequences of linear fractional transformations*, In. Continued fractions and orthogonal functions (eds.: S. Clement Cooper and W. J. Thron). Lecture Notes in Pure and Appl. Math., Marcel Dekker, vol **154** (1994), 281–304.

[Lore94b] L. Lorentzen, *Divergence of Continued Fractions Related to Hypergeometric Series*, Math. Comp. **62** (1994), 671–686.

[Lore94c] L. Lorentzen, *Properties of limit sets and convergence of continued fractions*, J. Math. Anal. Appl. **185**(2), (1994), 229–255.

[Lore03a] L. Lorentzen, *General convergence in quasi-normal families*, Proc. Edinburgh Math. **46** (2003), 169–183.

[Lore03b] L. Lorentzen, *A priori truncation error bounds for continued fractions*, Rocky Mountain J. Math. (2003), 409–474.

[Lore06] L. Lorentzen, *Musikk og kjedebrøker. 5-toners, 12-toners og 41-toners skala*, Det Kgl. Norske Videnskabers Selskab. Årbok 2005. (2006), 97–103.

[Lore07] L. Lorentzen, *Lisa Möbius transformations mapping the unit disk into itself*, The Ramanujan J. Math. **13**(1/2/3)(2007), 253–264.

[Lore08a] L. Lorentzen, *Continued fractions with circular value sets*, Trans. Amer. Math. Soc., to appear.

[Lore08b] L. Lorentzen, *Convergence and divergence of the Ramanujan AGM fraction*, Ramanujan J. Math., to appear.

[Lore08c] L. Lorentzen, *An idea on some of Ramanujan's continued fraction identities*, Ramanujan J. Math., to appear.

[LoWa92] L. Lorentzen and H. Waadeland, *Continued Fractions with Applications*, Studies in Computational Mathematics **3**, Elsevier Science Publishers B.V. 1992.

[McWy07] J. Mc Laughlin and N. Wyshinski, *Ramanujan and extensions and contractions of continued fractions*, Ramanujan J. Math., to appear.

[Mind69] F. Minding, *Über das Bildungsgesetz der Zähler und Nenner bei Verwandlung der Kettenbrüche in gewöhnliche Brüche*, Bull. Acad. Sc. Imp. St. Petersburg **13** (1869), 524–528.

[Muir77] T. Muir, *A Theorem in Continuants*, Phil. Mag., (5) 3 (1877), 137–138.

[Over82] M. Overholt, *A class of element and value regions for continued fractions*, In: Analytic Theory of Continued Fractions (eds.: W. B. Jones, W. J. Thron and H. Waadeland), Lecture Notes in Math vol **932**, Springer-Verlag (1982), 194–205.

[PaWa42] J. F. Paydon and H. S. Wall, *The Continued Fraction as a Sequence of Linear Transformations*, Duke. Math. J. **9** (1942), 360–372.

[Perr05] O. Perron, *Über die Konvergenz periodischer Kettenbrüche*, S. B. München **35** (1905).

[Perr11] O. Perron, Einige Konvergenz- und Divergenzkriterien für alternierende Kettenbrüche, Sb. Münch., 1911.

[Perr13] O. Perron, "Die Lehre von den Kettenbrüchen", Leipzig, Berlin 1913.

[Perr29] O. Perron, "Die Lehre von den Kettenbrüchen", Chelsea 1929.

[Perr54] O. Perron, "Die Lehre von den Kettenbrüchen", Band **1** 3. Aufl., B. G. Teubner, Stuttgart (1954).

[Perr57] O. Perron, "Die Lehre von den Kettenbrüchen" Band **2**, B. G. Teubner, Stuttgart (1957).

[Poin85] H. Poincaré, *Sur les equations linéaires aux differentielles ordinaires et aux differences finis*, Amer. J. Math. **7** (1885), 203–258.

[Prin99a] A. Pringsheim, *Über die Konvergenz unendlicher Kettenbrüche*, S.-B. Bayer. Akad. Wiss. Math. - Nat. Kl. **28** (1899), 295–324.

[Prin99b] A. Pringsheim, *Über ein Konvergenzkriterium für Kettenbrüche mit komplexen Gliedern*, Sb. Münch. **29**, 1899.

[Prin00] A. Pringsheim, *Über die Konvergenz periodischer Kettenbrüche*, S. b. München **30** (1900).

[Prin05] A. Pringsheim, *Über einige Konvergenzkriterien für Kettenbrüche mit komplexen Gliedern*, Sb. Münch. **35**, 1905.

[Prin10] A. Pringsheim, *Über Konvergenz und funktionentheoretischen Charakter gewisser limitärperiodischer Kettenbrüche*, S. b. Münch. (1910).

[Prin18] A. Pringsheim, *Zur Theorie der uendlichen Kettenbrüche*, Sb. Münch. 1918.

[Rama57] S. Ramanujan, "Notebooks", Vol. **2** Tata Institute of Fundamental Research, Bombay (1957). Now distributed by Springer-Verlag.

[ScWa40] W. T. Scott and H. S. Wall, *A Convergence Theorem for Continued Fractions*, Trans. Amer. Math. Soc. **47** (1940), 155–172.

[Seid46] L. Seidel, *Untersuchungen über die Konvergenz und Divergenz der Kettenbrüche*, Habilschrift München (1846).

[Seid55] L. Seidel, *Bemerkungen über den Zusammenhang zwischen dem Bildungsgesetze eines Kettenbruches und der Art des Fortgangs seiner Näherungsbrüche*, Abh. der Kgl. Bayr. Akad. der Wiss., München, Zweite Klasse, **7**:3 (1855), 559.

[Śles89] J. V. Śleszyński, *Zur Frage von der Konvergenz der Kettenbrüche* (in Russian), Mat. Sbornik **14** (1889), 337–343, 436–438.

[Smith60] H. J. S. Smith, *Report on the theory of numbers*. Report to the British Association, (1860), 120–172.

[Steen73] A. Steen, *Integration af Lineære Differentialligninger af Anden Orden ved Hjælp af Kjædebrøker*, Københavns Universitet, København (1873), 1–66.

[Stern32] M. A. Stern, Theorie der Kettenbrüche und ihre Anwendung, J. f. Math. **10/11**, 1832.

[Stern48] M. A. Stern, *Über die Kennzeichen der Konvergenz eines Kettenbruchs*, J. Reine Angew. Math. **37** (1848), 255–272.

[Stie94] T. J. Stieltjes, *Recherches sur les fractions continues*, Ann. Fac. Sci. Toulouse **8** (1894), J, 1–122; **9** (1894), A, 1–47; Oeuvres **2**, 402–566. Also published in Memoires Présentés par divers savants à l'Académie de sciences de l'Institut National de France **33** (1892), 1–196.

[Stir30] J. Stirling, *Methodus differentialis: sive tractatus de summatione et interpolatione serierum infinitarum*, London 1730.

[Stolz86] O. Stolz, "Vorlesungen über allgemeine Arithmetic", Teubner, Leipzig (1886).

[Sylv69] J. J. Sylvester, *Note on a New Continued Fraction Applicable to the Quadrature of the Circle*, Philos. Mag. Ser. **4** (1869), 373–375.

[Thie79] T. N. Thiele, *Bemærkninger om periodiske Kjædebrøkers Konvergens*, Tidsskrift for Mathematik (4)**3** (1879).

[Thie09] T. N. Thiele, "Interpolationsrechnung" Teubner, Leipzig (1909).

[Thron43] W. J. Thron, *Two families of twin convergenvce regions for continued fractions*, Duke Math. J. **10** (1943), 677–685.

[Thron48] W. J. Thron, *Some Properties of Continued Fraction* $1+d_0z+\mathbf{K}(z/(1+d_nz))$, Bull. Amer. Math. Soc. **54** (1948), 206–218.

[Thron49] W. J. Thron, *Twin convergence regions for continued fractions* $b_0+\mathbf{K}(1/b_n)$*, II*, Amer. J. Math. **71** (1949), 112–120.

[Thron58] W. J. Thron, *On Parabolic Convergence Regions for Continued Fractions*, Math. Zeitschr. **69** (1958), 173–182.

[Thron59] W. J. Thron, *Zwillingskonvergenzgebiete für Kettenbrüche* $1+\mathbf{K}(a_n/1)$*, deren eines die Kreisscheibe* $|a_{2n-1}| \leq \rho_2$ *ist*, Math. Zeitschr. **70** (1958/59), 310–344.

[Thron74] W. J. Thron, *A Survey of Recent Convergence Results for Continued Fractions*, Rocky Mountain J. Math. **4**(2) (1974), 273–282.

[Thron81] W. J. Thron, *A priori truncation error estimates for Stieltjes fractions*, In: Orthogonal polynomials, continued fractions and Padé approximation, Birkhäuser verlag (1981), 203–211.

[Thron89] W. J. Thron, *Continued Fraction Identities Derived from the Invariance of the Crossratio under Linear Fractional Transformations*, "Analytic Theory of Continued Fractions III", Proceedings, Redstone 1988, (L. Jacobsen ed.), Lecture Notes in Mathematics **1406**, Springer-Verlag, Berlin (1989), 124–134.

[ThWa80a] W. J. Thron and H. Waadeland, *Accelerating Convergence of Limit Periodic Continued Fractions* $\mathbf{K}(a_n/1)$, Numer. Math. **34** (1980), 155–170.

[ThWa80b] W. J. Thron and H. Waadeland, *Analytic Continuation of Functions Defined by Means of Continued Fractions*, Math. Scand. **47** (1980), 72–90.

[ThWa82] W. J. Thron and H. Waadeland, *Modifications of Continued Fractions, a Survey*, "Analytic Theory of Continued Fractions", Proceedings 1981, (W. B. Jones, W. J. Thron and H. Waadeland eds.), Lecture Notes in Mathematics **932**, Springer-Verlag, Berlin (1982), 38–66.

[VanV01] E. B. Van Vleck, *On the convergence of continued fractions with complex elements*, Trans. Amer. Math. Soc. **2** (1901), 215–233.

[VanV04] E. B. Van Vleck, *On the convergence of algebraic continued fractions, whose coefficients have limiting values*, Trans. Amer. Math. Soc. **5** (1904), 253–262.

[Vita35] G. Vitali, "Moderna teoria delle funzoni d variabile reale", 1935.

[Waad66] H. Waadeland, *A Convergence Property of Certain T-fraction Expansions*, Det Kgl. Norske Vid. Selsk. Skr. **9** (1966), 1–22.

[Waad67] H. Waadeland, *On T-fractions of certain functions with a first order pole at the point of infinity*, Norske Kgl. Vid. Selsk. Forh. **40** (1967), No 1.

[Waad83] H. Waadeland, *Differential Equations and Modifications of Continued Fractions, some Simple Observations*, "Padé Approximants and Continued Fractions", Proceedings 1982, (H. Waadeland and H. Wallin eds.) Det Kongelige Norske Videnskabers Selskabs Skrifter, Trondheim **1** (1983), 136–150.

[Waad84] H. Waadeland, *Tales about tails*, Proc. Amer. math. Soc. **90**, (1984), 57–64.

[Waad89] H. Waadeland, *Boundary versions of Worpitzky's theorem and of parabola theorems*, In: Analytic Theory of Continued Fractions III (ed.: L. Jacobsen), Lecture Notes in Math., Springer-Verlag vol **1409** (1989), 135–142.

[Waad92] H. Waadeland, *On continued fractions $K(a_n/1)$ where all a_n are lying on a cartesian oval*, Rocky Mountain J. Math., **22**(3) (1992), 1123–1137.

[Waad98] H. Waadeland, *Some probabilistic remarks on the boundary version of Worpitzky's theorem*, Lecture Notes in Pure and Appl. Math., Marcel Dekker, **199** (1998), 409–416.

[Wall45] H. S. Wall, *Note on a certain continued fraction*, Bull. Amer. Math. Soc. **51** (1945), 930–934.

[Wall48] H. S. Wall, "Analytic Theory of Continued Fractions", Van Nostrand, New York (1948).

[Wall56] H. S. Wall, *Partially bounded continued fractions*, Proc. Amer. Math. Soc. **7** (1956), 1090–1093.

[Wall57] H. S. Wall, *Some convergence problems for continued fractions*, Amer. Math. Monthly **64**(8), part II (1957), 95–103.

[Wallis85] J. Wallis, "Tractatus de Algebra", 1685.

[Weyl10] H. Weyl, *Über gewöhnliche Differentialgleichungen mit Singularitäten und die zugehörigen Entwicklungen willkürlicher Funktionen*, Math. Ann. **68** (1910), 220–269.

[Wimp84] J. Wimp, "Computation with Recurrence Relations", Pitman Advanced Publishing Program, Pitman Publishing Inc., Boston, London, Melbourne (1984).

[Worp62] J. Worpitzky, *Beitrag zur Integration der Riccatischen Gleichung*, Greifswald (1862), 1–74.

[Worp65] J. Worpitzky, *Untersuchungen über die Entwickelung der monodromen und monogenen Funktionen durch Kettenbrüche*, Friedrichs-Gymnasium und Realschule Jahresbericht, Berlin (1865), 3–39.

[Wynn59] P. Wynn, *Converging Factors for Continued Fractions*, Numer. Math. **1** (1959), 272–320.

Index